D1085290

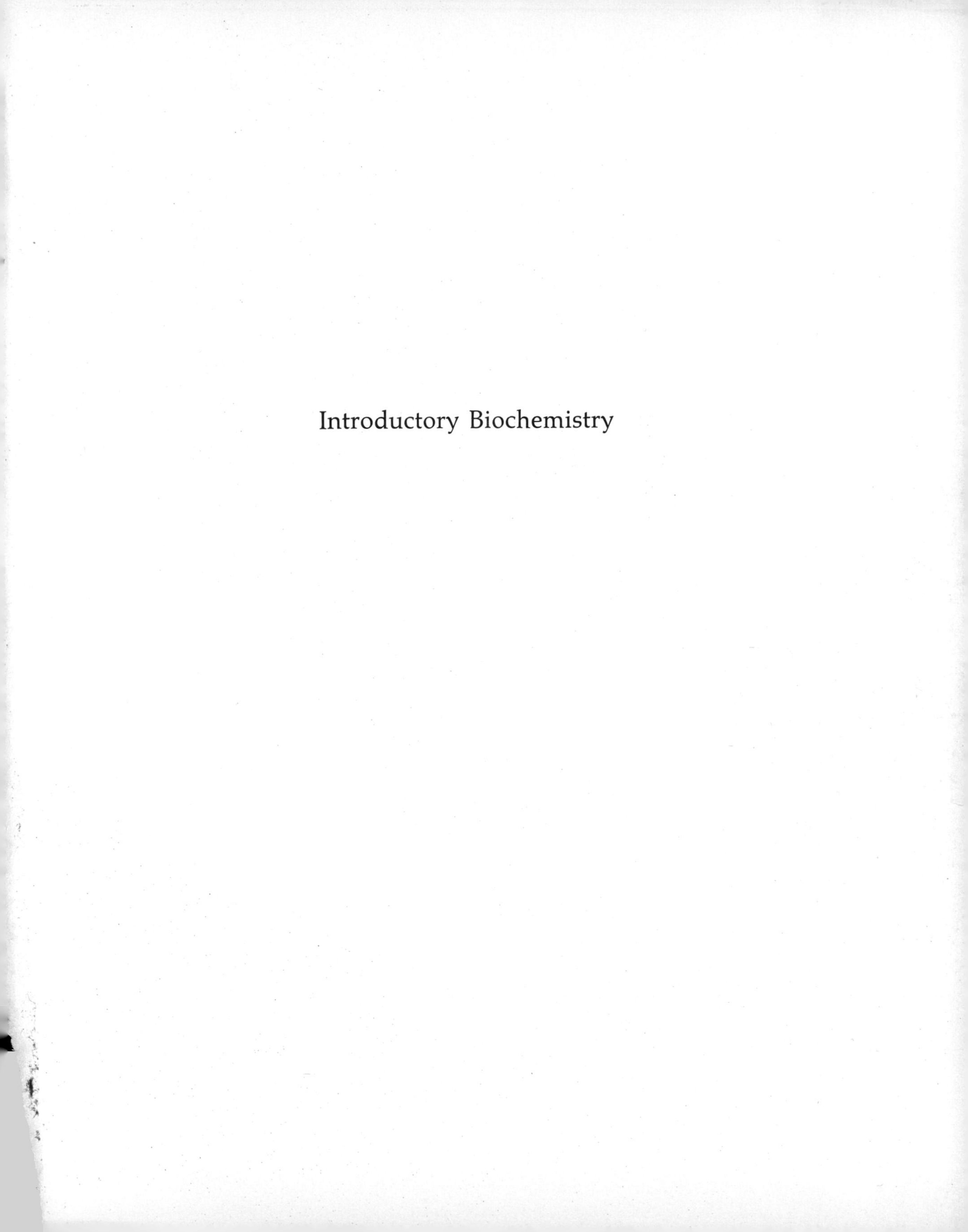

Introductory Biochemistry

Introductory Biochemistry

Fundamentals of Cellular Metabolism and Molecular Biology

Stuart J. Edelstein
Cornell University

HOLDEN-DAY, INC. SAN FRANCISCO Düsseldorf Johannesburg London
Panama Singapore Sydney Toronto

Editors: Gerald T. Papke and Sally Anderson

Designer: Michael A. Rogondino

Illustrator: Mark Schroeder

Production: Charles A. Goehring

Composition: York Graphic Services, Inc.
(Fototronic Elegante)

Printing and binding: Banta Levison Company

INTRODUCTORY BIOCHEMISTRY

500 Sansome Street, San Francisco, California 94111

Library of Congress Catalog Card Number: 72-88624
ISBN: 0-8162-2530-3

234567890 BL 8079876543

Printed in the United States of America

To My Teachers

PREFACE

This book developed from material prepared for teaching the one-semester biochemistry course required of all Cornell biology majors. I am grateful to Dr. Efraim Racker for providing me with the opportunity to teach the course and arranging for a year of preparation time for me to organize the material. During this period I developed the conviction that biochemistry would be more interesting to students if taught with a biological perspective. The principal themes of living systems could be emphasized with structural details interlaced into the main thrust of the book instead of placed in separate chapters. The experiences of teaching biochemistry in this format for several years have strengthened this conviction and convinced me that other schools might appreciate the opportunity to try this approach if a suitable textbook were available. I would now welcome reactions to the book from students and teachers, including comments on the general contents, specific details, or possible omissions.

I am also grateful to Dr. Gottfried Schatz for generously offering to give the lectures on metabolism in the Cornell course. Although the format in the metabolism section of this book differs from his approach, many topics presented here reflect insights gained from his lectures. Additional thanks go to him for carefully reading large portions of the text and offering many valuable comments. A number of his major recommendations for reorganizing the chapters on metabolism were not followed, however, and any faults found in the final product are my own.

Several other valuable reviews were arranged by the publisher. Thanks go to Dr. R. D. Cole for his page-by-page comments, to Drs. R. L. VanEtten, L. P. Hager, and B. H. Weber for their critical evaluations, and to Dr. C. Sagan for his comments on the last chapter. Discussions with Drs. H. K.

Schachman and P. Berg on several aspects of the text were very helpful and greatly appreciated. The help and encouragement of Fred Murphy and his staff at Holden-Day are gratefully acknowledged, especially the advice of Mr. Murphy to "Keep it simple." Special thanks go to Sally Anderson of Holden-Day for her skillful and tireless handling of editing and production matters.

I would also like to thank Jean Shriro for typing the manuscript and contributing many helpful comments and even a number of timely rejections, sending drafts back to me for more work before they were typed. Her judgments were very astute and saved a lot of wasted effort. I thank William Phillipson for convincing me that if I did not have the time to do the book now, I never would, and arranging the fruitful meeting with the publisher. Thanks especially go to my wife Lynn for showing great patience while I was preoccupied with work on the book, carefully reading the manuscript at various stages, providing many valuable comments, and saving the English language from my frequent incursions. In particular she convinced me that to frequently split infinitives is a bad practice. Final thanks go to the Cornell students and teaching assistants who provided much valuable feedback during the formulation and presentation of the material.

Stuart J. Edelstein
Ithaca, New York

CONTENTS

Part I

Life is an expression of evolution. Any form of life—such as a blossom, a deer, or an oak tree—has evolved by genetic variation and natural selection. Cells, plants, and animals reproduce. The offspring, which may differ slightly from their ancestors, compete in the environment. The most successful flourish, and favorable changes proliferate.

For all of the complexity of living things and their evolution, biochemistry has found a simple mechanism. While unable to explain each exact detail, the mechanism does provide a general outline and will serve as the focus for our study. It involves deoxyribonucleic acid (DNA)—the long double helix which stores the genetic information; ribonucleic acids (RNA), relatives of DNA, which are concerned with transmitting and translating the information in DNA into the structure of proteins; and proteins which mediate the actions and reactions of biochemical life.

This view of biochemistry emphasizes DNA as the fundamental vehicle of evolution. Three important features are built into its helical structure:

1 The stability for safeguarding the genetic contents of life.

2 The facility for reliable replication of the information stored with just enough "error" to provide the mutational variation necessary for evolution.

3 A system for coding the genetic information and readily decoding it when needed.

The chemical basis for storing the genetic information in the molecular structure of DNA will be described. A simple system of *complementary* nucleotide pairing through weak hydrogen bonds provides the basis for the genetic system. The same hydrogen-bond-directed pairing controls the synthesis of proteins. We will stalk the hydrogen bond through the interactions involving DNA and the various forms of RNA involved in protein synthesis. Proteins, built of 20 kinds of amino acids, are synthesized according to the order of the nucleotides found in DNA.

MACROMOLECULES

Together these steps provide the basic pattern for information flow: DNA ⟶ RNA ⟶ protein.

With the synthesis of proteins outlined, we will turn to an analysis of their structure and function. Cells require a multitude of proteins, most serving as catalysts called *enzymes,* to carry out the reactions in the biochemical world. Some enzymes catalyze reactions of nutrients to provide the cell with energy; others direct the degradation and synthesis of the precursors of nucleic acids, proteins, and other cellular components needed for growth. Many possess feedback sites for regulating their own activity. Small numbers of proteins act directly in protein synthesis and govern DNA replication. Changes in proteins, arising from mutations in DNA, may improve the performance of a reaction; the cell line or organism with this change multiplies more rapidly, and evolution is on its way.

The study of the two important classes of macromolecules—nucleic acids and proteins—will comprise Part I of this text. In Part II we will proceed to the design of metabolism among living cells. A great similarity is found in all living organisms as they tap the electron flow derived from sunlight or the oxidation of foodstuffs to yield energy trapped in the versatile adenosine triphosphate (ATP) molecule. The energy can then be harnessed to drive the synthesis of precursors for cell components and other important reactions. The essential features of the metabolic pathways will be outlined, as well as the highly specialized organelles such as mitochondria and chloroplasts that have evolved to house the machinery. Particular stress will be placed on the regulation and integration of the molecular steps of life and the interplay with the environment. We will emphasize the critical role of control effects generated by certain metabolites at key biochemical crossroads in changing properties of enzymes. The selective permeability and pumping properties of membranes will also be explored, as well as their organizational role in limiting that continuous, watery structural matrix known as the cell. Finally we will consider the signals that feed back to DNA itself and influence protein synthesis. The study of these mechanisms for holding the cell poised at optimal efficiency will permit a glimpse into higher-order biochemical integration—cells and structures approaching the stages of life that we see with our eyes. In sum, we will approach biochemistry with the same flow of information used by living systems: from DNA to proteins to cellular metabolism and intercellular activities, with feedback controls regulating the events at each level.

Chapter 1 Frontispiece *Electron micrograph of E. coli. Magnification 90,000×. Courtesy J. Waterbury.*

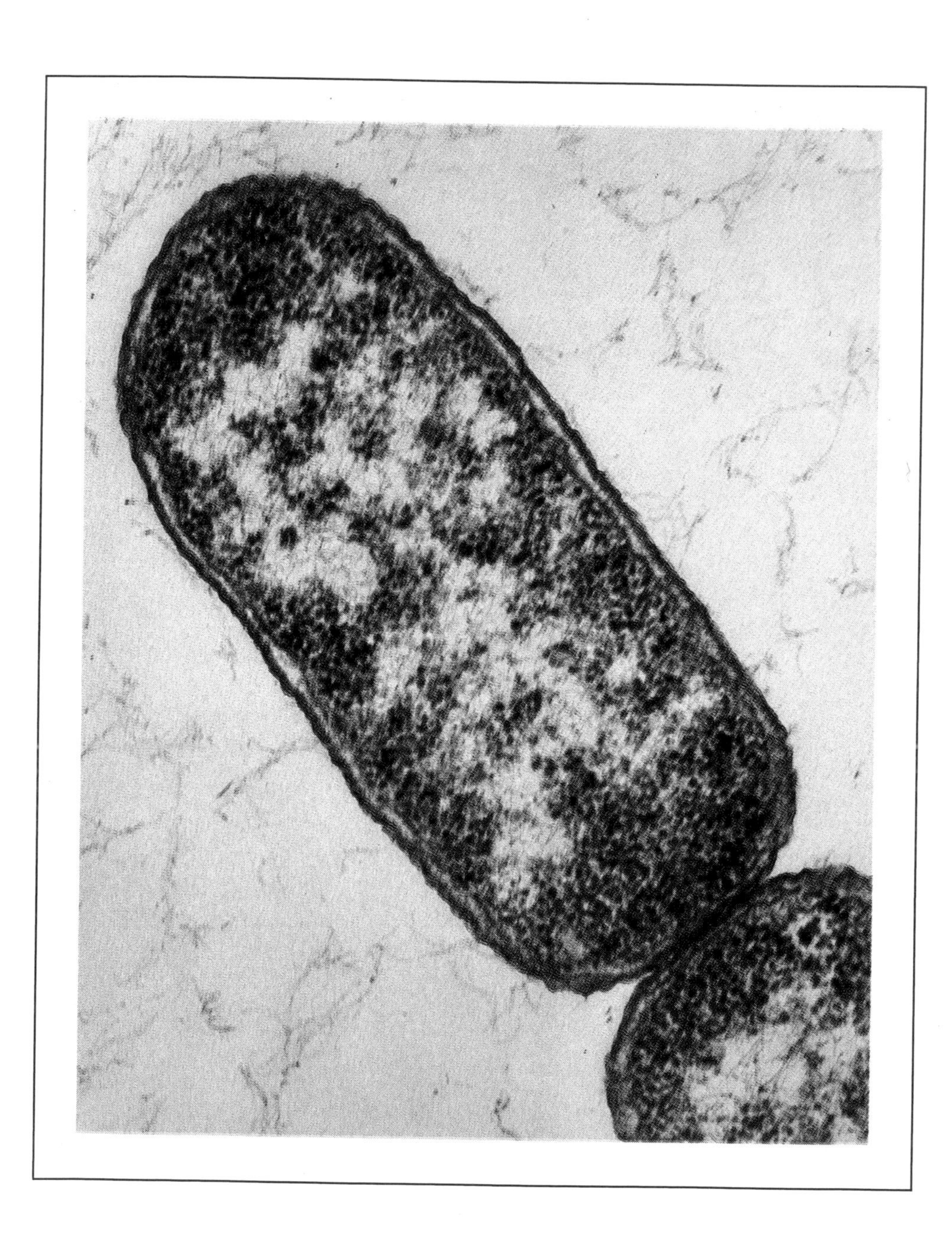

COLINEARITY

1

Diversity is a hallmark of life. Enormous varieties of sizes, shapes, colors, and patterns of behavior have evolved among living things. Yet, as suggested in the introduction, biochemistry has discovered a unity that accounts for the great diversity in life. This unity involves a single, relatively simple molecular mechanism centered around the information in DNA and its passage into RNA and proteins. To appreciate how the splendid details of living things radiate from the contents of DNA, we must first recall the concept of the gene. We will then go on to outline the fundamental relationship of DNA and proteins, the linear correspondence of the nucleotide base sequence of DNA with the amino acid sequence of proteins. One message is the translated form of the other, word for word. This *colinearity* of the genetic information with the proteins that direct the action of metabolic life is the central vein of molecular biology and is formulated in this chapter. The following five chapters will be occupied with unfolding its details.

"One Gene–One Enzyme"

Genetics (see Appendix I for a brief review of the basics of genetics) has revealed that many of the traits and properties of living organisms can be altered by *mutation*. Mutations are departures from the normal, or a "wild" type caused by a change in the DNA sequence, that is, a change in the gene. Thus the contents or structure of the chromosome—the *genotype*—controls the appearance of certain visible or measurable properties of living things—the *phenotypes*. Here again the types of mutations are as diverse as life itself. Well-studied mutations are the color changes in the eyes of fruit flies, the

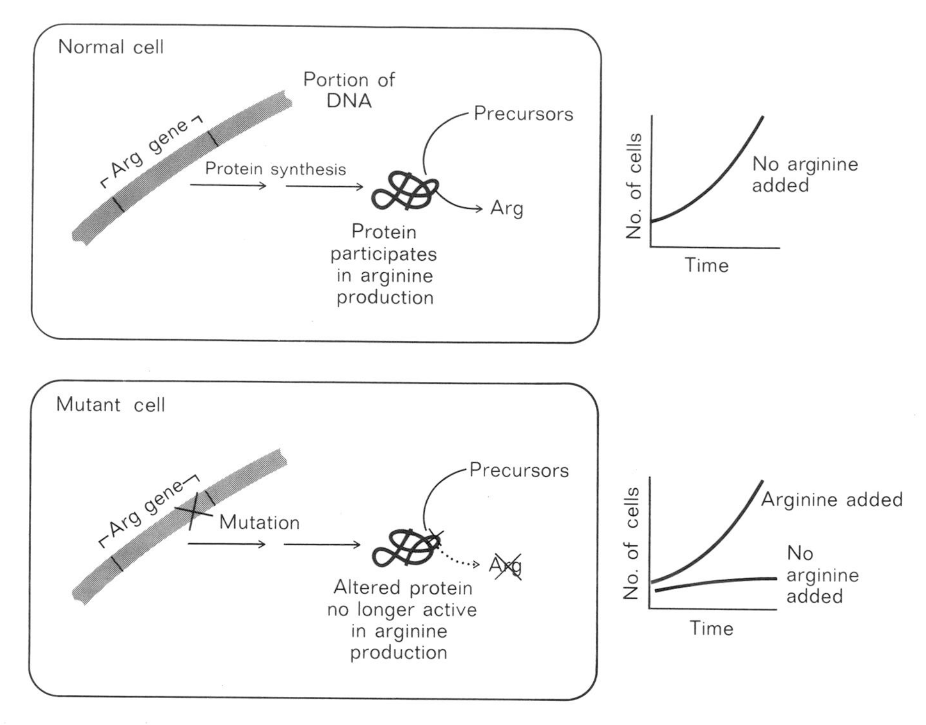

FIG. 1-1 *Genotype and Phenotype.* A cell makes arginine (Arg), an amino acid essential for its growth. A mutation in the DNA (new genotype) alters a protein involved in arginine biosynthesis. The mutants require added arginine for growth (new phenotype).

protruding Hapsburg lip of European royalty, the pea shape, the loss of the ability of mold to grow on arginine, to mention just a very few. Mutations *link* certain traits with certain genes, but further knowledge is needed to understand the mechanism of this connection. This knowledge involves proteins.

Proteins are responsible for not only the biochemical interactions of most cells but probably even the more complex interactions governing pea shape or the embryological development of the lower lip. Many of these complex phenotypes are undoubtedly controlled by a host of proteins acting as enzymes, and the designation of a single protein responsible for, as an example, eye color in man, may not be possible. However, in a sizable number of cases, one enzyme or protein can be isolated and characterized as the product of one gene. Such work, originally led by Beadle and Tatum in the 1940s, gave rise to the adage, "one gene–one enzyme." We now recognize that these terms are somewhat imprecise. Some genes govern the production of several proteins and some proteins are complexes composed of several subunits specified by several genes. Nevertheless the original terminology will be a useful focus until more precise terms are introduced toward the end of this chapter.

The first evidence for the one gene–one enzyme mechanism came from *Neurospora*, a mold. Certain strains, that is, specific genetic lines, with a mutation in a single gene were unable to synthesize the amino acid arginine (see Fig. 1-1). These modified strains were also deficient in a single enzyme in the arginine biosynthetic pathway. Other early evidence that mutations alter proteins was provided by sickle cell anemia in humans. This anemia affects oxygen transport and arises from an alteration of hemoglobin, the protein in the blood that carries oxygen.

Genetic studies with yeast and mammals provided important evidence, but the work suffered from certain limitations. The organisms are complex and take a relatively long time to multiply. Also, their cells are large and intricate (see Fig. 1-2). Known as *eucaryotic* cells, they contain many small cell-like structures or organelles, such as nuclei, mitochondria

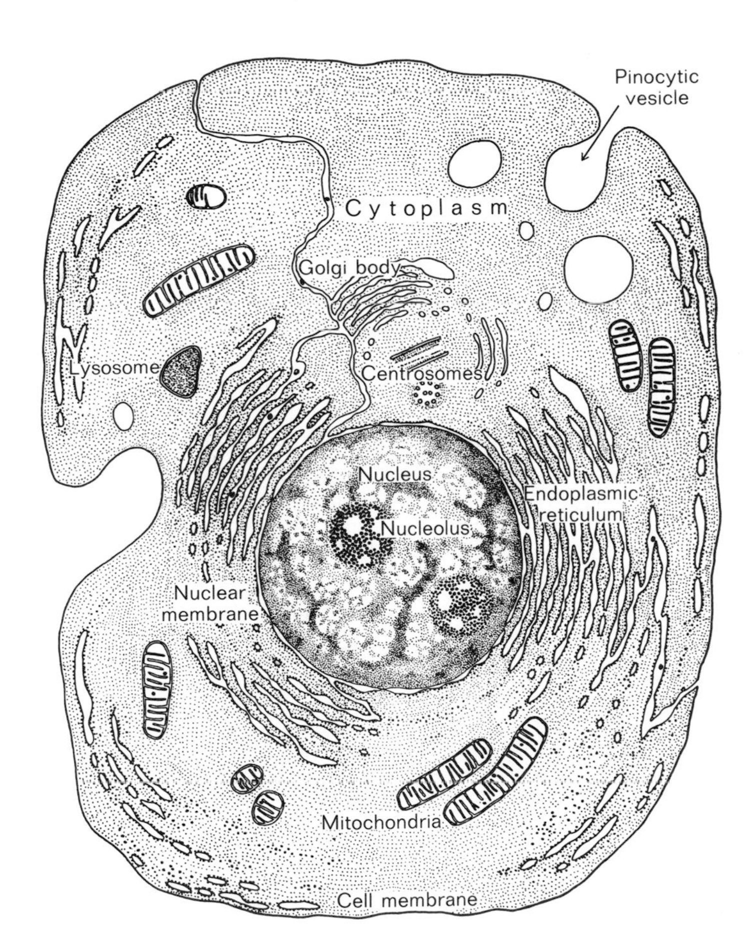

FIG. 1-2 *Eucaryotic Cell.* Chromosomes are located on the nucleus along with the nucleolus, site of ribosome synthesis. The mitochondria are organelles of oxidative reactions which provide energy for the cell. The network of membranes known as the endoplasmic reticulum is dotted with ribosomes, the sites of protein synthesis. The centrosomes, also known as centrioles, participate in separation of chromosomes during cell division. Lysosomes are vesicles containing hydrolytic enzymes. The Golgi body takes part in the formation of the endoplasmic reticulum and lysosomes. From J. Brachet, "The Living Cell," Copyright © *Scientific American*, Inc., September 1961. All rights reserved.

or chloroplasts, Golgi bodies and other storage systems. The term eucaryotic refers to the presence of a true nucleus with chromosomes separated from the rest of the cellular cytoplasm by a full nuclear membrane. More important, their genetic apparatus complicates analysis. The chromosomes of eucaryotic organisms are paired (with the exception of sex-determining chromosomes). Therefore most mutations are not revealed in the phenotype since they are not present on both sets of genes. In addition, the genes are frequently dispersed on many chromosomes (46 in humans) and the chromosomes are packed in a complicated structure that makes isolation difficult. For these reasons the important recent progress in biochemical genetics has come from the simple unicellular organisms, especially the intestinal bacterium *Escherichia coli*. These organisms are classified as *procaryotic* since they lack a nucleus and possess a simplified chromosomal structure.

Escherichia coli is a small, single cell about 1 micron across (abbreviated 1 μ), with one circular chromosome (see Fig. 1-3). Cultures of the cells grow quickly (doubling the population in about half an hour) and populations of cells in the billions can be grown easily, a great advantage for genetic experiments. Several thousand genes are present on the *E. coli* chromosome. Conceivably the protein or other product specified by each gene will someday be known. Already hundreds of genes have been identified (see the genetic map in Fig. 1-4), and scores of the proteins they specify have been isolated and characterized.

Structure and Colinearity of DNA and Protein

Work with *E. coli* has also produced many of the clues concerning how the gene "specifies" its corresponding protein.

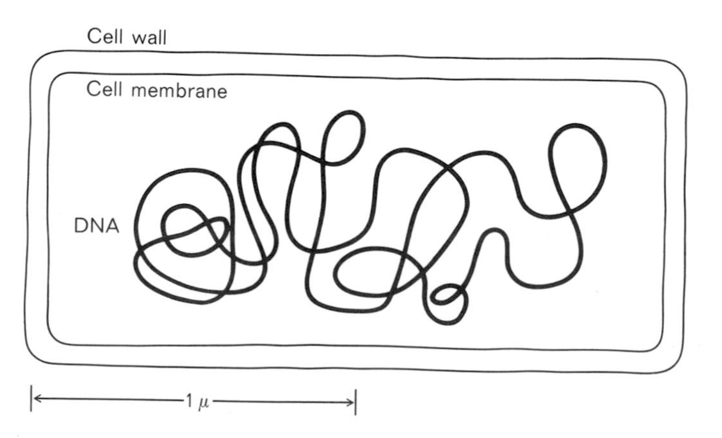

FIG. 1-3 *Procaryotic Cell: E. Coli.* A procaryotic cell is much smaller and simpler than a eucaryotic cell. The procaryotic cell is typically about the size of a mitochondrion of a eucaryotic cell and contains about 1000 times less DNA. See also Chapter 1 frontispiece.

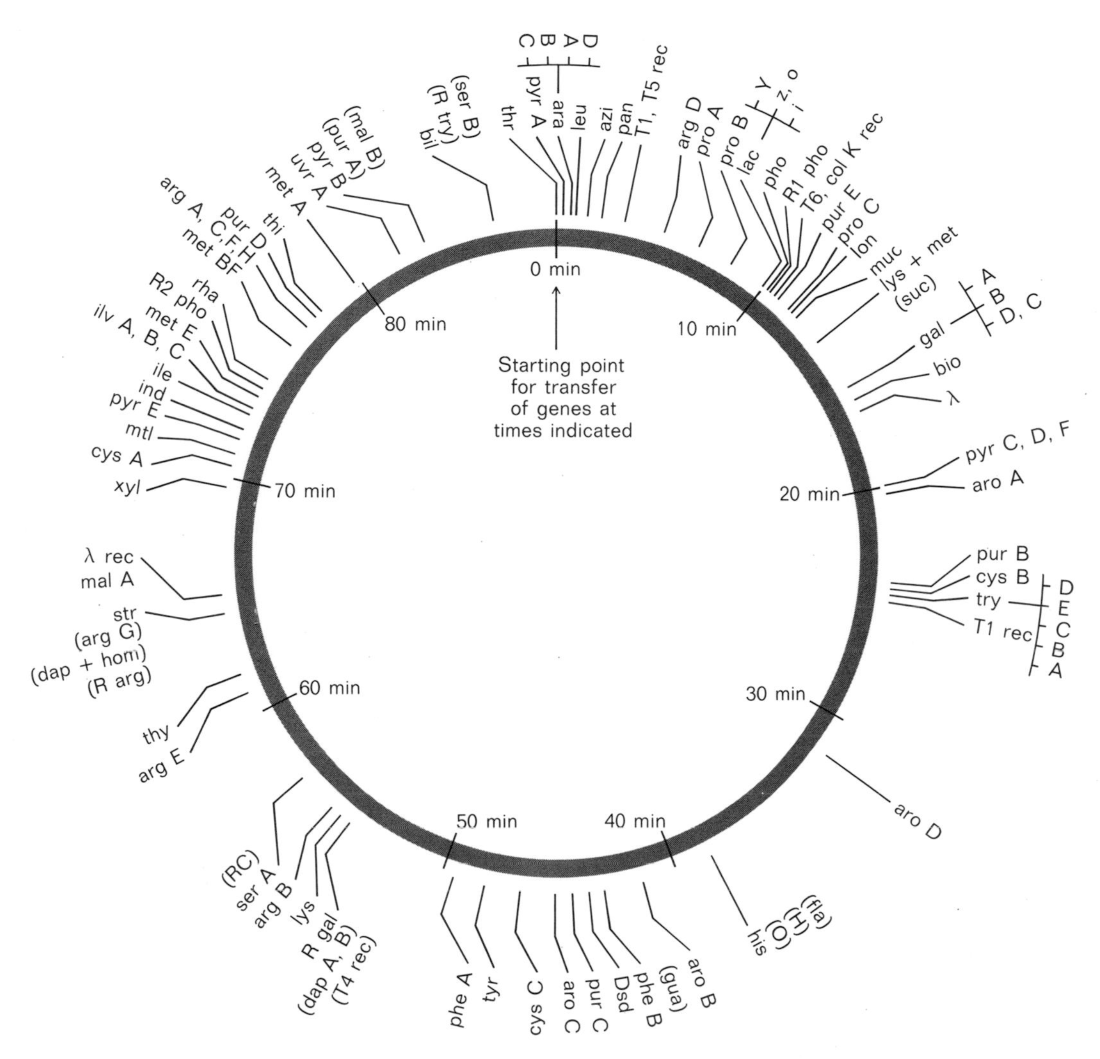

The genetic material, DNA, is a polynucleotide chain composed of four nucleotide bases:

Adenine (A)
Guanine (G)
Cytosine (C)
Thymine (T)

FIG. 1-4 *Genetic Map of E. Coli.* Certain strains of *E. coli* donate genetic material beginning at a fixed point on the chromosome and injecting it slowly into a recipient cell. In about 90 minutes a full complement can be transferred. By interrupting the process at various times and analyzing for transfer, the locations of the various genes can be determined. Adapted from A. L. Taylor and M. S. Thoman, *Genetics,* **50:** 667 (1964).

A hypothetical nucleic acid sequence is shown in Fig. 1-5. The essential features of the structure are a backbone of alternating sugar (deoxyribose) and phosphate units, joined in 3′,5′-phosphodiester linkage. Each sugar is attached to a base (A, T, G, or C) and the combination of sugar, phosphate,

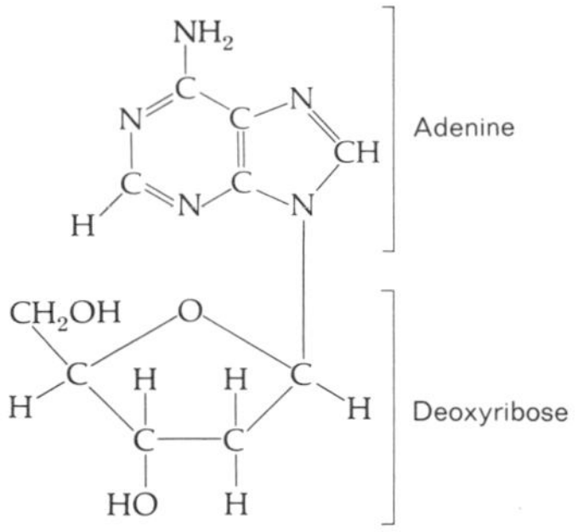

Adenine

Deoxyribose

Deoxyadenosine, a deoxynucleoside composed of a base (adenine) and a sugar (deoxyribose)

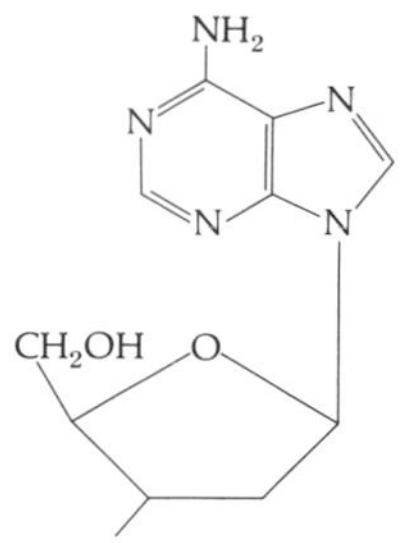

Deoxyadenosine, abbreviated form

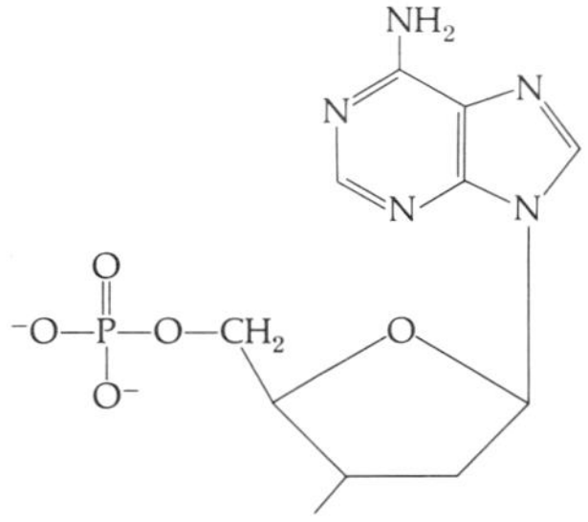

Deoxyadenosine monophosphate, a deoxynucleoside plus phosphate (also called a deoxynucleotide or deoxynucleoside monophosphate)

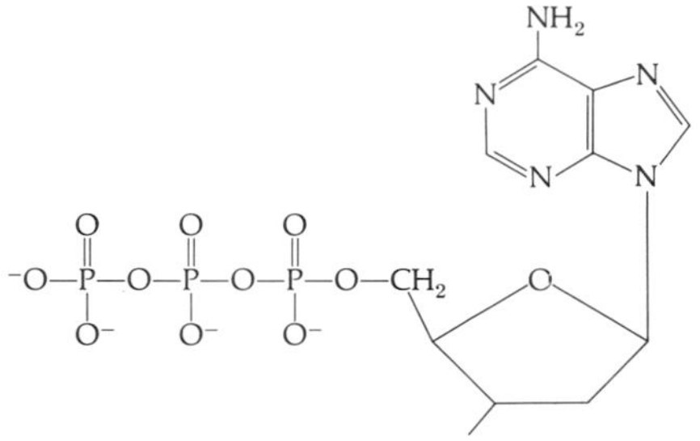

Deoxyadenosine triphosphate (deoxyadenosine diphosphate is also widespread in cells)

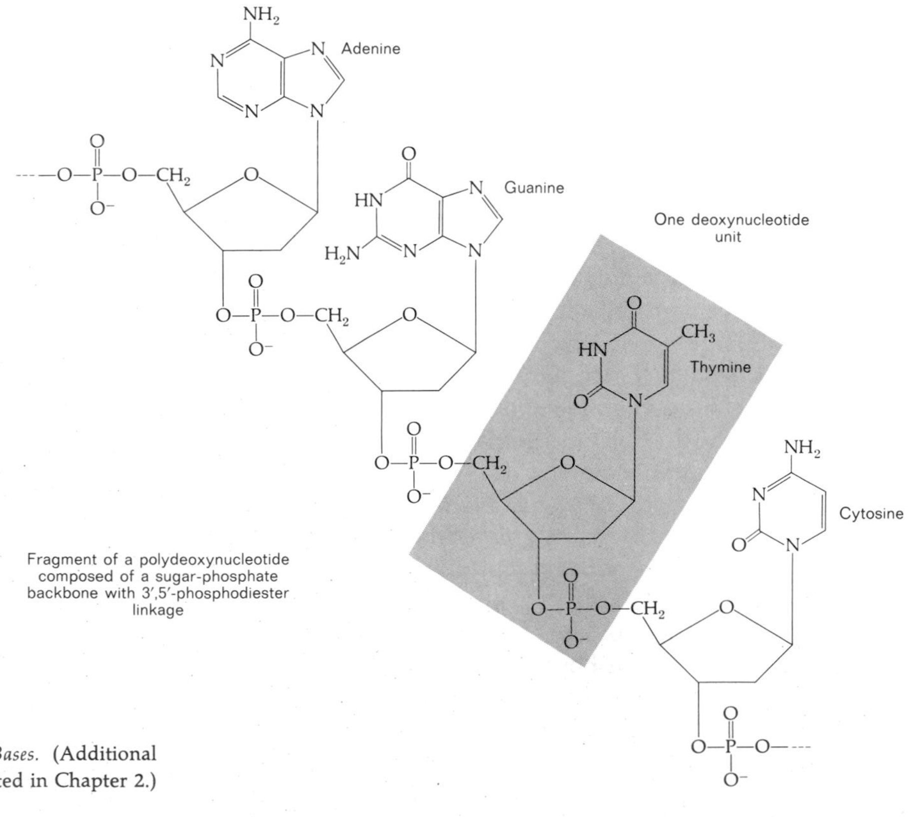

Adenine

Guanine

One deoxynucleotide unit

Thymine

Cytosine

Fragment of a polydeoxynucleotide composed of a sugar-phosphate backbone with 3′,5′-phosphodiester linkage

FIG. 1-5 *DNA and Its Bases.* (Additional structural details are presented in Chapter 2.)

and base is called a *deoxynucleotide*. Polydeoxynucleotides are typically millions of bases in length, as in *E. coli,* or even an order of magnitude longer, as in eucaryotic cells.

Proteins are constructed as a linear array of amino acids joined through peptide bonds. Amino acids possess the general structure

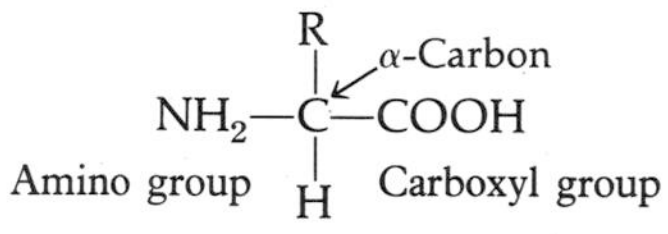

$$\begin{array}{c} \mathrm{R} \\ | \\ \mathrm{NH_2{-}C{-}COOH} \\ | \\ \mathrm{H} \end{array}$$

α-Carbon

Amino group

Carboxyl group

where R is one of 20 possible side chains. Peptide bonds are

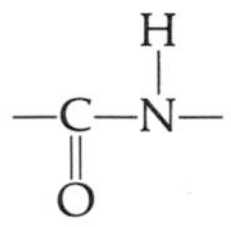

$$\begin{array}{c} \quad\ \mathrm{H} \\ \quad\ | \\ -\mathrm{C}-\mathrm{N}- \\ \| \quad\ \\ \mathrm{O} \quad\ \end{array}$$

units joining carboxyl and amino groups with the elimination of water. Thus proteins are composed of *polypeptide chains.* Only these 20 different amino acids are encountered in proteins, often in strings of several hundred. The 20 amino acids and their common abbreviations are:

Alanine (Ala)
Arginine (Arg)
Asparagine (Asn)
Aspartate (Asp)
Cysteine (Cys)
Glutamate (Glu)
Glutamine (Gln)
Glycine (Gly)
Histidine (His)
Isoleucine (Ile)
Leucine (Leu)
Lysine (Lys)
Methionine (Met)
Phenylalanine (Phe)
Proline (Pro)
Serine (Ser)
Threonine (Thr)
Tryptophan (Trp)
Tyrosine (Tyr)
Valine (Val)

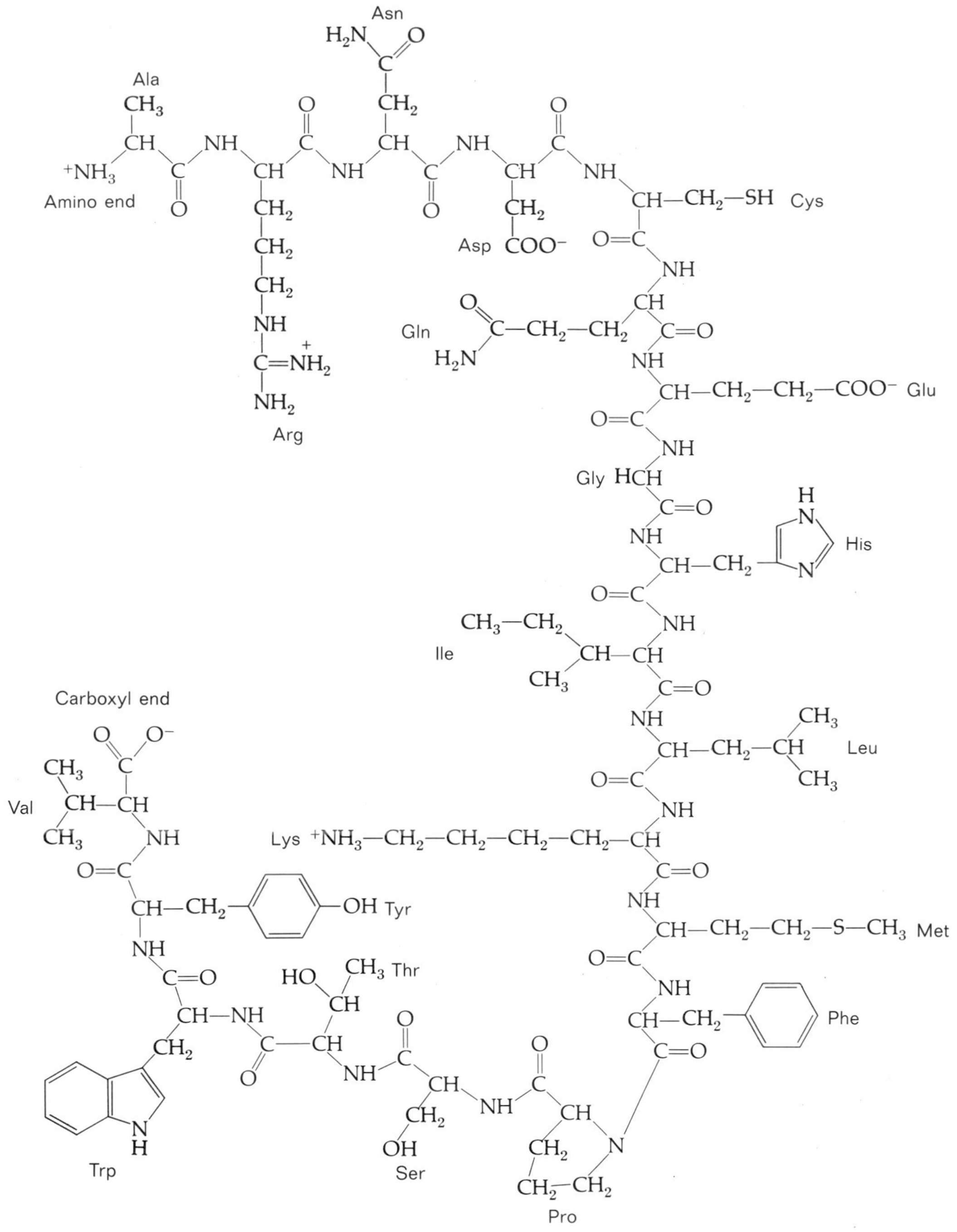

FIG. 1-6 *The Amino Acids.* The 20 amino acids regularly occurring in proteins are presented in a polypeptide composed of the amino acids in alphabetic order. The side chains are shown in color with the charges present at pH 7; the polypeptide backbone is black.

The abbreviations are generally reserved for amino acid residues in peptide linkage. A peptide of the amino acids joined in alphabetic order (Fig. 1-6) presents their structures. A wide variety of chemical entities are encountered in the side chains of the amino acids.

By an intricate and fascinating process, which we will study in detail, the order of the nucleotide bases of DNA determines the order of the amino acids in the polypeptide chain of a protein. One linear message determines another linear message. Thus the messages are *colinear*.

A clear example of this DNA-protein colinearity comes from elegant experiments conducted by Charles Yanofsky and his colleagues with the A polypeptide chain of the *E. coli* enzyme tryptophan synthetase and its gene. This enzyme catalyzes the last step in the biosynthesis of the amino acid tryptophan. The active form of the enzyme contains two A polypeptide chains as well as two B polypeptide chains. The A protein has been studied in greater detail. It contains 267 amino acids arranged in a specific order. The complete sequence of this protein is presented in Fig. 1-7. Many mutant strains of the A gene, or, more precisely, A cistron, were isolated. One *cistron* is the unit of the chromosome that codes for one polypeptide chain. Thus the original formulation "one gene-one enzyme" is more accurately phrased as "one cistron-one polypeptide chain." The mutations can arise naturally (at each nucleotide position an incorrect base is inserted during DNA replication about once in a billion times) or they can be induced at higher frequencies by chemical mutagens or ultraviolet light. In order to locate the position of each A-cistron mutant on the genetic map, the various mutants were crossed against one another in pairs. Due to a recombination process of the genetic material, known as *crossing-over*, a small fraction of the population of each cross will obtain a normal chromosome as long as the two mutations are not overlapping (see Fig. 1-8). The frequency of such normal recombinants is related to the distance between mutations: the farther apart two mutations, the "more room" for recombination and the higher the frequency of normal recombinants. By crossing mutants in various combinations, the order and position of each mutant on the genetic map is achieved.

In order to relate the positions of each mutation with a change in the A polypeptide chain or A protein, the corresponding A polypeptide chain produced by each mutant strain is also mapped or *sequenced* to determine which amino

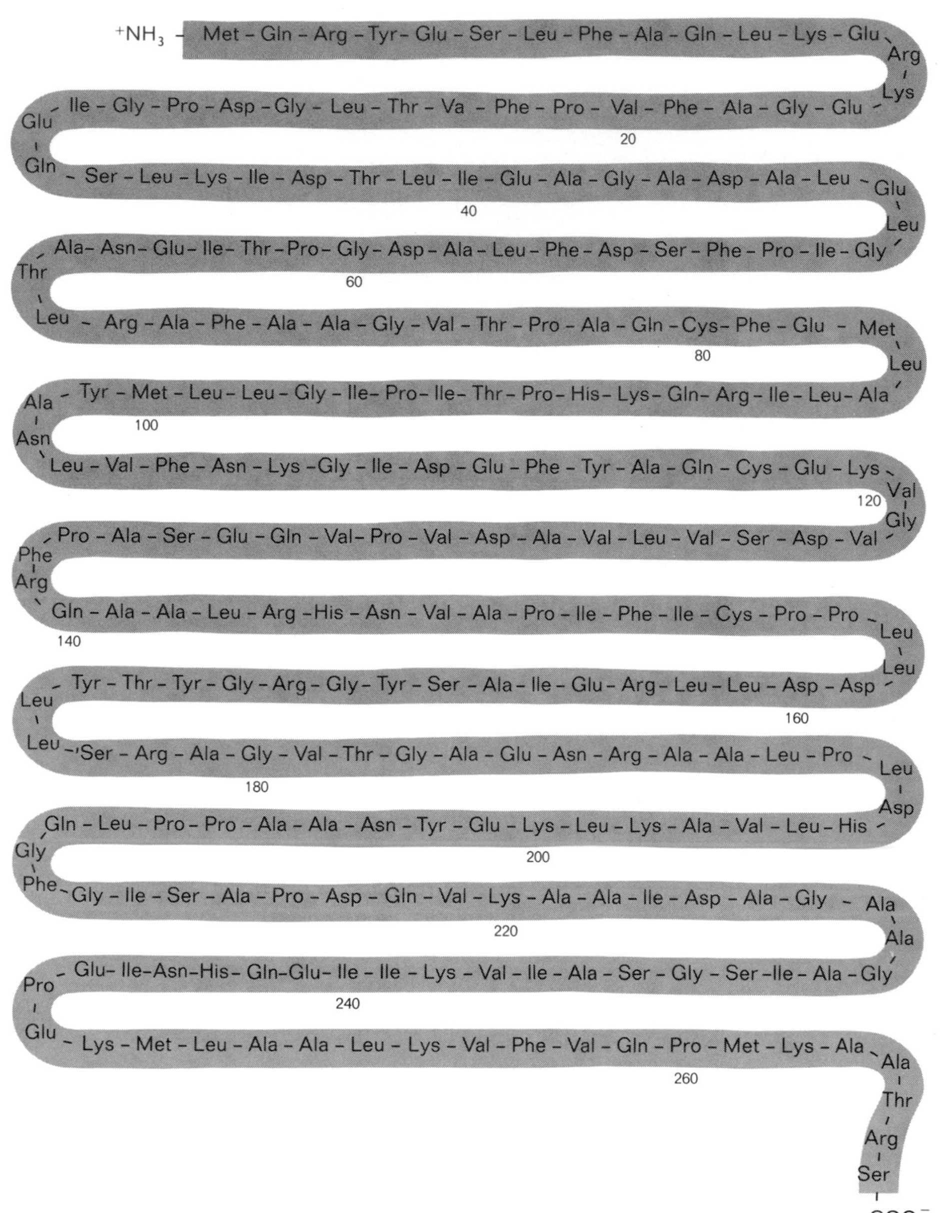

FIG. 1-7 *The Amino Acid Sequence of the A Protein.* From C. Yanofsky, G. R. Drapeau, J. R. Guest, and B. C. Carlton, *Proceedings of the National Academy of Sciences,* **57:** 296 (1967).

acid in the structure is altered. Each of the mutants discussed here is a *point mutation* and leads to a change in a single amino acid in the sequence. (Other types of mutations can alter many amino acids.) The exact methods employed for sequencing the normal and mutant polypeptide chains will be described in the section on details of protein structure in

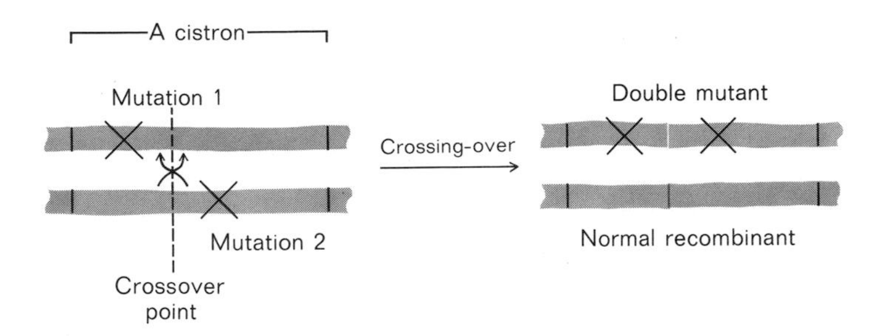

FIG. 1-8 *Crossing-over.* The greater the distance between two mutations on the cistron, the greater the number of potential crossover sites and the higher the frequency of appearance of recombinants with normal phenotypes. (See also Appendix I.)

Chapter 8. In the case of the tryptophan synthetase A protein, the experimenters were especially fortunate in that the mutant proteins, while functionally altered, still carried out a portion of the normal reaction. This partial activity could be used to identify and isolate the mutant proteins. Thus chemical analysis of the amino acid changes was achieved and the positions of the changes at the protein level were compared to the sequence of mutations on the chromosome. As seen in Fig. 1-9, a parallel between gene and protein is observed. Mutations at the left end of the A cistron give rise to changes at the left end of the A protein, that is, the end near the free α-amino group. Mutations near the center of the cistron produce changes near the center of the protein. The changes in the protein near the right end correspond to mutations at the right end of the cistron. More precisely, the fractional distance of a cistron or the gene traversed from left to right to reach a mutation is the same as the fractional distance of the corresponding polypeptide chain traversed from left to right (or amino end to carboxyl end) to reach the changed amino acid. In sum, the gene and the protein are *colinear.*

The colinearity of the genetic sequence with protein structure squarely frames the problems that must be solved to understand gene action and protein synthesis. In the introductory comments we noted the three major properties centered around DNA: information storage, replication, and coding. On the basis of the studies with tryptophan synthetase, the exact requirements in these three areas are pinpointed more clearly and will be discussed in the next five chapters:

> *DNA Structure.* How is the genetic information in the form of four nucleotides encoded in the structure of DNA?
>
> *DNA Replication.* What in this mechanism ensures that the gene for tryptophan synthetase A protein is duplicated with high fidelity and passed on from generation to generation?

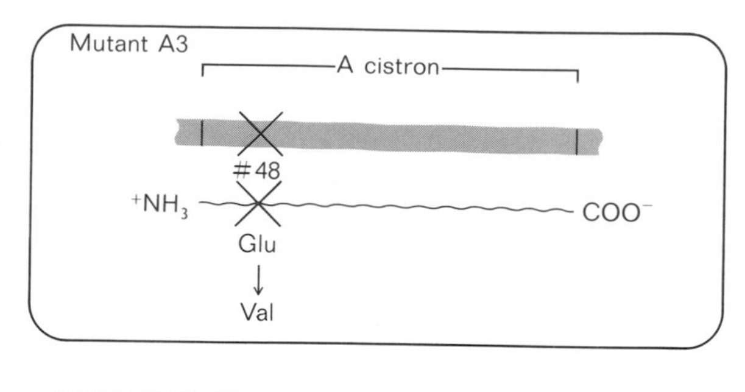

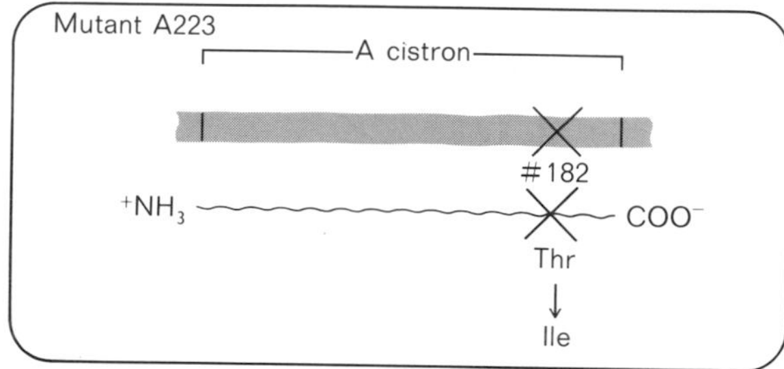

Summary of several mutants:

A cistron

A3 A487 A223 A46 A169

#48 #176 #182 #210 #234

$^{+}NH_3$ COO^-

Glu → Val; Leu → Arg; Thr → Ile; Gly → Glu; Ser → Leu

FIG. 1-9 *Colinearity.* The locations of individual mutations on the genetic map and in the amino acid sequence form two parallel linear arrays. Hence they are colinear. From C. Yanofsky, G. R. Drapeau, J. R. Guest, and B. C. Carlton, *Proceedings of the National Academy of Sciences,* **57:** 296B (1967).

Machinery of Protein Synthesis. What machinery permits the information stored in the cistrons of DNA to appear as a sequence of amino acids linked in a polypeptide chain?

The Genetic Code. What is the precise chemical configuration or array on DNA that specifies a particular amino acid for a certain location along a polypeptide chain? In other words, what are the chemical details of the genetic dictionary?

Coordination of Protein Synthesis. What are the exact steps that permit a protein such as tryptophan synthetase with 267 amino acids to be completely synthesized in a time estimated as short as 10 to 20 seconds?

PROBLEMS

1 Are the following examples of genotype or phenotype?
 (a) A change in one base of DNA
 (b) A change in one amino acid of a protein

2 Are the following features found in procaryotic or eucaryotic cells?
 (a) Mitochondria
 (b) A single circular chromosome
 (c) A nuclear membrane
 (d) Diploid chromosome

3 (a) What are the three kinds of chemical components in DNA?
 (b) One of the chemical components of DNA comes in four varieties. What are their names?

4 How many different amino acids are encountered in proteins?

5 The A protein of tryptophan synthetase has 267 amino acids. If we assume that each is specified by three nucleotides in the A cistron and an error rate of 10^{-9} prevails for replication of each nucleotide, what is the probability that a mutation in the A cistron will arise in any single replication of the chromosome?

6 Define colinearity.

Chapter 2 Frontispiece *Electron micrograph of E. coli bacteriophage T2, osmotically shocked. Magnification 84,000×. Courtesy A. K. Kleinschmidt. Reproduced from Biochimica Biophysica Acta 61: 861 (1962).*

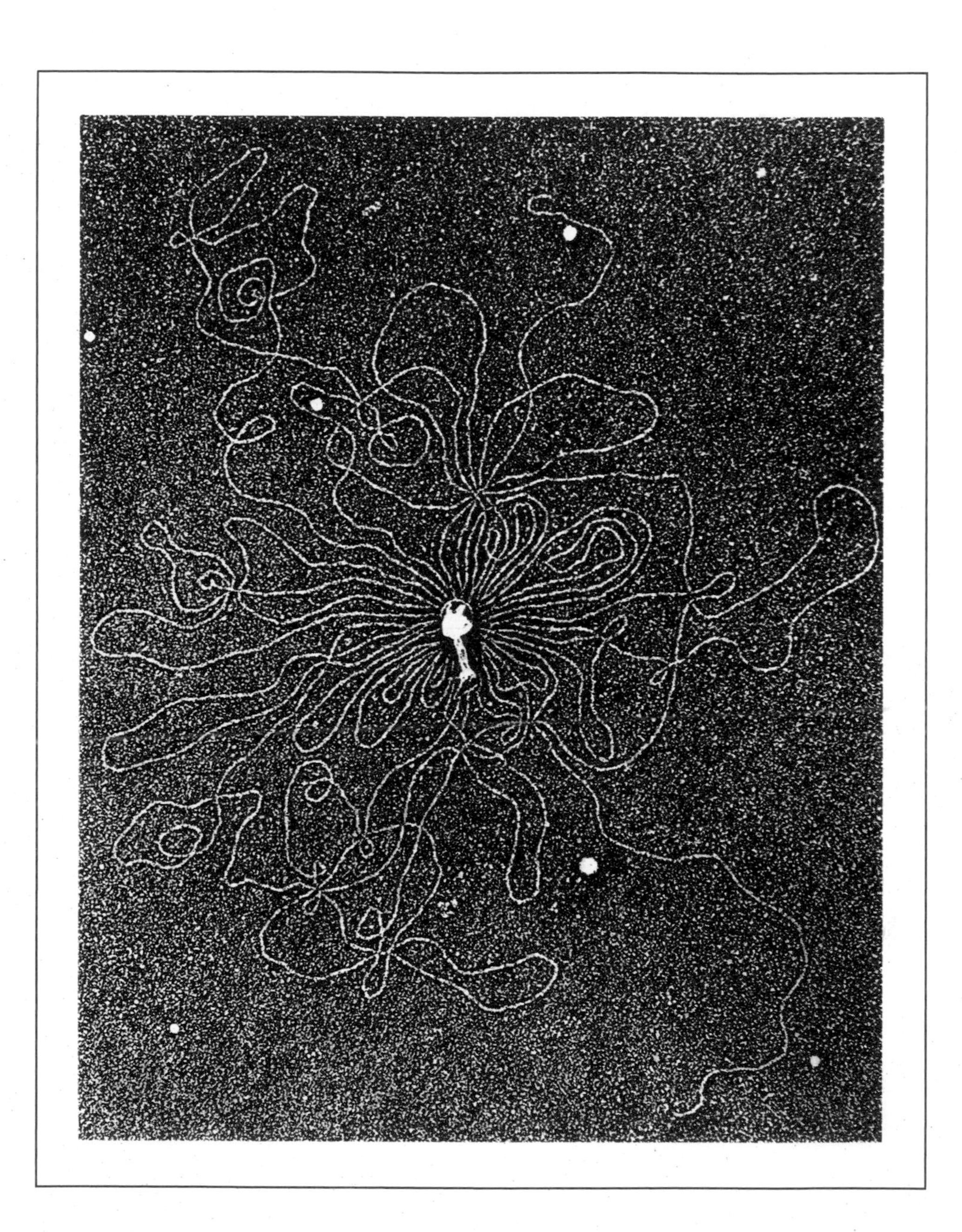

DNA STRUCTURE

2 At an early stage of biochemistry, little was known or even imagined about the details of gene action. The structure of the genetic material, DNA, was only vaguely resolved. Yet from the clues of what was known of the chemistry of DNA, pieced together with some aspects of chromosome function, a pivotal model for DNA emerged. The model encompassed both the three-dimensional structure of DNA and its replication. The model was described by James Watson and Francis Crick in 1953, and it set biochemistry on a new course. We will consider the contents and consequences of the DNA model, first recalling the proof for DNA as the genetic material, and then reviewing the chemical and physical background as a setting for the model itself.

DNA as Genetic Material

Early evidence for DNA as the hereditary material came from an analysis of the amount of DNA in plants and animals. In every diploid cell examined for a particular species, a nearly identical amount of DNA was observed. Furthermore, the corresponding haploid germ cells, such as sperm, contained half the amount of DNA located in the somatic cells (see Table 2-1). These distributions are just what would be expected for the genetic material.

Somewhat later work with microorganisms and viruses provided more compelling evidence for the role of DNA in the genetic machinery. Experiments with pneumococcal bacteria demonstrated that DNA alone could cause an inheritable *transformation* of cell shape. DNA isolated from "smooth" cells could, when incubated with normally "rough" cells, induce a transformation to the smooth character. The new smooth cells would then continue to grow

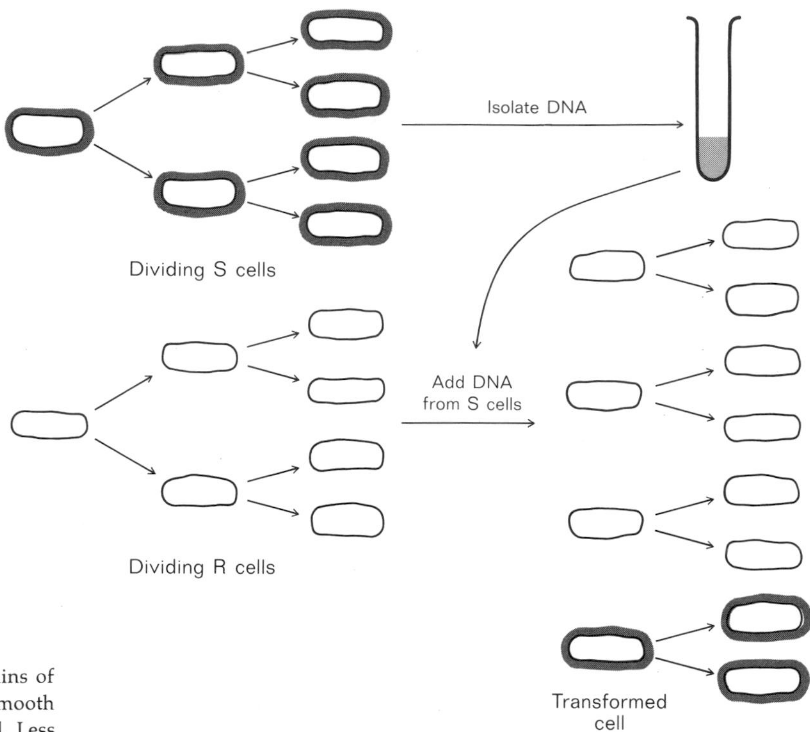

FIG. 2-1 *Transformation*. Virulent strains of *Pneumococcus*, designated S, possess a smooth polysaccharide capsule around the cell. Less toxic strains, designated R, have lost capsule-producing ability due to a mutation inactivating an enzyme needed to produce the polysaccharide. A small fraction of a culture of R cells can be transformed to S cells by addition of S-cell DNA. Transformed S cells grow and divide like normal S cells.

and divide, perpetuating the smooth character (Fig. 2-1). Since the transformation process showed all the qualities of a genetic alteration, DNA was identified as the genetic material. Further evidence in support of a genetic role for DNA came from experiments on virus reproduction. With radioactive elements to label DNA (^{32}P) and protein (^{35}S), experiments were conducted to demonstrate that only the DNA of the virus enters the host bacterium. The viral protein remained at the surface of the bacterium. It could be removed by vigorous agitation in a blender and separated from the bacterium. The DNA entered the cell and virus reproduction proceeded.

TABLE 2-1 DNA Content of Chicken Cells
(1 picogram = 10^{-12} gram)

Tissue	DNA (picograms per cell)
Heart	2.45
Kidney	2.20
Liver	2.66
Pancreas	2.61
Sperm Cells (haploid)	1.26

Apparently then only the DNA is needed to specify the next generation of viruses. These studies left little doubt that DNA is the genetic material.

Chemistry of DNA

Although the function of DNA in heredity was strongly suggested, nothing could be decided about its actual mode of action without knowledge of the chemistry of DNA. The detailed structure of DNA is complex, but fortunately it is composed of simple building blocks: the sugar deoxyribose (Fig. 2-2), phosphate, and four nitrogenous bases. Of the four bases, two are purines [adenine (A) and guanine (G)] and two are pyrimidines [cytosine (C) and thymine (T)] (Fig. 2-3). The bases are attached to the sugar through *β-glycosidic linkage.*

A major key to the structural model of DNA concerns the four nitrogenous bases A, C, G, and T. Careful analysis of the occurrence of these bases by Erwin Chargaff in the late 1940s revealed a most intriguing pattern. Although the relative amounts of the bases differ widely among different species, the ratios A:T and G:C are always unity. Accommodating these ratios was one of the major challenges for any structural model of DNA.

The final ingredients for the model of DNA came from x-ray analysis of DNA fibers. Crystalline forms of matter diffract x-rays in patterns that are characteristic of the smallest repeating units in the crystals. Long helical structures, although only semicrystalline, behave similarly, and in this case the x-ray patterns reveal the distances of periodically repeating units along the helix axis. Recorded patterns for DNA suggested a helical structure with a major repeating unit at 3.4 Å and a secondary unit at 34 Å. These distances provided the major geometric mold that any model of DNA would be required to fit.

The Double Helix

The identification of DNA as the genetic material had only modest impact on biochemistry since there was no physical model to relate gene and molecule. The contribution of Watson and Crick was to provide just such a physical model of DNA, one that not only accounted for the A:T and G:C ratios of unity and the spacing observed by x-ray measurements, but at the same time provided a simple and intelligible view of the gene and its replication.

The model is a double helix. Two long continuous polymers wind gradually, one around the other (see Fig. 2-4). The two polydeoxynucleotide strands run in opposite directions. Hence the structure is termed *antiparallel*. One strand

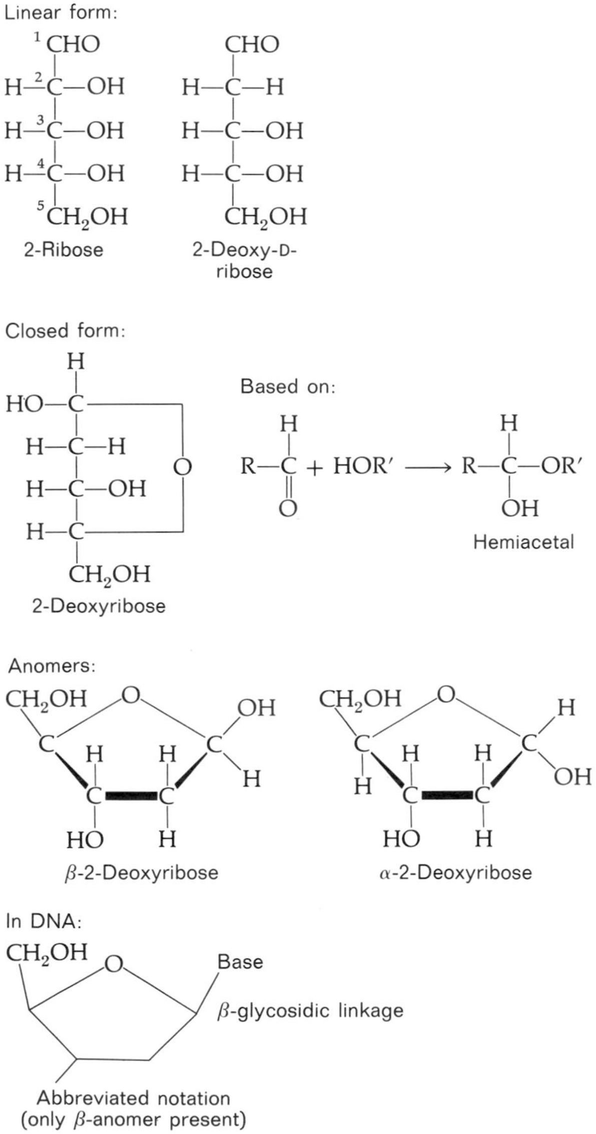

Linear form:

^{1}CHO, H—^{2}C—OH, H—^{3}C—OH, H—^{4}C—OH, $^{5}CH_2OH$

2-Ribose

CHO, H—C—H, H—C—OH, H—C—OH, CH_2OH

2-Deoxy-D-ribose

Closed form:

H, HO—C—, H—C—H, H—C—OH, H—C—, CH_2OH, O

2-Deoxyribose

Based on:

$$\text{R—C(H)=O} + \text{HOR}' \longrightarrow \text{R—C(H)(OH)—OR}'$$

Hemiacetal

Anomers:

CH_2OH, O, C, H, H, C, OH, H, C, C, HO, H

β-2-Deoxyribose

CH_2OH, O, C, H, H, H, C, H, OH, C, C, HO, H

α-2-Deoxyribose

In DNA:

CH_2OH, O, Base

β-glycosidic linkage

Abbreviated notation (only β-anomer present)

FIG. 2-2 *Deoxyribose: Linear and Closed Forms.* For the linear form, D refers to the configuration at the penultimate carbon (OH to the right on carbon 4). The closed form is based on the hemiacetal formation. Closing the linear structure makes carbon 1 a center of asymmetry, and α- and β-anomeric structures can be distinguished. In DNA, only the β-anomer is present. (See Appendix III for additional background on sugar chemistry.)

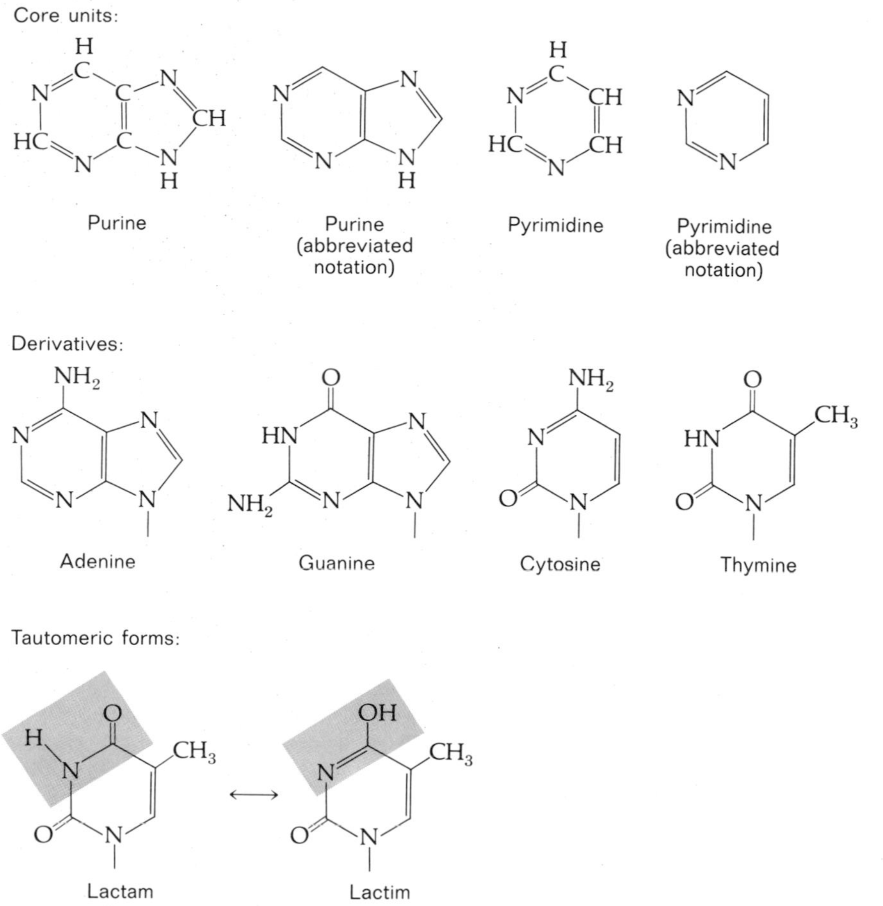

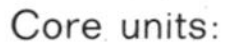

FIG. 2-3 *Structure of the Bases.* The four bases of DNA are either purines or pyrimidines. The bases can exist in two tautomeric forms, lactam and lactim. The lactam forms predominate at neutral pH. (See also Appendix II.)

appears in the 5′ ⟶ 3′ sense (phosphodiesters link the 5′-carbon of one sugar to the 3′-carbon of the next sugar); the other strand is inverted compared to the first since its phosphodiester linkage progresses in the 3′ ⟶ 5′ sense. Of the two possible helical types, left- and right-handed (see Fig. 2-5), the DNA double helix is the right-handed variety. The polynucleotide chains are arranged with the sugar-phosphate backbone on the outside and the bases aligned facing the center. The planar bases are all perpendicular to the long axis of the double helix and the bases from one strand are in contact with the bases from the other strand. Thus the overall construction of the double helix resembles a ladder coiled around an imaginary center pole with each rung consisting

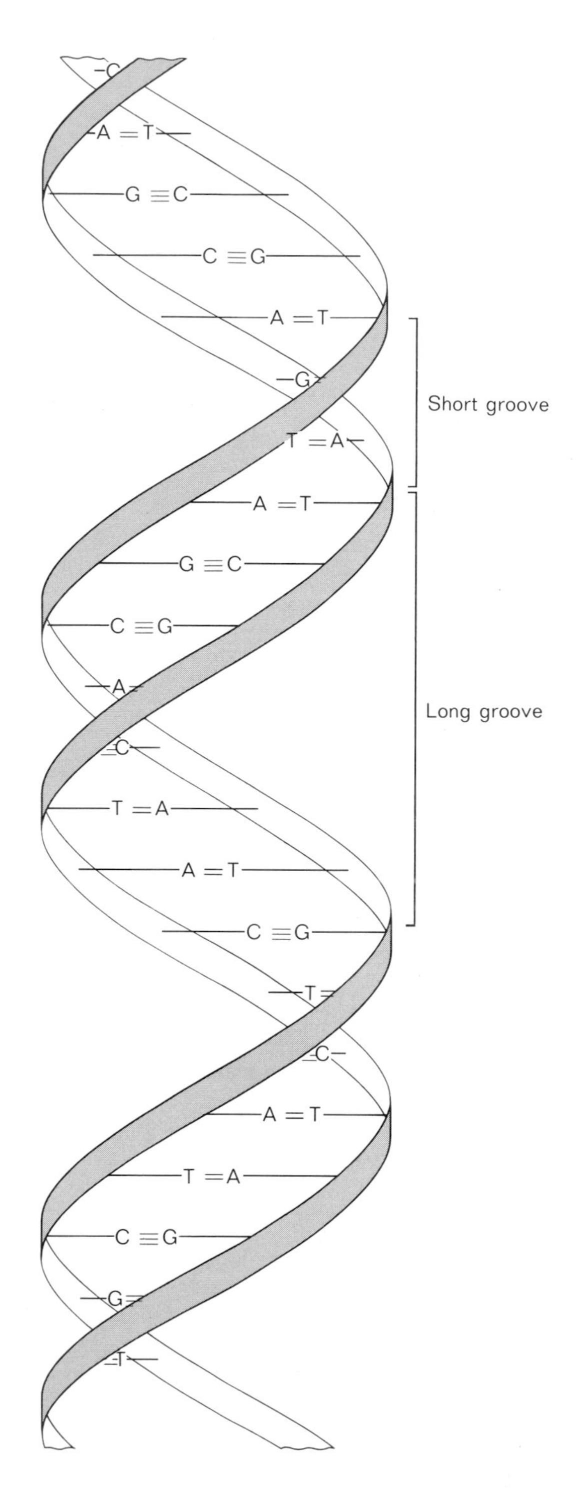

FIG. 2-4 *DNA Double Helix*. The two strands of DNA run in opposite directions (antiparallel) and the structure displays long and short grooves. Based on J. D. Watson and F. H. C. Crick, *Nature*, **171:** 964 (1953).

of a pair of bases 3.4 Å from the next. This 3.4 Å spacing accounts for the principal repeating distance deduced from x-ray diffraction measurements. The secondary repeating unit at 34 Å arises from the completion of one full turn of the helix—every 10 bases.

Since the double helix is a coiled structure, the bases do not pass through the center of the molecule, and the strands appear to be staggered. The helix displays a narrow groove where the strands are in close proximity for pairing of the bases and a wide groove in the gap between paired strands (see Fig. 2-4). In biological terms, the most important feature of the double helix is the interaction of the bases, uniting the two strands and ultimately holding the genetic information. The geometry of the structure permits only purine-pyrimidine pairing. And the chemistry of the bases allows only the pairing suggested by the ratios found by Chargaff. The purine A from one strand pairs with the pyrimidine T from the other strand. The purine G pairs only with the pyrimidine C. The other combinations do not occur.

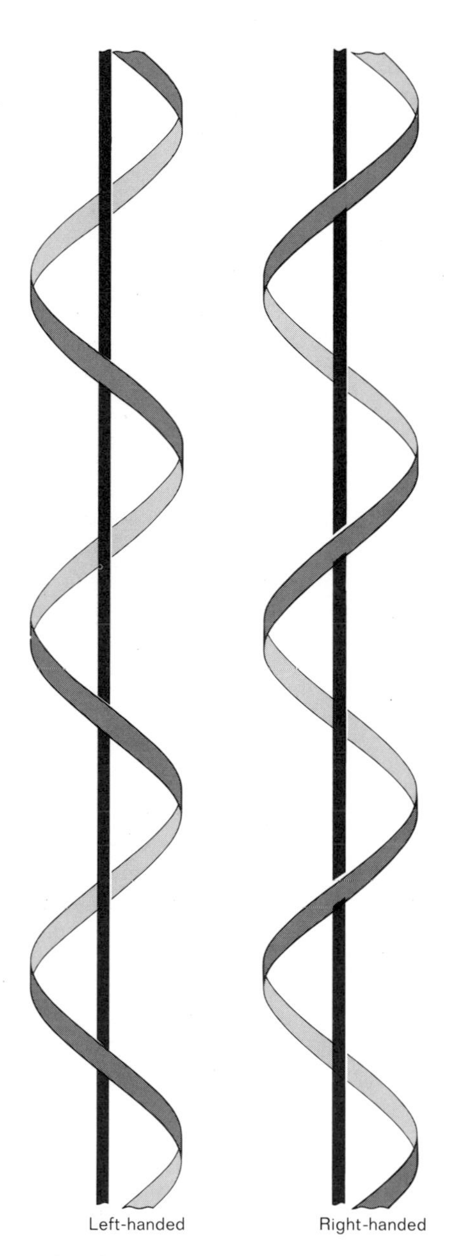

FIG. 2-5 *Left- and Right-handed Helices.*

Hydrogen Bonds

Base pairing is thus the key to the double helix, and the key to base pairing is *hydrogen bonding.* As a molecular jigsaw puzzle, hydrogen bonds occur in just the right places to permit pairing between A and T, and G and C, but none of the other purine-pyrimidine combinations "fit" (see Fig. 2-6). The base in one strand appears to be inverted or "flipped" compared to its complementary partner in the other strand due to the antiparallel structure of the helix. In addition to one of the bases of each pair being inverted, both bases are tilted toward the center of the helix by approximately 40° to achieve the proper alignment.

The A-T pair provides two hydrogen bonds. Three hydrogen bonds occur in the G-C pairing. Thus the genetic system of life on this planet is based on five weak bonds. The hydrogen bond, along with van der Waals interactions and coulombic forces, is one of the weakest chemical forces known to exist between atoms. Just how weak the hydrogen bond is cannot be stated precisely, but the strength of the hydrogen bonds in DNA probably lies somewhere in the vicinity of 1 kcal/mole. In contrast, the carbon-carbon bond has a stability of over 80 kcal/mole. Yet, in spite of being held together by weak forces, DNA has an enormous stability.

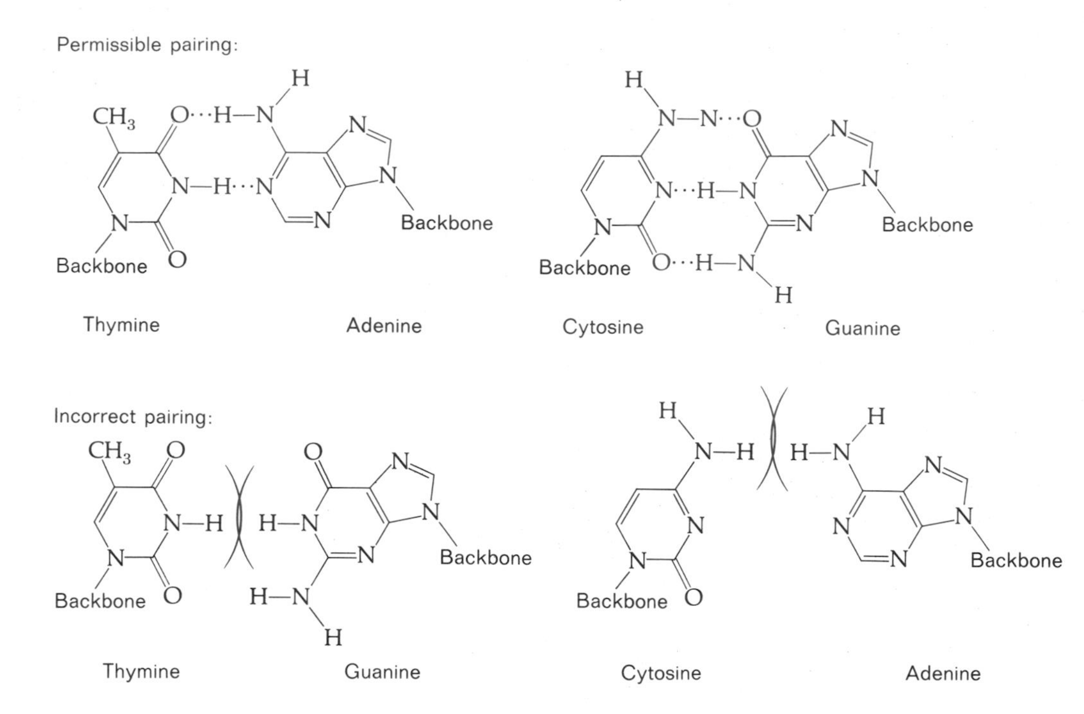

FIG. 2-6 *Base Pairing.* The permissible purine-pyrimidine pairs are presented above and the incorrect pairs are below. In each example the base on the left is "inverted" in accord with the antiparallel structure of the double helix.

The reason for the stability is that the hydrogen bonds cooperate, like the many strings of the Lilliputians. For example, consider the pairing of two nucleotides A and T free in solution. Let the complex A-T be favored over the free forms A and T by a factor of 10. A factor of 10 in a reaction of this type would be considered very weak by most chemical standards. Now join two A's and two T's together covalently through diester linkage to form the dinucleotides AA and TT. All other things being equal, the dinucleotides will pair to form A_2-T_2, but the new complex is now favored over the free forms by a factor of 10×10, or 100. If 10 units are present in each strand of A and T, the bonded form is favored by 10,000,000,000 to 1. Thus the hydrogen bonds are coupled when present on covalently linked nucleotides and their effects are multiplicative or cooperative. Clearly then, if many, many hydrogen bonds are present—as in the pairing of DNA strands—their association will be strong, approaching that of covalent chemical interactions.

Van der Waals Interactions

The hydrogen bonds are not completely alone in stabilizing the double helix. As seen in the depictions of Fig. 2-4, the stacking of the flat base pairs along the helix axis results in some contact between adjacent bases on the same strand. This stacking is shown more clearly in Fig. 2-7. Since each base is one-tenth of a revolution farther around the helix axis than its neighbor, the contact is incomplete. However, it is sufficient for some contact of electron clouds, and van der Waals interactions do occur. Van der Waals interactions are caused by attractive forces between transient dipoles of uncharged atoms. The forces operate at relatively long range, approaching maximum stability as two atoms approach contact. If the distance between atoms is closer than the optimal spacing, strong repulsive forces occur. A typical van der Waals profile is illustrated in Fig. 2-8 and is generally described as a 6–12 potential. The stabilizing force depends on the interatomic distance r, to the sixth power; the repulsive force depends on this distance to the twelfth power. Therefore the apparent free-energy change $\Delta G°$ can be expressed by the equation

$$\Delta G° = \Delta G°_{\text{min}} \left[\left(\frac{r_{\text{min}}}{r} \right)^{12} - 2 \left(\frac{r_{\text{min}}}{r} \right)^{6} \right]$$

The terms $\Delta G°_{\text{min}}$ and r_{min} refer to the free-energy change and interatomic distance corresponding to maximum stability, respectively. In general van der Waals interactions are on the order of only several tenths of a kcal per mole. Thus they are generally very weak, quite likely even weaker than hydrogen bonds. Nevertheless, their importance can be illustrated qualitatively. Certain homopolymers of nucleotides, such as poly A and poly C (polymers of the type shown in Fig. 1-5, with only one type of base, and ribose in place of deoxy-

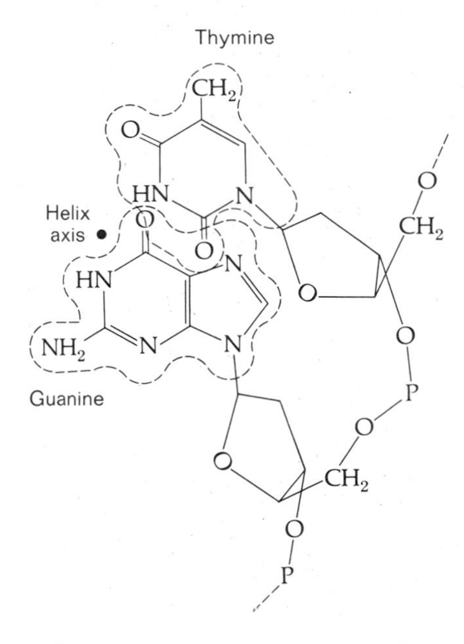

FIG. 2-7 *Overlap of Stacked Bases.* The bases are shown as seen when looking down the helix axis with T above G on the same strand. From R. Langridge, H. R. Wilson, C. W. Hopper, M. H. F. Wilkins, and L. D. Hamilton, *Journal of Molecular Biology,* **2:** 19 (1960).

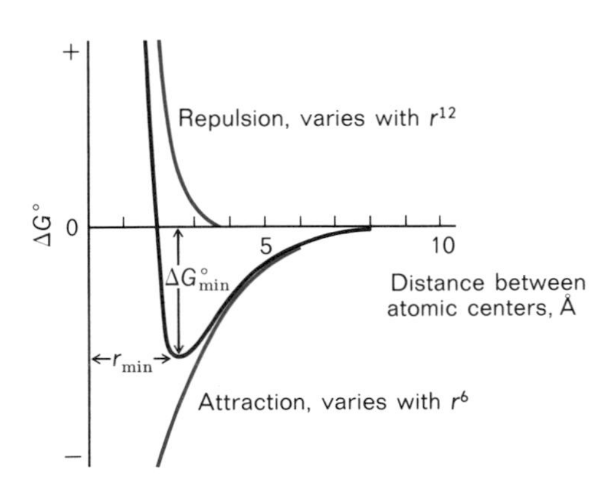

FIG. 2-8 *6–12 Potential.*

ribose) can be synthesized enzymatically. Under certain conditions poly A and poly C exist in a helixlike structure, but as a single-stranded form, presumably resembling one strand of the double helix left without its complementary strand. In this case no hydrogen bonds can possibly form. Therefore the stability of these helices—considerably less than duplex DNA, but appreciable—must be derived entirely from van der Waals stacking forces.

Some small additional stability for the DNA structure may also be attributed to "liberating" water. A few molecules of water would be associated with each base if it were not in contact with another base inside the double helix. This freeing of the water molecules enhances the tendency of the bases to associate together. Factors of this type, largely based on entropy changes and called *hydrophobic interactions,* will be considered at length in the discussion of proteins, an area where they have been more thoroughly explored. Hydrophobic interactions would contribute to the stacking tendency of the bases, but their exact contribution is difficult to distinguish from van der Waals interactions.

In conclusion, some contribution to the DNA double helix from van der Waals and hydrophobic interactions must be acknowledged. However, only hydrogen bonds are likely to be responsible for the specificity of the complementary base pairing. Hydrogen bonds—three in the G-C pair and two in the A-T pair—thus provide the foundation for the structural complementarity that specifies the base sequences and encodes the genetic wealth of all species. As we will see, the same complementary base-pairing relationships send the information through the machinery of protein synthesis to define amino acid sequences. Finally, these amino acid sequences specify the active proteins which carry out the manifold actions and reactions of the biochemical world. It is awesome to ponder the frailty of this mechanism based only on five hydrogen bonds and yet so evidently successful.

Size of DNA

Most molecules of biochemical interest are too small to be seen, even with the most powerful microscope. Therefore establishing their size requires the application of physical-chemical methods. DNA, however, is so exceptionally large and so abnormally behaved because of its long rodlike shape that it virtually defied physical-chemical analysis. Fortunately DNA is large enough to be seen at least in broad outline with the electron microscope. Autoradiographic methods can also be applied. An autoradiograph of DNA in the process of replication is shown in Fig. 2-9. From pictures of this type the contour length of the DNA can be measured. Since a length of 3.4 Å corresponds to one base pair, the contour length can be converted into numbers of base pairs or total molecular weight. In the case of *E. coli*, the single chromosome has a contour length of 1.2 mm. This length corresponds to about 4 million base pairs or a molecular weight of over 2000

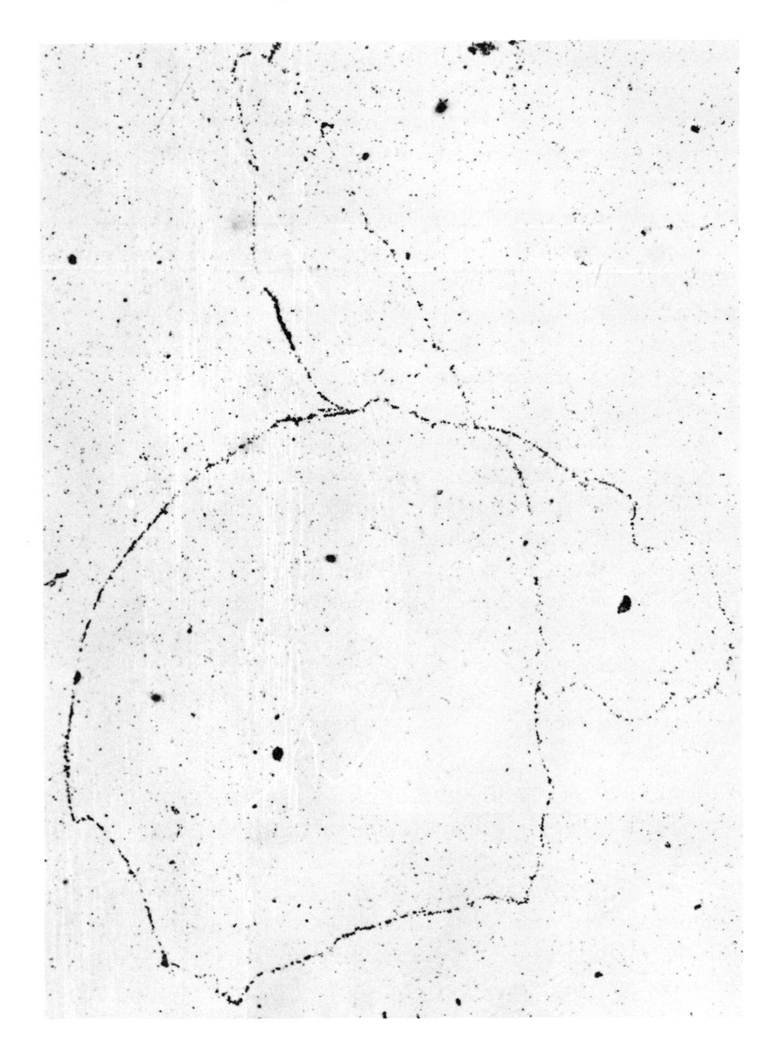

FIG. 2-9 *Autoradiograph of E. Coli DNA*. The DNA molecule is undergoing replication. Courtesy of J. Cairns. Prior publication in *Scientific American*, January 1966.

million (2×10^9). A chromosome of this size accommodates many thousands of genes. Smaller chromosomes from viruses have also been measured. The chromosomes of diploid organisms, however, are much more difficult to study. In addition to the DNA, large amounts of associated proteins known as histones, as well as other proteins, are present, and the general structure of the chromosome is not stretched out but folded and convoluted in some as yet unsolved way. It is clear, however, that the total number of base pairs is very great. Careful chemical analysis reveals that each set of chromosomes in a human being contains about 6×10^{-12} g of DNA. Since each base pair has a weight of about 600, dividing by Avogadro's number (6×10^{23}) yields a value of about 10^{-21} g per base pair. Hence a total of 6×10^9 base pairs must be present in each cell—a large number indeed. Well over a million genes may be present. If this DNA were extended as in the chromosome of *E. coli,* a value for the contour length of slightly more than 2 m would be obtained. One of the major unsolved problems of biochemistry is how such incredibly long molecules are "packaged" in cells of microorganisms, cell nuclei of higher forms of life, or the extremely compact virus structures. The 1.2-mm length of *E. coli* DNA, for example, resides inside a bacterial cell with a diameter of about $1\ \mu$ ($1000\ \mu = 1$ mm).

Some appreciation of the enormous length of DNA and the vast amount of information it can encode may be achieved with the following analogy. Imagine that we are storing the genetic contents of one human in terms of a string of small beads, each 1 cm in diameter. Four different colors of beads would be used to represent the four bases. In order to represent the necessary genes, a string of beads long enough to go around the world would be needed.

Stability of DNA

The stability of the DNA double helix, or duplex as it is sometimes called, is substantial considering that it is derived largely from weak hydrogen bonds. Estimates of DNA stability come from an experimental approach called *melting*. The ordered structure of DNA undergoes a transition somewhat like the melting of ice. In the melting of DNA, the hydrogen bonds break, the bases become unstacked, the strands separate, and when the melting process is complete the molecule takes on the properties of a random coil (two random chains

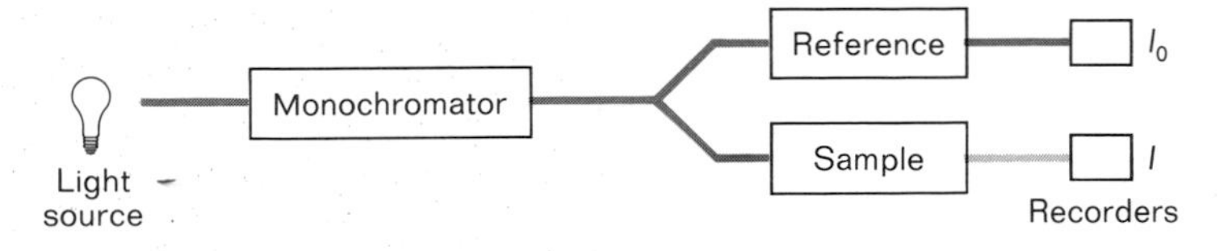

FIG. 2-10 *Measurement of Absorbance.* Light through "reference" is unabsorbed and gives level of incident light intensity I_0. Light through "sample" is absorbed, and intensity of light transmitted is diminished to a value I. The ratio I_0/I depends on absorbance: the more light absorbed by "sample," the larger I_0/I. Absorbance is defined by $A = \log(I_0/I)$; thus a value of $A = 1$ corresponds to $I_0/I = 10$ or the transmittance by the sample of only 10% of the light.

per duplex). Melting is a major structural change and is easily monitored by physical methods. The viscosity of DNA solutions, for example, falls dramatically with melting, from the very high values characteristic of rodlike DNA to the moderate viscosity of the melted coils. However, the melting is most conveniently followed spectrally. The nitrogenous bases of nucleic acids absorb light in the ultraviolet (uv) region of the electromagnetic spectrum, with a maximum absorption in the region near 260 nanometers (1 nm = 10 Å = 10^{-9} m). Absorption spectra are often important distinguishing characteristics for biochemical molecules and will be referred to often. Also, since absorbance or optical density (OD) is proportional to the amount of material present, spectral measurements are often used to measure concentration (see Fig. 2-10). Absorbance is related to concentration by the equation

$$A = \varepsilon l c$$

where ε is the extinction coefficient, a constant for a particular structure at a given wavelength, l is the optical pathlength, and c is the concentration.

In the absorption of light by DNA, the close stacking of the bases depresses the degree of absorption, changing ε. Thus unstacked bases of melted DNA absorb nearly $1\frac{1}{2}$ times more than the bases in the double-helical structure. Because of this increase in absorbance or *hyperchromicity,* the melting of DNA can be easily monitored and the stability of DNA can be estimated by determining what conditions are needed to achieve melting. The conclusion reached is that the DNA is very stable, since for some species temperatures near the boiling point of water are needed for melting. The exact melting temperature or t_m (midpoint of the melting curve) depends on the pH and ionic strength and is very sensitive to chemicals that disrupt the double helix. Ions stabilize the helix by shielding the repulsive forces of the negative charge on the phosphate group. Extremes in pH favor melting by increasing charge, and many organic molecules interfere with stacking forces.

The cooperative nature of the bonding due to many weak

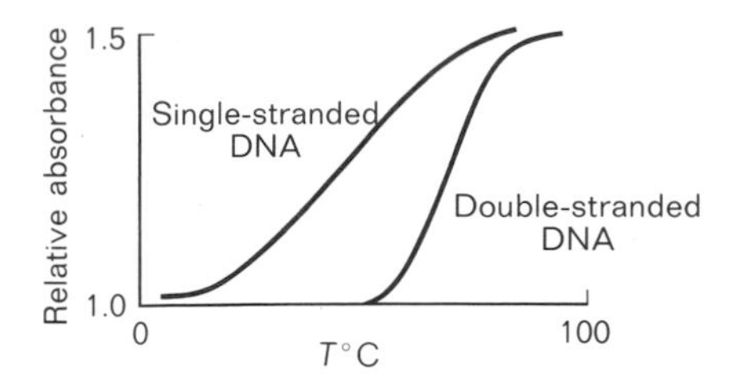

FIG. 2-11 *DNA Melting Curve.*

hydrogen bonds is indicated by the steep slope of the melting curves (Fig. 2-11). Single-stranded DNA exhibits some melting, due to unstacking, but with much less cooperativity. The diminished cooperativity is reflected by a more gradual melting curve. The role of hydrogen bonding is also illustrated by considering the significance of the third hydrogen bond in G-C pairing compared to A-T pairs with only two hydrogen bonds. As a consequence of just this one additional hydrogen bond (with some possible contribution from an especially strong stacking tendency of G), G-C pairs lend much more stability to the DNA helix than do A-T units. This conclusion is reached by examining the t_m of a variety of DNA molecules from different species and with varying proportions of G-C and A-T. The more G-C, the higher the t_m, and for one standard set of conditions, the t_m is given by

$$t_m = 69°\text{C} + 0.41(\text{mole \% G-C})$$

Thus the difference in stability between G-C-rich and G-C-poor molecules can approach 40°.

PROBLEMS

1 Which bases pair in the DNA double helix? Draw the structures of the base pairs. Can the base pairs form in the lactim structure?

2 Distinguish between the transformation and virus growth experiments in establishing DNA as the genetic material.

3 According to the base ratios of DNA, [A]/[T] = [G]/[C] = 1. Is it also true that
(a) [purines]/[pyrimidines] = 1
(b) [A]/[G] = 1

4 Calculate the number of base pairs in chicken DNA from the weight per heart cell in Table 2-1. This number corresponds to what length for an extended molecule?

5 A solution of DNA transmits 50% of the incident light at 260 nm. What is the absorbance? What is a reasonable estimate for the absorbance of the same solution if it is heated above its melting temperature?

6 A solution of DNA is found to give a hyperchromic shift in absorbance as a function of increasing temperature, with a midpoint at 89°C. What is the approximate percentage of G-C pairs in the DNA?

7 What is the difference between a nucleoside and a nucleotide.

Chapter 3 Frontispiece *Electron micrographs of closed circular DNA obtained from HeLa cells. Courtesy J. Vinograd. For additional details see Journal of Molecular Biology 69: 193 (1972).*

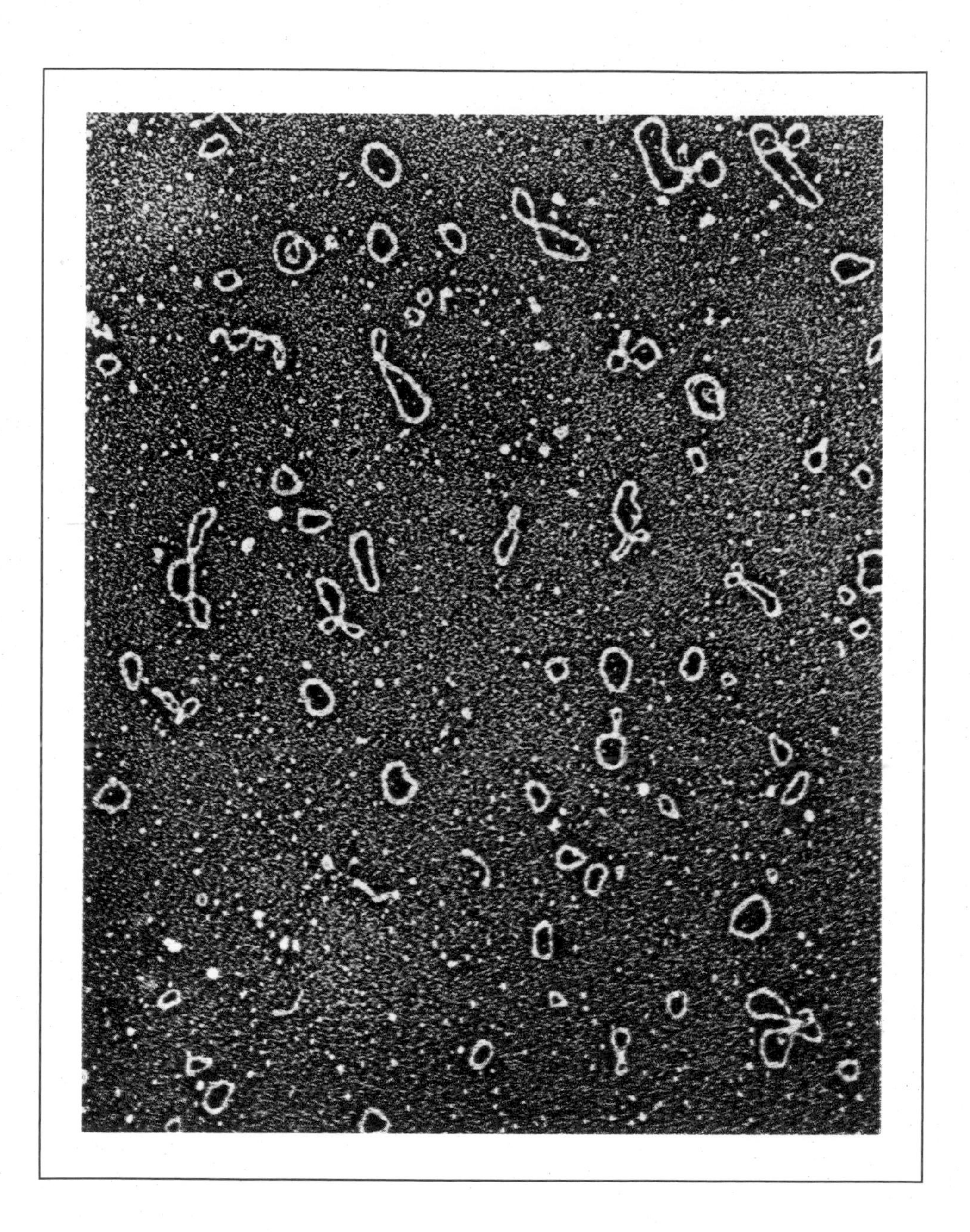

REPLICATION OF DNA

3

In the introduction to Part I, we noted three important qualities for genetic material. A stable structure is needed to ensure the preservation of genetic information, and in Chapter 2 we reviewed the structure of DNA and the factors that determine its stability. The second important quality concerns the means for replication. This aspect of the DNA molecule will be considered now. The third and more complex quality—information decoding—will be treated later.

Semiconservative Replication

A very striking feature of the double-helix model for DNA is the mechanism of replication it directly implies. Since base pairing follows strictly the A-T and G-C rules, one strand implies the other complementary strand. Thus if the two strands part and each missing partner is replaced with a new strand, two identical DNA molecules are generated from the original one. This scheme of replication is illustrated in Fig. 3-1. Each of the two complete DNA molecules formed contains one strand of the original parent molecule. This mode of replication is called *semiconservative.* Alternative patterns can also be imagined. In the *conservative* scheme both strands of the original DNA molecule would remain together and two entirely new strands would be generated for a second molecule. Another possibility is a *nonconservative* mechanism. Here parts of the original molecule are combined with some newly synthesized fragments and no single complete strand can be identified following replication.

Proof for the semiconservative mode of DNA replication was obtained by Matthew Meselson and Franklin Stahl in an elegant experiment based

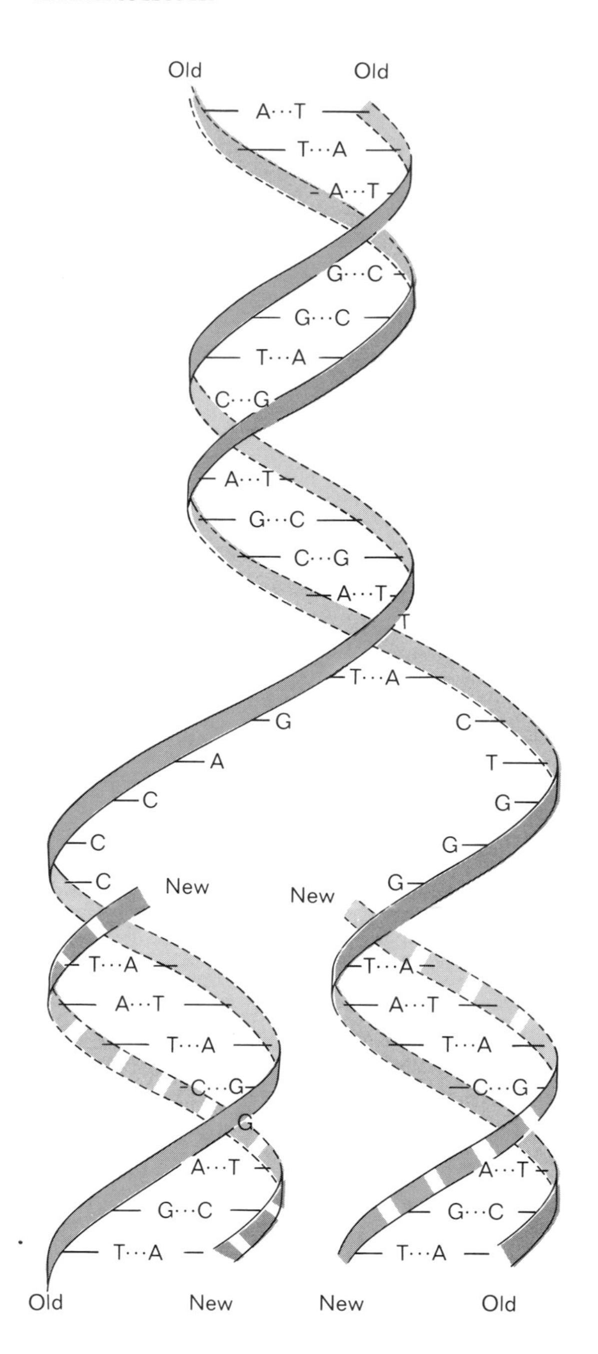

FIG. 3-1 *Model of DNA Replication.*

on the banding method of ultracentrifugation. Several techniques of ultracentrifugation are widely used in biochemistry. For work with macromolecules, especially proteins, sedimentation methods are especially useful. Usually either the rate of transport or the distribution at equilibrium of the material is studied in the centrifugal fields (both of these approaches will be discussed in other chapters). However, banding is based on a different principle: the response of small amounts of a macromolecule to an imposed density gradient generated by the redistribution of cesium chloride (CsCl) in the centrifugal field (see Fig. 3-2). With high concentrations of CsCl and high rotor speed, a stable gradient of CsCl will be generated that in turn creates a gradient of density. In a typical experiment a gradient of density of 1.6 g/cc at the meniscus to 1.8 g/cc at the bottom might be present. When DNA is also present in the solution it will tend to form a sharp band at its *buoyant density*, about 1.71 g/cc for *E. coli* DNA. DNA molecules at points in the solution at higher density will tend to float until they reach the buoyance point. Molecules in regions of lower density will sink or sediment. Thus density gradient centrifugation will conveniently form distinct bands of materials with different densities. The precise position of the band depends on the G-C content of the nucleic acid. The greater the percent of G-C, the higher the buoyant density. The buoyant density is approximated by

$$\rho_B = 1.66 + 0.1(\text{mole fraction G-C})$$

Banding experiments also provide information on macromolecule molecular weight (from band width) and homogeneity (from band symmetry). However, only the property of forming sharp density-dependent bands was needed to prove the semiconservative mode of replication.

To apply banding to the replication question, cells were grown on the heavy isotope of nitrogen, ^{15}N. The "heavy" DNA prepared from these cells banded at a point well below the position of normal (^{14}N) DNA (see Fig. 3-3). A growing population of the ^{15}N cells was quickly transferred to a normal (^{14}N) medium. Cells were allowed to continue growth for another generation and the DNA was then isolated. When examined in a banding experiment, the DNA was found to form a single band located *between* the positions of the heavy and normal bands (Fig. 3-3). This finding is just what would be predicted by the semiconservative mode of replication. A hybrid of old and new or heavy and normal DNA was then

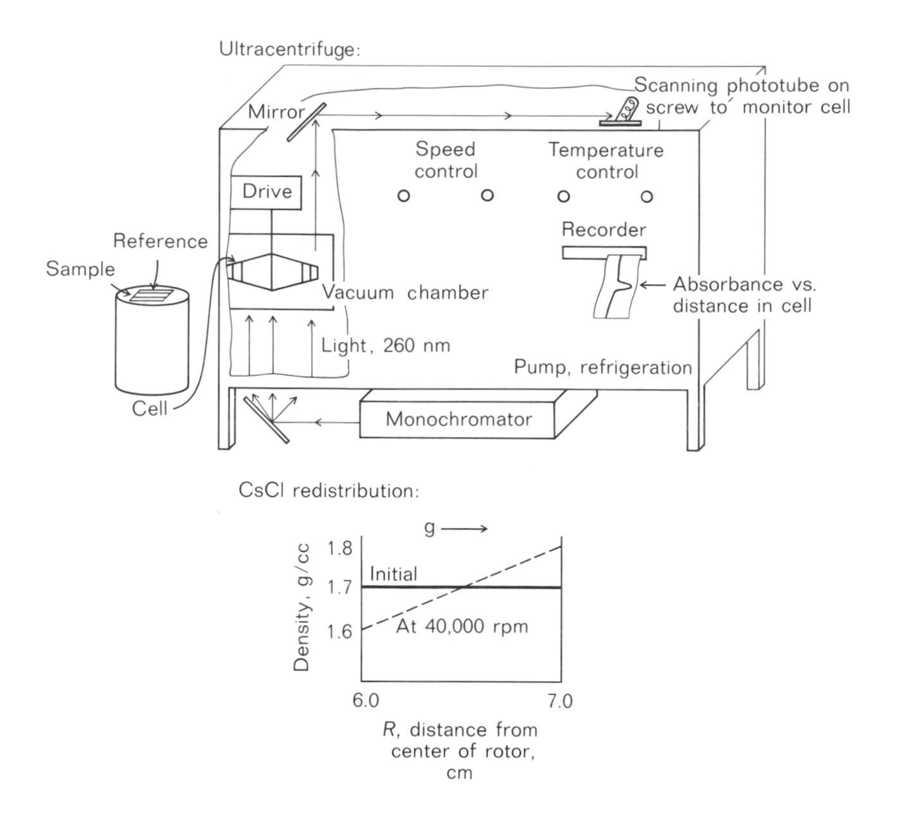

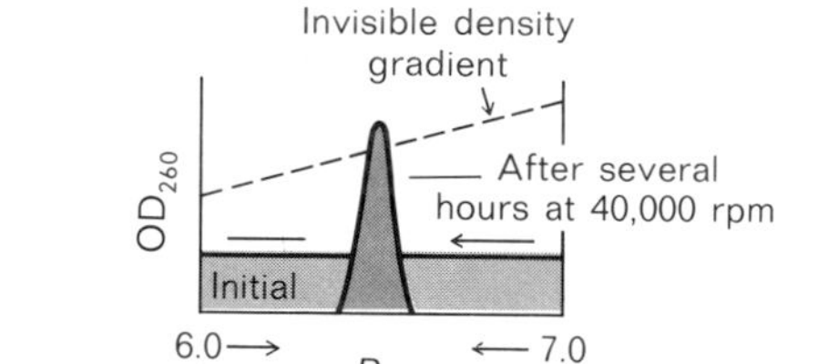

FIG. 3-2 *Banding.* A DNA solution in 6 *M* CsCl is placed in an ultracentrifuge cell, spun at 40,000 rpm, and examined with light of 260 nm. The CsCl (invisible at 260 nm) redistributes due to the high centrifugal force and concentrates toward the bottom of the cell, creating a density gradient. The DNA redistributes in the CsCl density gradient to a band at a position where the CsCl density is equal to the buoyant density of the DNA.

formed. Also as expected, further growth through additional generations "diluted out" the intermediate or hybrid DNA.

With the general scheme of replication established as semiconservative, the details of the process were investigated. Did the replication proceed from one growing point on the circular chromosome of *E. coli?* Or, were there several growing points or different directions of growth for the two strands being replicated? Autoradiography provided the answers to the questions. As seen in Fig. 2-9, the chromosome has been caught in the act of replicating, and part of the circular structure is duplicated. Autoradiographs from samples at different stages of replication indicated clearly that the process begins at a particular initiation point with a "replicating fork" con-

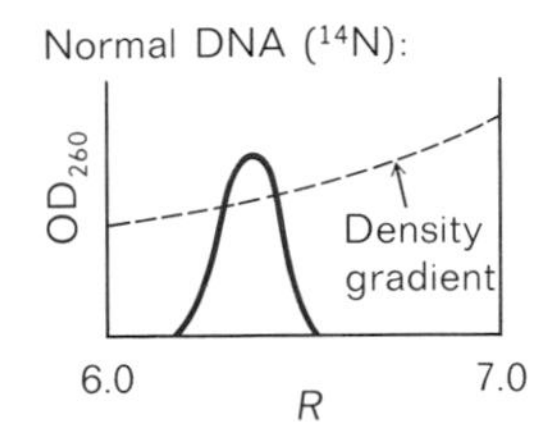

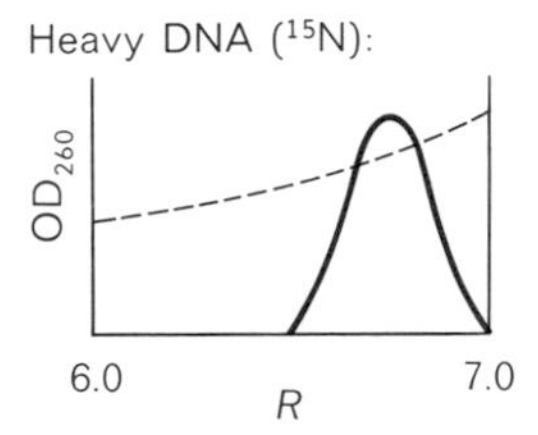

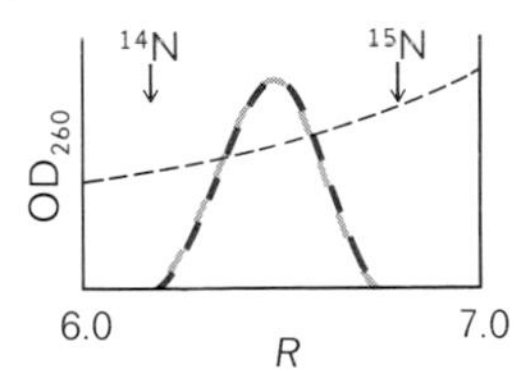

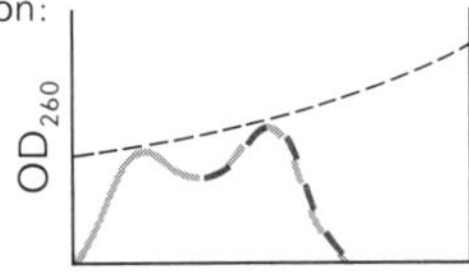

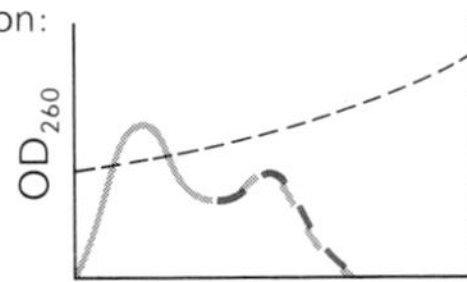

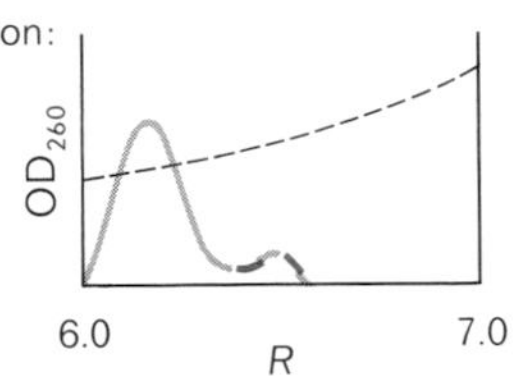

FIG. 3-3 *Meselson-Stahl Experiment.* Based on M. Meselson and F. W. Stahl, *Proceedings of the National Academy of Science,* **44:** 671 (1958).

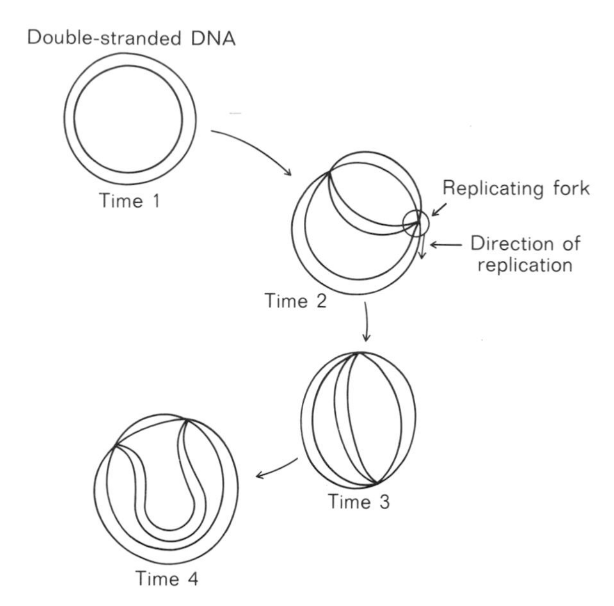

FIG. 3-4 *Replicating Fork.* DNA isolated at various times during replication.

tinuing around the chromosome, leaving a duplicated chromosome behind it (see Fig. 3-4). Each of the two new chromosomal fragments of DNA contains one old and one new strand. Under favorable conditions of growth, the newly synthesized DNA is formed with the addition of about 1500 nucleotides per second.

Enzymatic Aspects of DNA Replication

The DNA of the chromosome is replicated at speeds so fast that a nucleotide is added almost every half of a millisecond. The strands are unwound and parted, new bases are brought into place according to the hydrogen-bond complementarity, and new covalent linkages are formed. There is a wondrous amount of chemistry taking place in a very short time, and the analysis of the enzymes that govern replication is certain to unveil a molecular mechanism of extraordinary delicacy and coordination. However, at the present moment it is not at all clear how this process occurs or even which enzymes are involved in replication. Nevertheless, work on this topic has already provided a wealth of information on the structure of DNA and the various enzymatic activities directed toward DNA.

The enzymatic synthesis of DNA has been studied for many years by Arthur Kornberg and his colleagues. Their first success began with an incubation of deoxynucleoside triphosphates of the four bases known to occur in DNA (dATP, dCTP, dGTP, and dTTP), a purified preparation of the DNA to serve as a template for replication, and finally an extract of *E. coli*. Nucleoside triphosphates were selected since this form is active in many biosynthetic events in the cell, most notably ATP. Efforts with the crude extracts of *E. coli* were encouraging. Small amounts of ^{14}C-labeled triphosphate nucleosides could be incorporated into a polydeoxynucleotide fraction with the properties of DNA. Moreover, the new DNA had the same base composition as the template DNA; omission of just one of the triphosphates or the DNA template resulted in no new synthesis. The factor that catalyzed this reaction was purified from *E. coli*, and eventually a homogeneous protein was obtained. This protein is known as *DNA polymerase*.

Work with the enzyme DNA polymerase is of special interest in studying two aspects of DNA: the mechanism of replication and the structure of DNA itself. The exact role of DNA polymerase in replication is not presently clearly defined. However, studies with DNA polymerase have provided some very important results concerning the structure of DNA, particularly from the work on *nearest-neighbor analysis*. The triphosphate nucleosides used to synthesize *de novo* DNA are all 5′-triphosphate nucleosides (5′-dNTP), as shown in Fig. 3-5. The α-phosphate of the triphosphate nucleoside (the phosphate closest to the dNTP deoxyribose ring) is attached via an ester linkage to its carbon number 5′. As a triphosphate adds onto a growing chain of DNA, the α-phosphate provides a bridge to the carbon number 3′ of the terminal nucleotide residue on the chain. Eventually a 3′,5′-phosphodiester is formed. Splitting off the terminal phosphates yields energy and helps to drive polynucleotide synthesis.

To follow the steps in nearest-neighbor analysis, first picture the addition of one nucleotide to a growing DNA chain. For example, the chain ends in T. A molecule of ATP then comes up and the 3′-hydroxyl group of the T attacks the phosphate attached to the A (at the 5′ position), and A is incorporated. Subsequently, additional bases after the A will be incorporated, each added in the 5′ $\longrightarrow$ 3′ direction. Now if the material is digested with the appropriate nuclease, the phosphodiester bond between the phosphate group and adenine will be cleaved. The phosphate itself will remain attached to T. If the phosphate had been labeled with ^{32}P,

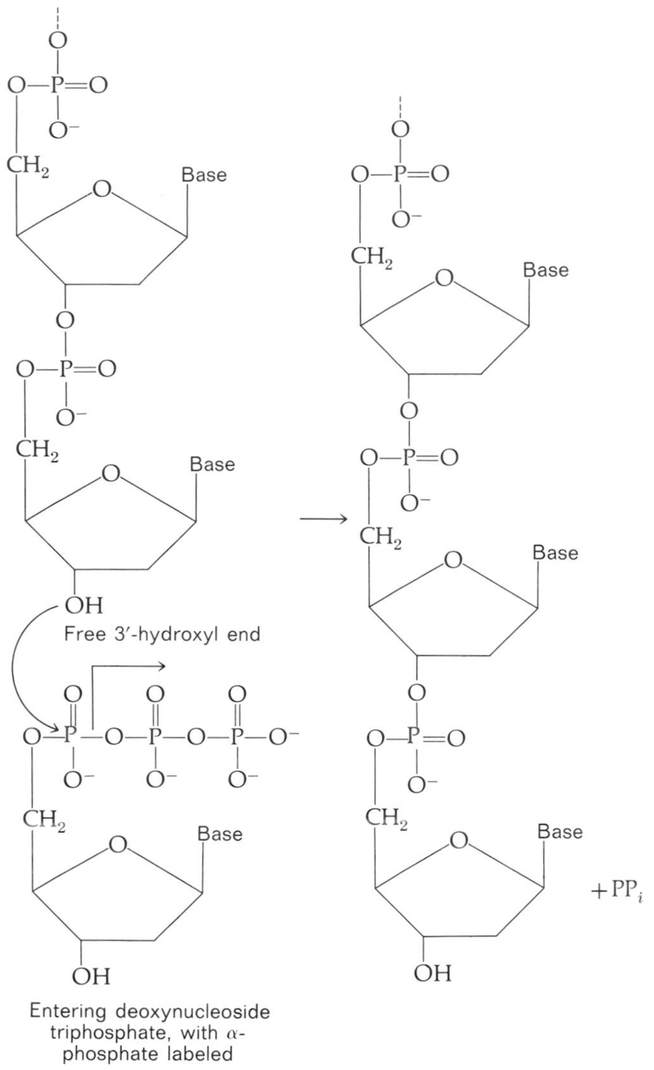

Free 3′-hydroxyl end

Entering deoxynucleoside triphosphate, with α-phosphate labeled

$+ PP_i$

FIG. 3-5 *Addition to 3′-Hydroxyl End.*

radioactivity would remain with T. Then if the entire fraction of newly synthesized DNA were digested after an incubation with ^{32}P-labeled ATP, the frequency of bases on the 5′ side of A would be revealed (see Fig. 3-6). When the digestion with nuclease was complete, the four bases could be fractionated and their specific activity in terms of ^{32}P measured. With dATP labeled initially, the frequency with which G, C, T, and A occur as neighbors of A would be obtained. By alternately incubating the synthesis mixture with ^{32}P in just one of the four nucleotides and analyzing each of the four nucleotides at the completion of the digestion, nearest-neighbor frequencies could be obtained for all the bases. A typical experiment of this type is reported in Table 3-1. Two important conclusions can be derived from the nearest-neighbor analysis. First, the nearest-neighbor pairs show a precise arrangement. For example, the frequency of occurrence of CC is identical to the occurrence of GG. Similarly, the occurrence of AA is just matched by the occurrence of TT. These reciprocal relationships are exactly what is predicted for DNA with two structurally complementary strands. Since one strand is the complement of the other, the fraction of CC must be mirrored by the fraction of GG.

The second conclusion derived from the nearest-neighbor analysis has even deeper significance for the structure of DNA. In the original model of the double helix, the two strands of DNA were postulated to run in opposite directions; however, other structures could be imagined that would satisfy many of the requirements for DNA structure even if the two strands run in the same direction. The two possibilities known as *parallel* and *antiparallel helices* are readily

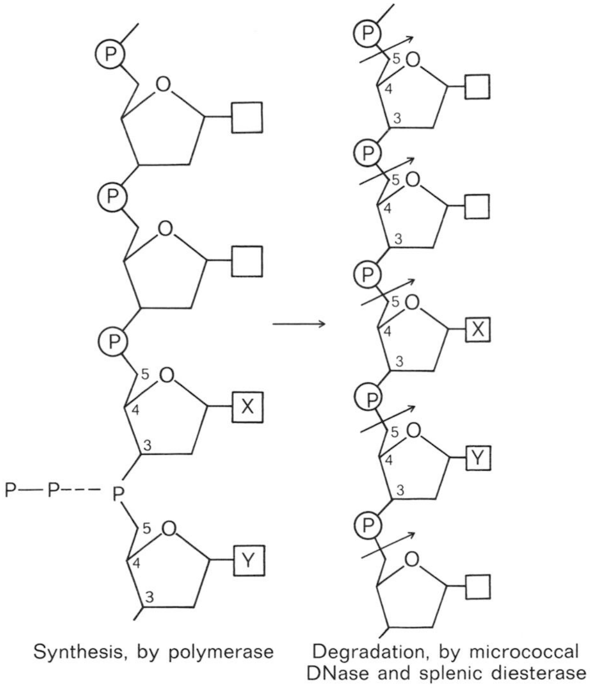

FIG. 3-6 *Basis of Nearest-Neighbor Analysis.* DNA is synthesized with ^{32}P-labeled nucleotides and degraded with the ^{32}P attached to the adjacent nucleoside at the 3′ position. The arrows on the right indicate the linkages cleaved by micrococcal deoxyribonuclease and calf-spleen phosphodiesterase. Modified from A. Kornberg, *Enzymatic Synthesis of DNA*, Wiley, New York, 1961.

distinguished by the nearest-neighbor analysis. As illustrated in Fig. 3-7, quite different results are predicted by the parallel and antiparallel possibilities. The data observed are compatible only with the antiparallel structure.

Thus DNA polymerase has provided much valuable information concerning DNA itself, but its precise role in DNA replication remains uncertain. It was noted that DNA polymerase operates to add nucleotides in the 5′ ⟶ 3′ direction. However, the autoradiography experiments on the growing chains of newly synthesized DNA suggest that both strands are replicated together. Therefore only one strand must be replicated in the 5′ ⟶ 3′ direction, while the other strand is apparently replicated in the 3′ ⟶ 5′ direction. DNA polymerase could fulfill this requirement if it synthesized the 3′ ⟶ 5′ chain by taking short segments in the 5′ ⟶ 3′ direction that would then be linked by another

TABLE 3-1 Nearest-Neighbor Frequencies of *Mycobacterium phlei* DNA*

Reaction No.	Labeled Triphosphate	Isolated 3′-Deoxyribonucleotide			
		Tp	Ap	Cp	Gp
1	$dATP^{32}$	*a* TpA 0.012	*b* ApA 0.024 I	*c* CpA 0.063 II	*d* GpA 0.065 III
2	$dTTP^{32}$	*b* TpT 0.026 I	*a* ApT 0.031	*d* CpT 0.045 IV	*c* GpT 0.060 V
3	$dGTP^{32}$	*e* TpG 0.063 II	*f* ApG 0.045 IV	*g* CpG 0.139	*h* GpG 0.090 VI
4	$dCTP^{32}$	*f* TpC 0.061 III	*e* ApC 0.064 V	*h* CpC 0.090 VI	*g* GpC 0.122
	Sums	0.162	0.164	0.337	0.337

Identical Roman numerals designate those sequence frequencies that should be equivalent in an antiparallel double helix with A-T, G-C pairing; identical lowercase letters designate sequence frequencies that should be equivalent in a model with strands of similar polarity. The symbol TpA stands for deoxythymidyl-(5′-3′)-deoxyadenine.

*Chemical analysis of the base composition of the primer DNA indicated molar proportions of T, A, C, and G of 0.165, 0.162, 0.335, and 0.338, respectively.

Source: A. Kornberg, *Enzymatic Synthesis of DNA,* Wiley, New York, 1961.

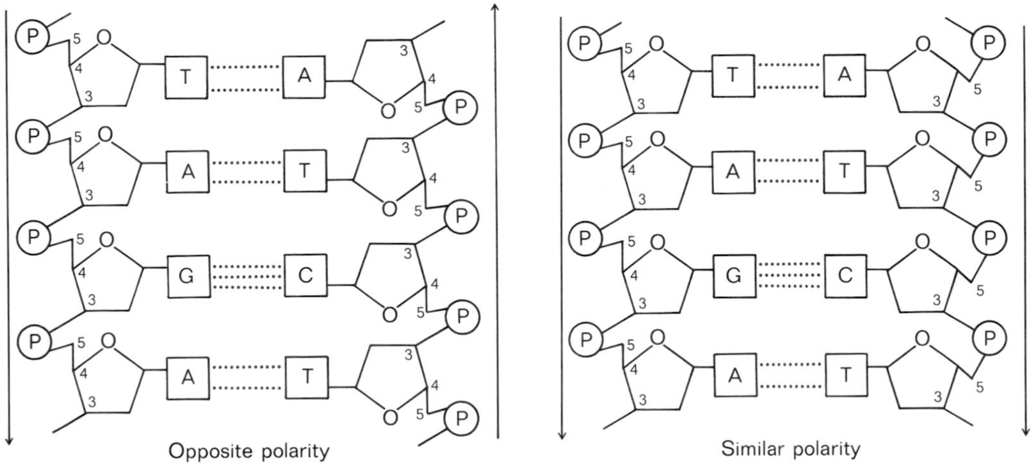

Opposite polarity

TpA(0.012) = TpA(0.012)
ApG(0.045) = CpT(0.045)
GpA(0.065) = TpC(0.061)

Similar polarity

TpA(0.012) = ApT(0.031)
ApC(0.045) = TpG(0.061)
GpA(0.065) = CpT(0.045)

FIG. 3-7 *Predictions for Parallel and Antiparallel Structures.* Values in parentheses are the dinucleotide frequencies expected with *M. phlei* DNA. From A. Kornberg, *Enzymatic Synthesis of DNA,* Wiley, New York, 1961.

enzyme to form complete strands. Just such a linking enzyme has been isolated; it is called polynucleotide ligase and has lent strength to the argument that DNA polymerase could operate to synthesize both strands of DNA.

Nevertheless, an even more serious and probably fatal objection remains to the assignment of DNA polymerase as the principal enzyme of DNA replication. A mutant of *E. coli* has been isolated by John Cairns, after an extensive screening process, which appears to lack the enzyme DNA polymerase entirely. Moreover, new enzymatic activity in replication, associated with the membrane of *E. coli,* has been identified in these mutants. The new enzyme(s) is a strong candidate for designation as the true "replicase." The role of the originally designated DNA polymerase in the cell may only involve repair (filling in of gaps in one of the strands). This DNA polymerase does, in fact, operate more readily on single-stranded DNA. According to this hypothesis, the enzyme would function to repair missing sequences in one chain by catalyzing synthesis of nucleotides complementary to the remaining chain.

General Aspects of Replication

We have just considered some of the problems in the detailed enzymatic mechanism for the synthesis of DNA, that is, the way in which a series of individual nucleotides is added to growing strands of DNA. The pattern of DNA synthesis in terms of the overall configuration of the molecule may also be considered. We have noted that in many cases DNA is present in a circular form, and the replication pattern is very intimately related to this circularity. For example, autoradiograph studies (summarized in Fig. 2-9,) indicate that the circular structure of DNA is maintained during replication. Yet the circularity presents certain problems. Independent evidence suggests that DNA for a given species begins its replication at a fixed point on the chromosome. Moreover, DNA synthesis involving base pairing must take place on separated strands.

These requirements, at first sight, seem incompatible with the circular structure for the DNA molecule since there is no obvious starting point and the circular structure makes strand separation difficult. A possible solution to these difficulties came with the discovery that DNA can exist either in the closed-circular form or in the open-linear form. Work on these forms began with DNA from the bacteriophage λ. Its DNA molecule has a molecular weight of slightly greater than 10^7 and appears to be cleaved by a very specific enzyme at two points on its structure. The enzyme is a sequence-recognizing endonuclease (i.e., an enzyme that splits nucleic acids at interior sequences, not at the ends) and makes a nick at points on each of the two strands separated by about 12 nucleotides (see Fig. 3-8). These breaks permit the strands

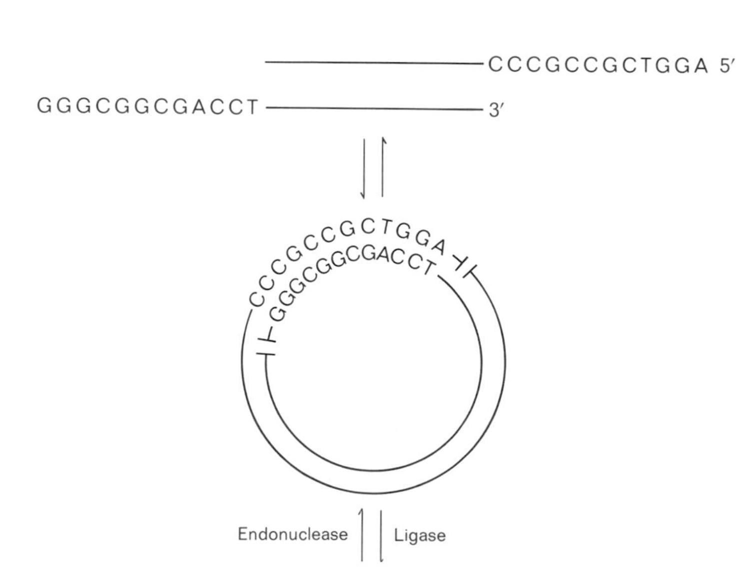

FIG. 3-8 *Linear and Circular Forms of λ DNA.* The region between the bases specified is not drawn to scale but is much reduced. Data from R. Wu and E. Taylor, *Journal of Molecular Biology,* **57:** 491 (1971).

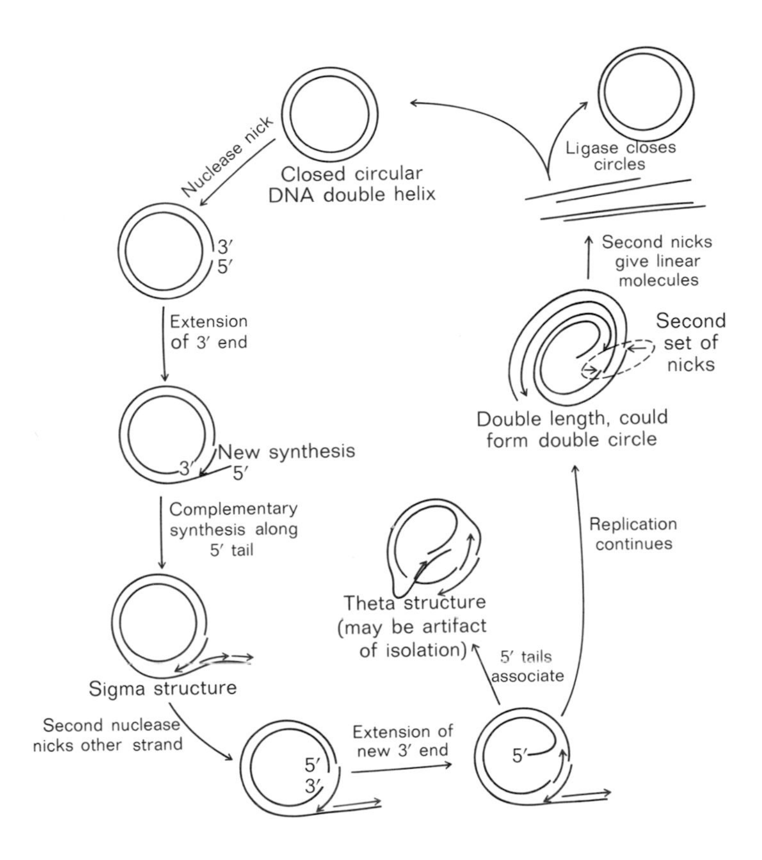

FIG. 3-9 *Rolling Circle Model of DNA Replication.* Extension of second 3′ end probably does not normally occur. Adapted from J. D. Watson, *Molecular Biology of the Gene,* 2nd ed., copyright © 1970 by J. D. Watson, W. A. Benjamin, New York, 1970.

to form a linear molecule in which each end contains an extra 12 nucleotides at the 5′ terminus, complementary to the opposite end. These ends are known as "sticky" ends since the molecule can re-form a circle through the base pairing of these nucleotides. When such pairing has been accomplished, another enzyme, the polynucleotide ligase already noted, is capable of forming intact circles by resealing the loose ends, which were cut when the breaks in the stands were first created.

The transformations of circular λ DNA have important implications in the mechanism of replication of other DNAs. If a closed circle is nicked by an endonuclease similar to the enzyme that can open circular DNA, the free 3′ terminus of DNA which is now formed can serve as the initiation point for new synthesis of DNA (see Fig. 3-9). As new synthesis is extended from the 3′ terminus, the other end of the chain, the 5′ terminus, is displaced away from the circle. This end of the molecule can also be replicated by complementary synthesis even with a DNA polymeraselike enzyme that synthesizes in the 5′ ⟶ 3′ direction. Short sections of complementary synthesis on the 5′ tail could occur as the circle "unrolls." These sections could be linked by a ligase to form a continuous strand. Even the extension of the free 3′ end could be achieved by synthesis of short sections later linked by ligase. In fact evidence in favor of such a mechanism has been reported by R. Okazaki. When growing cells are incubated with ^{3}H-thymidine for short periods, the label appears to be incorporated into short polynucleotide chains (1000 to 2000 nucleotides in length). Longer incubations with ^{3}H-thymidine (approaching the time of one generation) result in the appearance of label in longer polypeptide chains.

While we can explain replication in general terms with a DNA polymerase–type enzyme and a ligase, up to this point we would have a mode of replication known as the σ pattern, since it resembles a Greek sigma. However, as noted in the autoradiograph studies, the replicating form of DNA looks much more like a θ. In order to develop a model that does include this θ-like behavior, a second nick must be introduced in the remaining intact strand. New synthesis of DNA can then, again take place to extend the second 3′ terminus, displacing the remaining second 5′ terminus, in a series of reactions analogous to the progress in the first stage of new synthesis after the initial nick. With the addition of the second growing point, two free 5′ terminal ends can pair with one another and form the loop of the θ structure. However, it is possible that the θ structure is formed only in isolation since the structure itself is "locked" and cannot continue to "unwind" as is needed for additional synthesis. Eventually, as synthesis in the σ mode continues, the entire molecule becomes duplicated with the parent molecule remaining attached in a linear fashion. Finally, the double-sized molecule can be cleaved, in a mechanism similar to the opening of the circle. The cleavage would yield two linear molecules with "sticky" ends to go on to form a circle with final closure achieved by action of the ligase. Thus the process yields two molecules where there had been only one.

PROBLEMS

1 A DNA sample was found to have a buoyant density of 1.685 g/cc in a CsCl banding experiment. What is the approximate mole percent G-C?

2 A hybrid DNA band was observed in the Meselson-Stahl experiment after one generation of growth of ^{15}N cells in

^{14}N medium. What would be expected for conservative replication?

3 A nearest-neighbor experiment was performed, and the frequencies of the 16 possible dinucleotides obtained. How many pairs of dinucleotides should have the same frequency only in the case of an antiparallel DNA structure?

4 Are the nicks observed in producing the linear form of λ DNA from the circular form adequate to explain all the structures observed by autoradiography during DNA replication?

Chapter 4 Frontispiece *Electron micrograph of nucleolar genes isolated from an oocyte of the spotted newt, Triturus viridescens, during transcription of ribosomal RNA. Magnification 25,000×. Courtesy O. L. Miller, Jr., and B. R. Beatty. For additional details see Science* **164**: *955 (1969).*

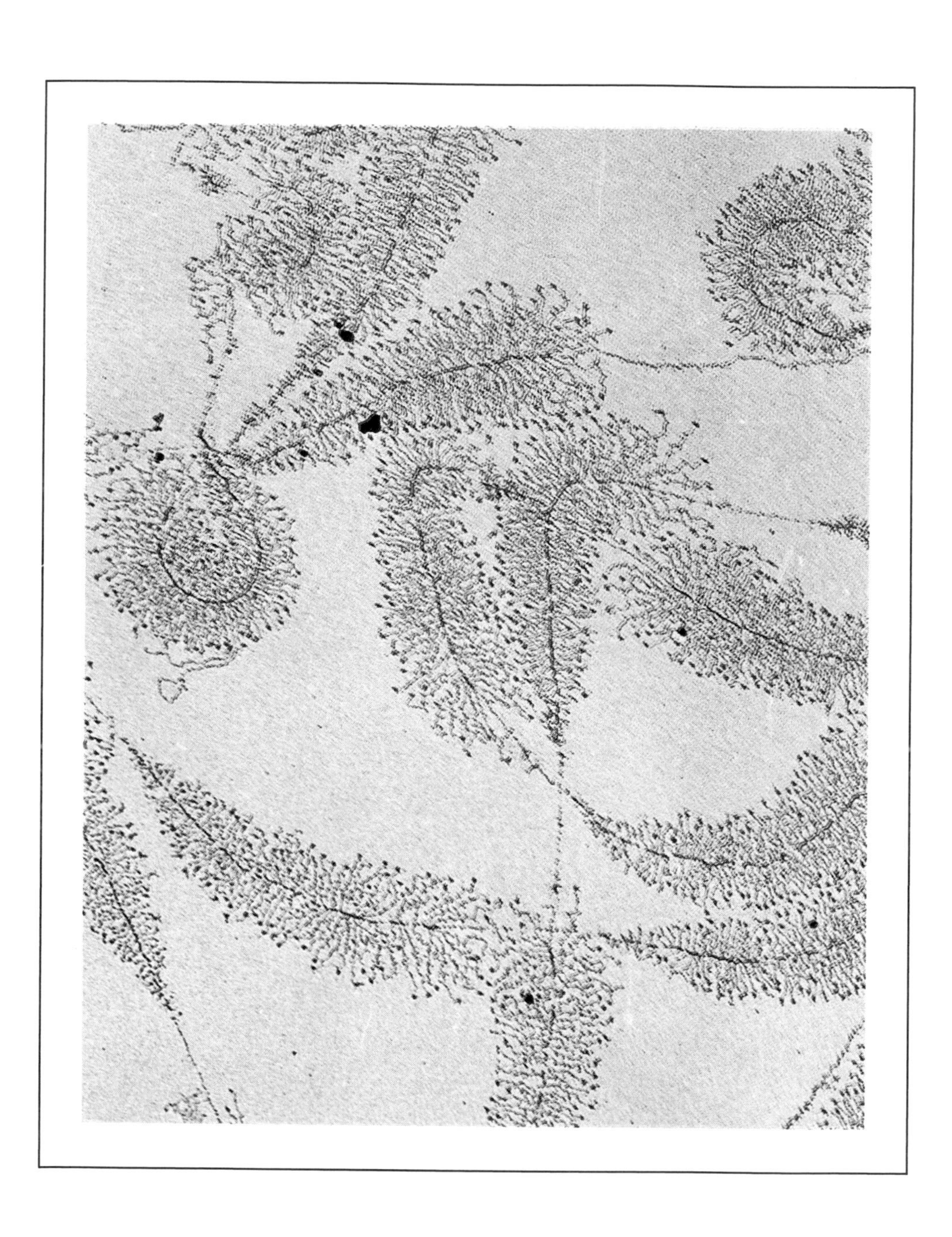

MACHINERY OF PROTEIN SYNTHESIS

4

DNA is a double helix composed of two polynucleotide strands united by complementary base pairs. As noted, the structure itself suggests a mode for its replication—the semiconservative pattern of chain synthesis. However, the mode of the other important process involving DNA, translation of nucleotide sequences into colinear amino acid sequences, is not readily apparent from the structure of DNA. Two major questions dominated early investigations in this area: (1) How many nucleotides are required to specify one amino acid? (2) How is the information contained in a polynucleotide sequence converted into a sequence of amino acids?

A minimum number of nucleotides needed to code for all 20 amino acids could be obtained from purely mathematical considerations. Individually the four nucleotides could code for only four amino acids. The various combinations of pairs of nucleotides would still only specify 16 amino acids (4^2). Thus nucleotide triplets with 64 possible combinations (4^3) are the minimum units required to provide a distinct nucleotide sequence for each of the 20 amino acids. As we will see in the next chapter, the minimum number of nucleotides per amino acid or "codon" size was clearly tested and in fact found to be three.

The question of possible mechanisms for translating nucleotide sequences into amino acid sequences is much more complex. Two general types of mechanisms could be imagined, direct and indirect. In the *direct* system each amino acid would bear a distinct stereochemical relationship to the atomic structure of the set of nucleotides that specifies that amino acid. Thus a complementary relationship, such as the hydrogen bonding and base size that define A-T and G-C pairing, would also prevail with bases and amino acids. With an *indirect* mechanism no direct complementary stereochemical correspondence between nucleotides and amino acids is needed. Rather an *adaptor* system is involved, providing a "third party" capable of recognizing nucleotide arrays and particular amino acids and relating the two in some appropriate way.

The possibility of direct nucleotide–amino acid recognition was quickly dismissed when coding was first pondered, due in large part to a failure to devise any suitable stereochemical relationships. While the A-T and G-C pairing system for replication is clear and well defined, no comparable nucleotide–amino acid pairing mechanism can be readily formulated. Not only would nucleotides be required to recognize amino acids of diverse shapes and sizes, but improper matches would have to be strictly guarded against since errors at this level would not benefit evolution and could be very detrimental. Therefore indirect mechanisms were favored; these are now fully documented.

In this chapter the indirect system for converting information in DNA into protein sequences will be described. It is a complicated system, involving a battery of proteins and nucleic acids. Chapter 5 will consider the precise dictionary of the genetic coding system, the codons; Chapter 6 will then discuss the overall coordination of protein synthesis. While all the stages involve indirect specification, we will see that the arrangement of the genetic dictionary does, nevertheless, hint at some direct stereochemical logic between amino acids and their codons. Some speculations on the perplexing question of the stereochemical basis of the genetic code will be included in our discussion.

The Adaptor System

An indirect mechanism of protein synthesis became evident as elements of an *adaptor system* were found in living cells. The need for specific interactions between nucleotides and amino acids is thus replaced by the adaptor system forming a bridge across the nucleotide–amino acid gap. The adaptors are *transfer RNA* (tRNA) molecules. There are 20 sets of transfer RNAs, a specific set for each amino acid. The transfer RNA molecules recognize nucleotide codons by complementary base pairing and carry an attached amino acid to be incorporated into a polypeptide sequence specified by a nucleotide sequence. The transfer RNA molecules are polynucleotides. They resemble one strand of DNA except that ribose is present in place of deoxyribose, and uracil (U) re-

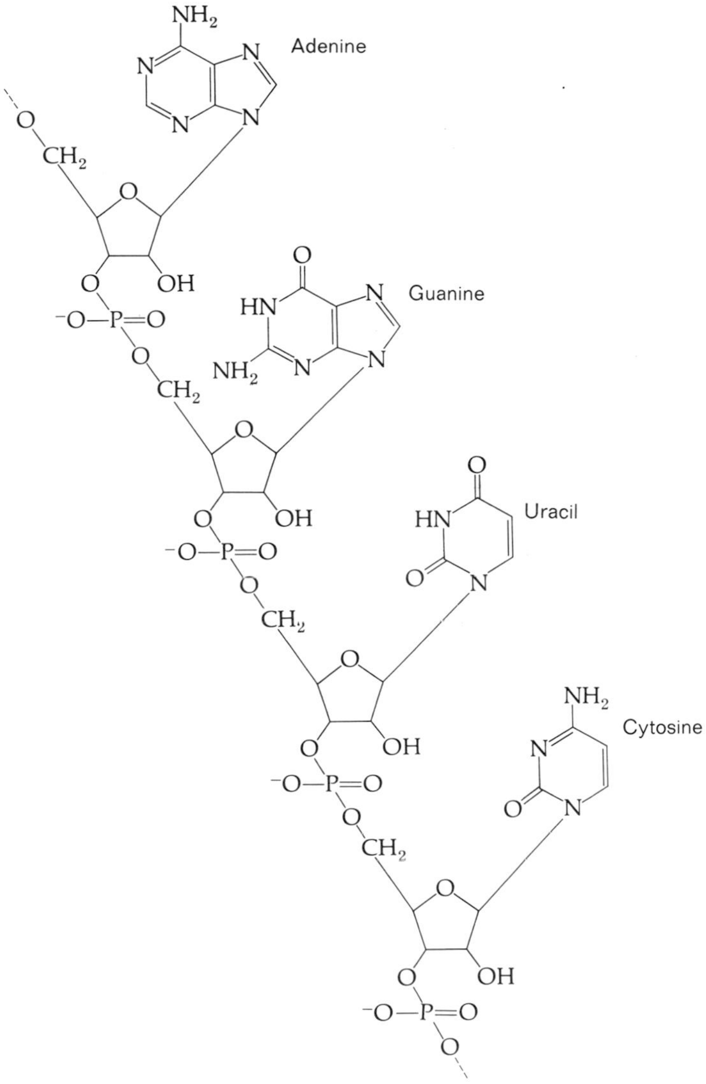

FIG. 4-1 *Structural Units of RNA.*

places thymine (see Fig. 4-1). Generally about 80 nucleotides are present in each structure, which contains many double-helix-like arrays formed by the molecule folding back on itself. The amino acid is attached at the 3′ end by ester formation between a ribose hydroxyl group and the amino acid carboxyl group (see Fig. 4-2).

The transfer RNA molecules provide the part of the adaptor system that recognizes nucleotides, using the same hydrogen-bonding mechanism that prevails in DNA. The

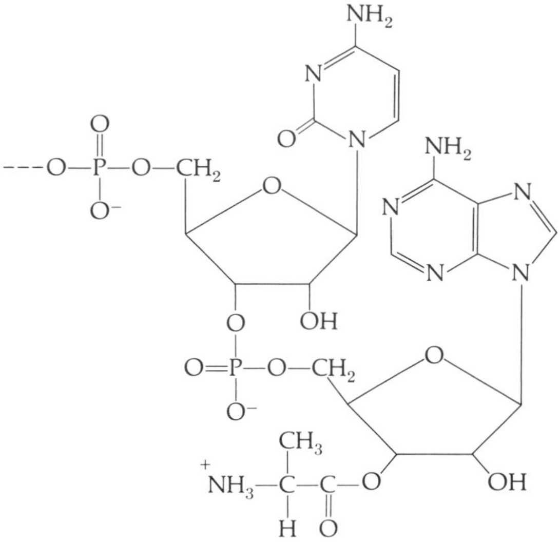

3′ end of tRNA with an amino acid (alanine) attached

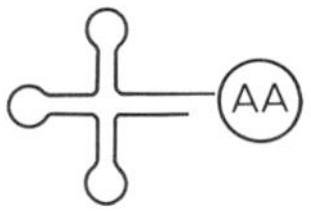

General structure of a tRNA with amino acid (AA) attached

FIG. 4-2 *Transfer RNA.*

connection to amino acids is provided by 20 *synthetase* enzymes. Each synthetase recognizes one amino acid by a highly specific three-dimensional structural complementarity. The amino acid is held on the surface of its enzyme and is "activated" by reaction with ATP to form aminoacyl adenylate (see Fig. 4-3):

$$\text{ATP} + \text{amino acid} \rightleftharpoons \text{aminoacyl AMP} + \text{PP}_i$$

Each synthetase also recognizes and binds just one of the 20 types of transfer RNA molecules. The transfer RNA is held in exactly the right position for the activated amino acid to react with the 3′-hydroxyl group of the free ribose at the end of the transfer RNA chain (Fig. 4-3):

$$\text{Aminoacyl AMP} + \text{tRNA} \rightleftharpoons \text{aminoacyl-tRNA} + \text{AMP}$$

The product formed, called aminoacyl-transfer RNA, is thus the link between nucleotides and amino acids.

The actual positioning of the aminoacyl–transfer RNAs on codons to align amino acids in the correct colinear fashion requires a great deal of cellular machinery, and DNA itself is not directly involved. First, a replica is made of each cistron to be read. In some cases clusters of cistrons that specify several enzymes of one pathway are read together. (These

First step, amino acid activation:

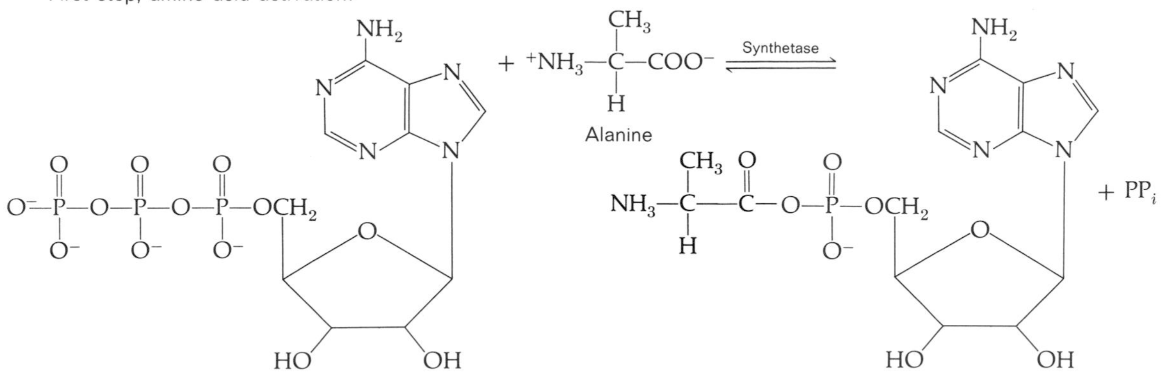

ATP + $^{+}NH_3{-}CH(CH_3){-}COO^-$ (Alanine) $\xrightleftharpoons{\text{Synthetase}}$ $NH_3{-}CH(CH_3){-}C(=O){-}O{-}P(=O)(O^-){-}OCH_2$-(ribose, adenine) (Aminoacyl AMP) + PP_i

Second step, aminoacyl transfer:

tRNA + Aminoacyl AMP $\xrightleftharpoons{\text{Synthetase}}$ Aminoacyl-tRNA (See Fig. 4-2 for structure)

FIG. 4-3 *Synthetase Reactions.* ATP and an amino acid (alanine, in this example) first react to form an aminoacyl AMP. The acyl bond is highly reactive and permits the activated amino acid to be attached to the correct transfer RNA in the second step.

clusters of cistrons are known as *operons* and will be discussed at length in Chapter 16.) For each cistron or operon the genetic information is transcribed as *messenger RNA* (mRNA), a single-stranded RNA. Molecules of messenger RNA are much longer than transfer RNA—sometimes several thousand nucleotides long. The synthesis of messenger RNA is analogous to the replication of DNA, although only one strand is copied. The copying process is known as *transcription* and quickly duplicates the number of copies of a given cistron in a cell. The enzyme RNA polymerase is responsible for the messenger RNA synthesis. For each gene or cistron in the double helix of DNA, only one strand contains information in the proper sense. The other strand contains the information as well, but in an inverted complementary form. It is the "inverted" strand to which nucleoside triphosphates (with ribose) pair according to the hydrogen-bond mechanism (G-C and A-U base pairs). Directed by RNA polymerase, the ribonucleotides are then covalently linked, via 3′,5′-phosphodiester bonds, to provide a replica of the DNA strand with the coding information in the proper sense. By a mechanism not yet clearly defined, the RNA polymerase recognizes an initiation site on the correct strand and begins transcription at the proper point. Transfer RNA and ribosomal RNA are also produced in a transcription process catalyzed by RNA polymerase.

The next element in protein synthesis is the ribosome, the actual site of protein synthesis. It is on the surface of ribosomes that transfer RNA molecules bearing their particular amino acids are aligned to messenger RNA codons and the appropriate peptide bonds are formed. A constellation of initiation, elongation, and termination factors participates with the ribosomes in this complicated task. Ribosomes are very large and complex, with a molecular weight of about 2.5 million and a large complement of RNA (known as ribosomal RNA) as well as many proteins in their structure.

The complete system for adapting nucleotide sequences to amino acid sequences thus invokes many individual components. Now we will look more closely at some of the individual components of protein synthesis.

Transfer RNA and the Synthetase System

Our perspective in the study of the biochemical events of the cell is oriented around the structure of DNA, the storage of information in the structure, and the expression of the information in proteins. All of these events take place by molecular mechanisms that rely on structural complementarity. Structural complementarity is clear in terms of base pairing in DNA, but is more subtly expressed in the adaptor system of protein synthesis. This system must unite codons with amino acids in a specific way. The task is accomplished by transfer RNA—composed of nucleotides—recognizing the nucleotides of messenger RNA, and an enzyme, the synthetase—composed of amino acids—recognizing the amino acid. The principle that emerges from this relationship is that structural complementarity often exists between similar structures. Nucleotide-nucleotide interactions provide one side of the adaptor system, and amino acid–amino acid interactions provide the other side. What happens when the two sides of the system are joined? How does the synthetase recognize the transfer RNA?

The synthetases are very complex enzymes—some contain almost 1000 amino acids linked in a single polypeptide chain. Thus the synthetase enzymes rank among the larger enzymes known. The great intricacy of structures of this size suggests that they would be capable of recognizing transfer RNA in a number of ways. The first possibility is that the enzyme itself forms some arrangement of amino acid side chains that "impersonates" a sequence of nucleotides by placing hydrogen-bond-forming groups in the right geometrical arrangement. This part of the enzyme could then recognize and bind by structural complementarity (involving a system of hydrogen bonds) to the same part of the transfer RNA (the anticodon) that recognizes the nucleotide codon in the messenger RNA. With this mechanism the enzyme's transfer RNA recognition step would occur by having the enzyme mimic the nucleic acid. The second possibility is that the transfer RNA molecule, also a sizable molecule, acquires a particular three-dimensional structure, a *tertiary structure*. In this case the precise tertiary structure of the RNA or the topography of a helical region from the "outside" is the principal feature recognized by the synthetase, and the part of the RNA that recognizes the messenger RNA, the anticodon, plays no role in the process. In this second type of mechanism the transfer RNA mimics a protein in the sense that it acquires a precise and presumably irregular three-dimensional structure that serves as the "password" for recognition by the appropriate synthetase.

Although the details of the binding of transfer RNA to its synthetase have yet to be elaborated, it seems likely that the second mechanism applies. The available evidence indicates that the synthetase recognizes the overall structure of the transfer RNA. Thus two distinct types of structural complementarity can be found in the transfer RNA–synthetase system, one based on hydrogen bonding and the other based on tertiary-structure recognition (see Fig. 4-4). For protein synthesis to succeed, both types of recognition must be nearly perfect. As noted in the general introduction, some error in the genetic mechanism is needed to provide variation, a prerequisite for selection of new characters and evolution. But mistakes at the level of the adaptor system provide no reproducible mutational variation since they are not propagated in successive generations.

The difficult discrimination problems that have been overcome by the adaptor system are especially apparent when amino acids of close similarity in structure are considered. Insights in this area were achieved with the fractionation of individual synthetases, particularly by Paul Berg and his collaborators. The amino acid isoleucine, for example, is structurally very similar to leucine and valine. Yet the isoleucine synthetase discriminates against these other amino acids and

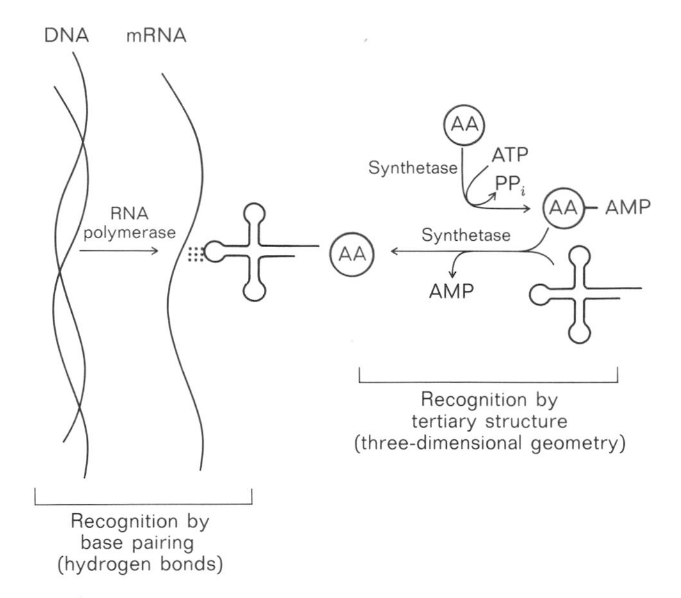

FIG. 4-4 *Adaptor System.*

only links isoleucine to the isoleucine transfer RNA. The fact that a process of this type would tax even the versatility of an enzyme is verified in the case of isoleucine synthetase, an extensively studied enzyme. When tested with the valine, some activated valyl AMP was observed. However, valine is activated to a much lower extent than isoleucine. Nevertheless, the activation is appreciable and could interfere with the fidelity of the protein-synthesis mechanism. Although the enzyme has not been able to discriminate fully against the valine, a second level of checking avoids charging the isoleucine transfer RNA with valine. Both isoleucine and valine are activated, but the valine is not linked to the transfer RNA. In fact the valyl AMP, which is normally stable on the enzyme, is actually hydrolyzed in the presence of isoleucyl-transfer RNA. Hence the genetic information is secure at least for the phase covered by the isoleucine synthetase.

Structural Details of Transfer RNA

The last section dealt with how the transfer RNA molecule is charged with an amino acid through a reaction involving the enzyme synthetase. The transfer RNA for each amino acid must possess enough structural specificity to distinguish it from the other 19 types of transfer RNA. In addition, the transfer RNA once charged must recognize some nucleotide-sequence codon on the messenger RNA. To gain some understanding of the way in which transfer RNA accomplishes these tasks, the structure will be considered in greater detail. One of the first important breakthroughs in the chemistry of transfer RNAs was the complete analysis of the nucleotide synthesis in alanine transfer RNA by Robert Holley in the mid-1960s. Since then many transfer RNAs have been sequenced. The strategy of sequencing yeast alanine transfer RNA involved cleavage with two specific ribonucleases (RNase): pancreatic ribonuclease, which cleaves the backbone on the 3′ side of pyrimidine units, and T1 ribonuclease, which cleaves the backbone on the 3′ side of the guanine nucleotide units (refer to Fig. 4-5). Digestion with each enzyme produces a distinct set of fragments, and the sequence of bases in the fragments can be determined by digestion with snake venom phosphodiesterase, which removes nucleotides one by one from the 3′ end. Eventually enough information was obtained from sequencing fragments to piece together the "linear jigsaw puzzle" to give the complete sequence.

Virtually all of the transfer RNAs studied fall into the pattern first described for the alanine transfer RNA, often referred to as a *cloverleaf* structure. The name cloverleaf stems from the possibility of arranging the sequence of nucleotides to give four regions of hydrogen bonding, three terminating with loops. Yeast alanine transfer RNA is expressed in this pattern in Fig. 4-6. Several features are significant. The molecule possesses the terminal sequence CCA. This same sequence has been found in all of the transfer RNAs studied so far, and the amino acid is always attached to the terminal A. Moving along the molecule, three loops are evident. The loop at the far end of the molecule from the terminal A contains the bases that physically recognize the genetic code on the messenger RNA molecules. (This anticodon region will be discussed in detail in the next chapter.) The other two loops may be involved in the recognition of the transfer RNA molecules by the synthetase enzymes and in the recognition of the transfer RNA molecules by the ribosomes. A third minor loop called the *lump* is also evident and varies in size among different transfer RNA molecules that have been studied.

A special characteristic of transfer RNAs is that they frequently contain unusual bases in addition to the four principal bases A, G, U, and C. These bases may have additional alkyl groups or structures that are quite distinct from the four

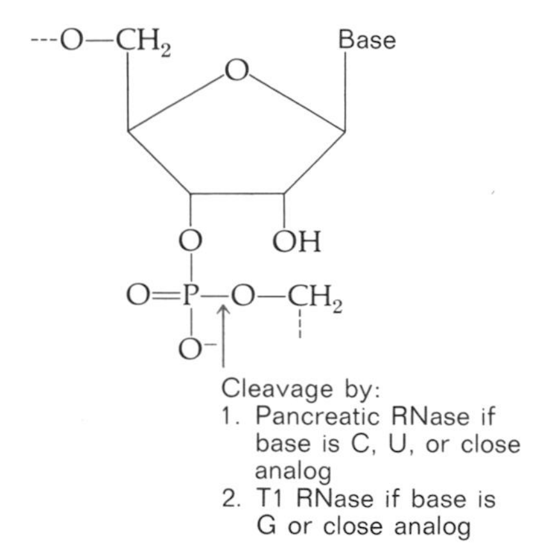

FIG. 4-5 *Nuclease Action on RNA.*

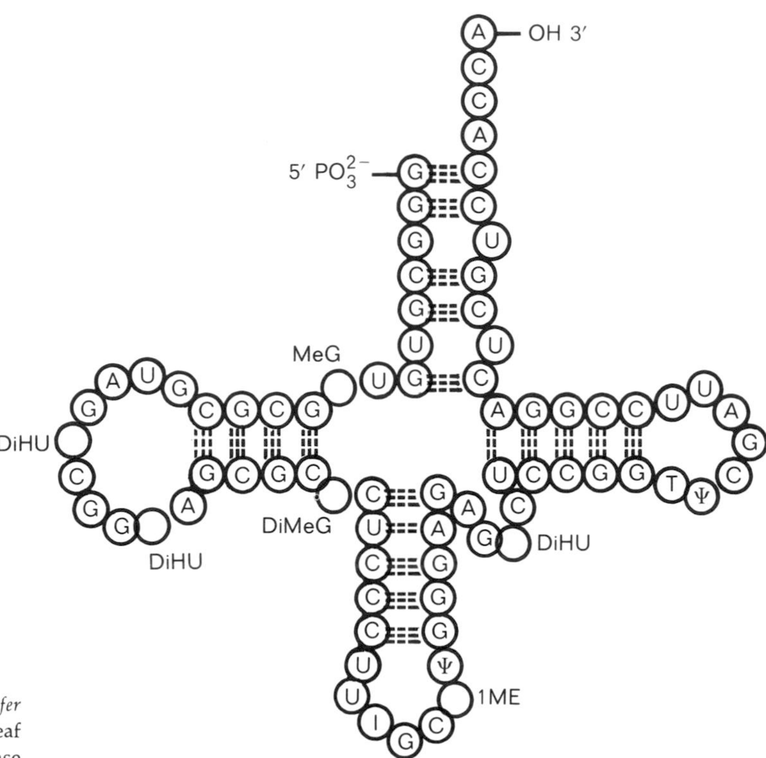

FIG. 4-6 *Structure of Yeast Alanine Transfer RNA*. The bases are arranged in the cloverleaf structure. Note high incidence of G-C base pairs. From R. W. Holley et al., *Science*, **147:** 3664 (1965). An additional G is included as suggested by C. A. Merril, *Biopolymers* **6:** 1727 (1968). For the structure of a transfer RNA, see S. H. Kim et al., *Science*, **179:** 285 (1973).

common ribonucleotides. The occurrence of these unusual bases is indicated in Fig. 4-6 by the special abbreviations, and the exact structure of these bases is shown in Fig. 4-7. As yet, complete knowledge of the atomic coordinates of a transfer RNA is unavailable. Several laboratories are working toward solving the structure by x-ray crystallography and the precise three-dimensional structure should be very revealing. The work of Alexander Rich and his colleagues indicates an L-shaped structure for yeast phenylalanine transfer RNA and now raises specific questions concerning the precise alignment of transfer RNA molecules on the ribosomes.

Ribosomes

Ribosomes with their enormous size and complexity are perhaps the most dexterous structures involved in protein synthesis. The process of transcription provides the essential information content of protein synthesis. But information alone would be useless without a system capable of working with and delivering the information. The genetic code without ribosomes would be like having FORTRAN and no computers. Other essential stages in protein synthesis such as messenger RNA formation and transfer RNA–amino acid coupling represent to the biochemist only small extensions of reactions he is already familiar with. The reactions involving ribosomes, on the other hand, involve a much higher degree of complexity and manipulation. Molecules must be oriented spatially, and very large molecules must be moved physically in a coordinated manner. The role of ribosomes in translation—coordinating the reading of messenger RNA, alignment of charged transfer RNA, and initiation and elongation of the polypeptide chain—will be discussed in Chapter 6, after a description of the detailed signals of the genetic code in Chapter 5. Here we will consider the structural properties of ribosomes.

Ribosomes were first characterized by their movement in centrifugal fields. One of the simplest forms of sedimenta-

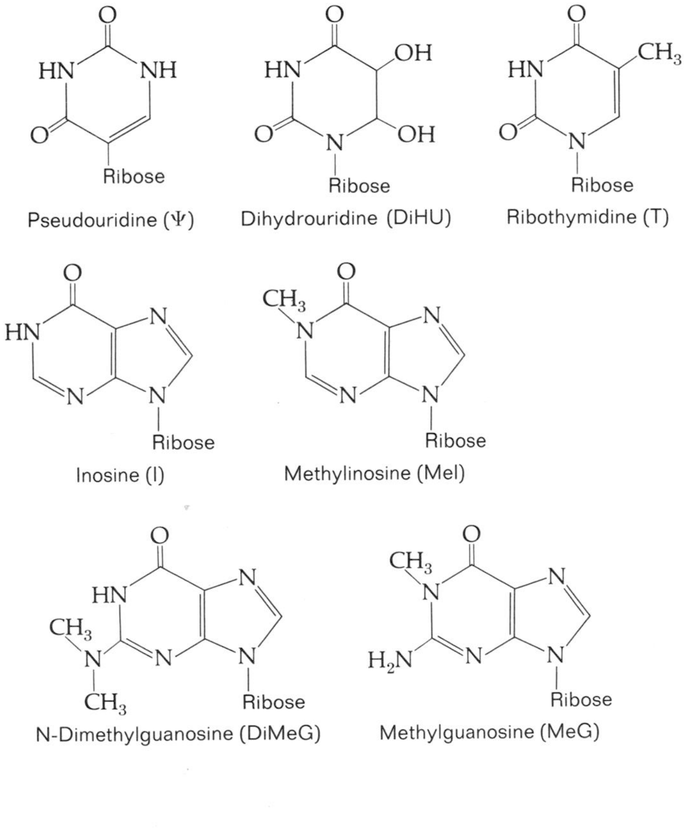

Pseudouridine (Ψ) Dihydrouridine (DiHU) Ribothymidine (T)

Inosine (I) Methylinosine (MeI)

N-Dimethylguanosine (DiMeG) Methylguanosine (MeG)

FIG. 4-7 *Unusual Bases Found in Transfer RNA.*

tion analysis involves measuring the rate of transport of material in a centrifugal field. The rate of movement through a centrifuge cell, dr/dt (r = distance from the center of rotation), varies with the strength of the centrifugal field, $\omega^2 r$ (ω = angular velocity of the rotor in radians per second). Therefore a simple normalized measure of the rate of sedimentation, called the *sedimentation coefficient* or s, is given by

$$s = \frac{dr/dt}{\omega^2 r} = \frac{d \ln r/dt}{\omega^2}$$

Ribosomes were first noticed in cell extracts because of their high rate of sedimentation, coupled with the apparent homogeneity of the sedimenting material. The complete ribosomal unit from *E. coli* is characterized by a sedimentation coefficient of s = 70S, corresponding to a molecular weight of over 2.5 million. (The capital S refers to Svedbergs, the units of sedimentation, named for Thé Svedberg who built the first ultracentrifuge in Sweden in the 1920s.)

If magnesium ions, essential for the structure of ribosomes, are not present, the 70S structure dissociates into two major ribosomal subunits (Fig. 4-8). These are the 50S and 30S ribosome subunits. Both are essential for complete protein synthesis, and it is now possible to delineate individual functions for these two units (see Chapter 6). Also, both possess a high degree of structural complexity (see Fig. 4-9). Each contains a very long single-stranded molecule of RNA, the ribosomal RNA. The RNA, in fact, represents over half the weight of each of these subunits and values of 500,000 (16S) and 1,100,000 (23S) are normally accepted for the 30S and 50S subunits, respectively. A smaller 5S ribosomal RNA molecule is also contained in the 50S subunit. At present the role of these ribosomal RNA molecules is unclear. They may have some functional importance in the recognition of messenger and transfer RNA molecules or may simply provide a structural matrix for the protein molecules in ribosome assembly.

In addition to these long RNA molecules, the ribosomes contain a large array of proteins. Approximately 30 proteins have been identified from the 50S ribosomes and 20 proteins have been characterized from the 30S ribosomes. More proteins have been found than can be accommodated by the molecular weights of the ribosomes. This observation implies that there is some heterogeneity in ribosomes, a characteristic that could play a significant role in protein synthesis.

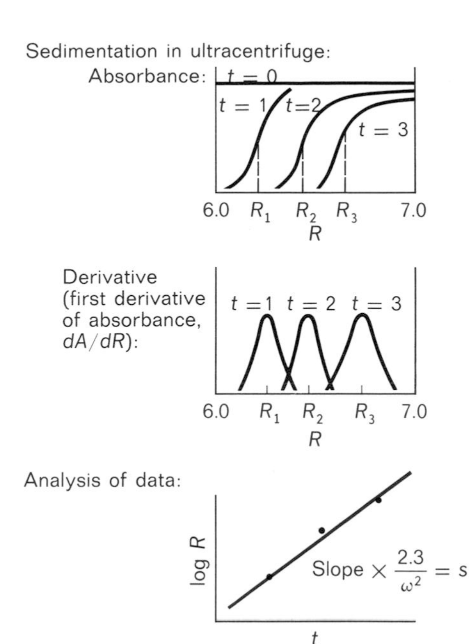

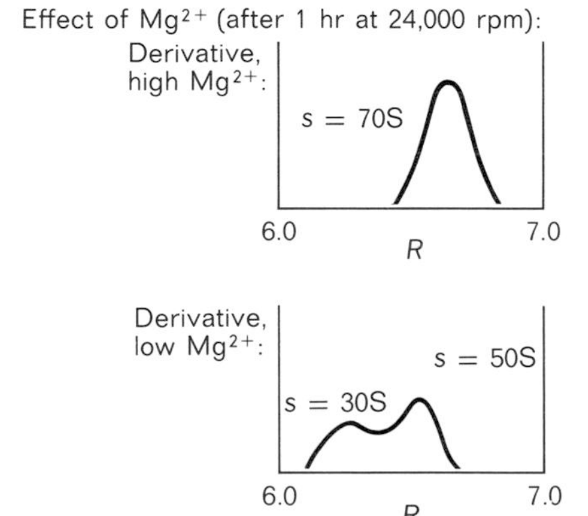

FIG. 4-8 *Characterization of Ribosomes.* When ribosome solutions are centrifuged, sedimentation occurs with time. The logarithm of the boundary position versus time gives the sedimentation coefficient s. The s value for ribosomes depends on the concentration of magnesium (Mg^{2+}) present. At high Mg^{2+}, a single boundary with s = 70S is found. At low Mg^{2+}, two components are seen, with s = 30S and 50S.

What is especially striking about the large number of proteins in ribosomes is that their functions are highly cooperative. Dissociation and reconstitution of ribosome proteins by Traub and Nomura has been surprisingly successful, especially with the 30S subunit. The parent particles can be fractionated into their constituent units and the individual components, when properly mixed, will spontaneously re-form active ribosomes. Reconstitution experiments have been performed with the omission of one or another of the proteins and the results have been carefully monitored. When one of

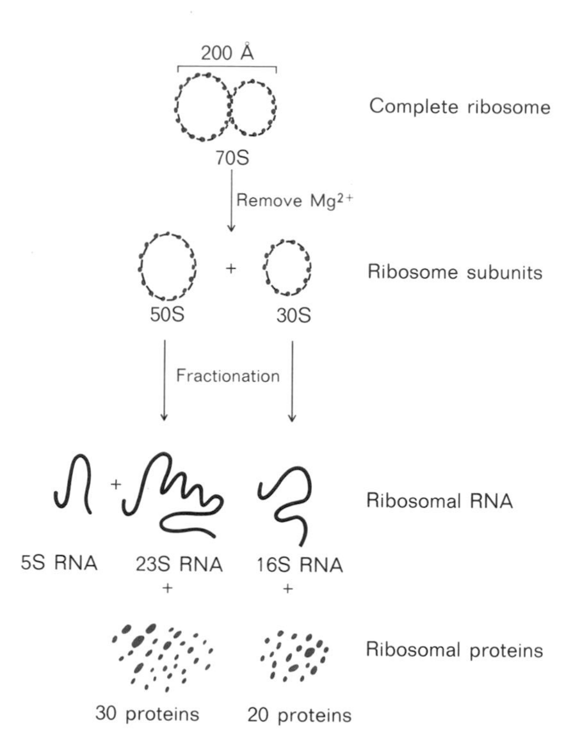

FIG. 4-9 *Ribosome Structure.* The 50S and 30S subunits have molecular weights of 1.6 million and 0.9 million, respectively. For further details, see C. G. Kurland, *Annual Review of Biochemistry,* **41:** 377 (1972).

a large class of proteins is eliminated from the reconstitution mixture, the reassembly into the 30S structure will not take place. The proteins are likely to assemble in a specific sequence and omission of just one key protein may interrupt the process. In addition there is another class of proteins that, when omitted from the reconstitution mixture one at a time, leads to ribosomes that appear to have the proper shape and structure, but that are inactive in protein synthesis.

Thus the ribosome emerges as a highly ordered and highly intricate structure in which the various proteins involved in the structure and function are linked in such a way that the precise arrangement of many of these proteins is indispensable for normal ribosome activity. Just how this activity is expressed will be discussed in Chapter 6.

PROBLEMS

1 Nucleotide triplets with four types of nucleotides can be arranged in 64 possible combinations. If the genetic code were based on only two types of nucleotides, how many combinations would be possible for nucleotide triplets?

2 If, instead of tertiary-structure recognition, synthetases recognized their transfer RNA molecules by complementary hydrogen bonding to the bases of the anticodons, which amino acid side chain could conceivably participate in such hydrogen bonding?

3 Which unusual nucleotide found in transfer RNA closely resembles a component of DNA?

4 The 50S ribosomal subunit is about 1.8 times the size of the 60S ribosomal subunit. For spherical structures, the sedimentation coefficient varies with $[MW]^{2/3}$. On the basis of this information, are the ribosomal subunits approximately spherical?

Chapter 5 Frontispiece *Photograph of a petri dish showing plaques of T2 bacteriophage on a lawn of E. coli. Three types of plaques are present: wild type (r^+), a variant with a mutation in the r gene (r), and a variant that arises when both r and r^+ types are grown jointly (mottled). From G. Stent, Molecular Biology of Bacterial Viruses, W. H. Freeman and Company, San Francisco, 1963.*

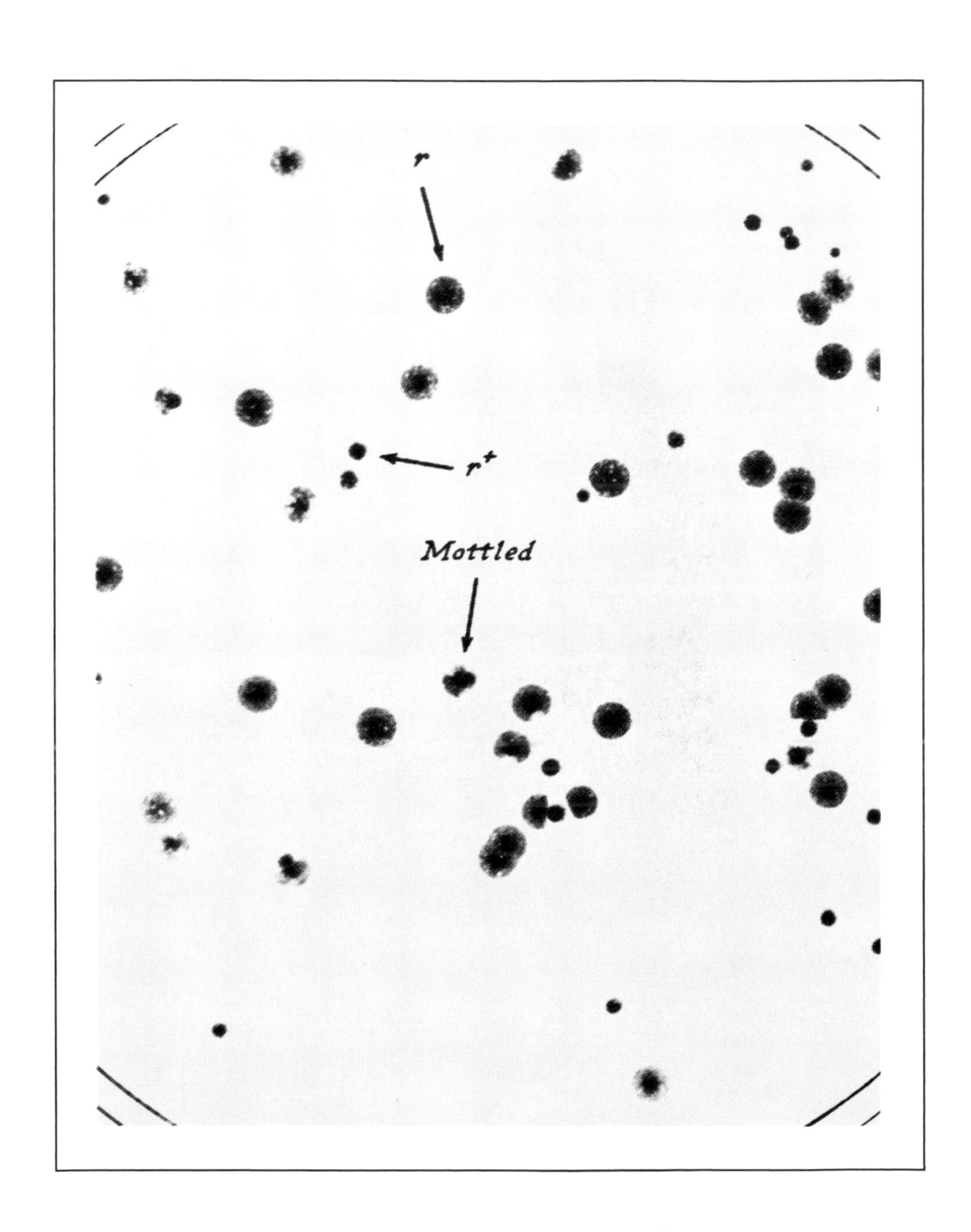

r
r+
Mottled

GENETIC CODE

5

Chapter 4 reviewed the machinery involved in the synthesis of proteins in living cells. This chapter deals with the *genetic code,* the precise information system that instructs the machinery to insert a particular amino acid into a polypeptide in response to a particular sequence of genetic material. The complete function of the machinery was only fully appreciated with the development of knowledge about the genetic code. Similarly, the genetic code could only be solved with a good appreciation for the machinery. Thus there is a very important reciprocal relationship between the machinery and the code, and after a discussion of the code in this chapter, a subsequent chapter will present an integrated view of the machinery and the code.

The Experimental Elucidation of the Code

There are four distinct stages of experimental work that must be considered in understanding the genetic code. The first stage was very biological and involved the conclusion that each amino acid is most likely specified by a set of three nucleotides. This conclusion was reached from the "phase-shift"- (also known as "frame-shift"-) mutation experiments with a bacteriophage of *E. coli,* T4. The second and third stages of work on the genetic code, which were realized principally through the efforts of Marshall Nirenburg, involved the priming of a reconstituted protein-synthesis system with synthetic messenger RNA, or a reasonable facsimile. The initial phase of these experiments, which really created the sense of a breakthrough in the analysis of the genetic code, was a demonstration that synthetic messenger RNA composed entirely of uridine (poly U) would give rise to the synthesis of a polymer of phenylalanine when introduced in the proper mixture of ribosomes, RNA, and other soluble factors. This second stage of experimental work on the code led to the assignment of codons for several other amino acids, but it was really with the third stage of development that the entire dictionary could be assigned. This third stage involved the discovery that transfer RNA molecules bind to ribosomes in the presence of ribonucleotide triplets. Moreover, the composition of the triplets determines which transfer RNA was bound. Thus, for example, the triplet UUU when incubated with ribosomes directs the binding of phenylalanine transfer RNA. Further examination of this system revealed that each transfer RNA molecule contains a triplet complementary to the codon in the messenger RNA that directs its binding to the ribosome. The fourth stage in the elucidation of the genetic code is in many ways the most elegant and complete. This concerned the actual synthesis of messenger RNA molecules of known sequence, and the determination of the amino acids incorporated into polypeptide chains in response to each synthetic message. These studies are due to the pioneering work of Har Gobind Khorana. Now each of the four stages in the elucidation of the genetic code will be considered in detail.

Phase-shift Mutations and a Triplet Code

Molecular biologists sometimes revel in their ability to unravel complicated problems without getting their hands wet. An outstanding example of this strategy in practice concerns the experiments with intact cells that led to the idea that the genetic code is based on nucleotide triplets. The leading parts in this drama were played by Francis Crick and Sydney Brenner, with supporting roles assigned to *E. coli* and its bacteriophage T4. A single bacteriophage can attach to an *E. coli* cell, inject its DNA, and quickly transform the bacterium into a factory for manufacturing more bacteriophages until hundreds are formed. The cell is then lysed and the many bacteriophages are released to attack other cells.

The T4 bacteriophage had been shown by Seymour Benzer to possess an especially interesting property, one that made genetic analysis very simple. When T4 was grown on *E. coli* strain B, it was possible to recognize bacteriophage mutants with an unusual property. Even, round spots are formed by these bacteriophages as they destroy regions of *E. coli*. These spots, known as *plaques,* form in regions of a petri dish layered with a lawn of *E. coli* to which a small amount of the bacteriophage is added. When the bacteriophages have attacked a cell and begin to multiply rapidly, the bacteria in the region will be destroyed and eventually a small plaquelike clear area will appear on the plate. The mutants produced uneven, ragged, "mottled" spots. When mutants that formed the mottled plaque obtained with *E. coli* B were grown on strains of *E. coli* K, no plaques were seen. In contrast, the wild-type T4 grew quite normally on strain K.

The gene responsible for this altered behavior—mottled plaques on strain B and no plaques on strain K—was called *rII*. These properties (summarized in Fig. 5-1) led to a very simple assay for wild-type genetic recombinants to map the *rII* gene. Two mutants of *rII* at distinct points on the chromosome are grown together on strain B and some recombinants form normal *rII* bacteriophages by crossing-over of their DNA. The frequency of these recombination events (proportional to the distance between the two mutants) could be easily monitored by replating the material on strain K, since only the "normal" recombinant would grow at all. Experiments of this type permitted an enormous number of mutations to be mapped with such a high resolution of mapping that point mutations only one nucleotide apart could be distinguished. These experiments in fact helped to substantiate the idea that individual nucleotides of DNA were important units in the coding of genetic information.

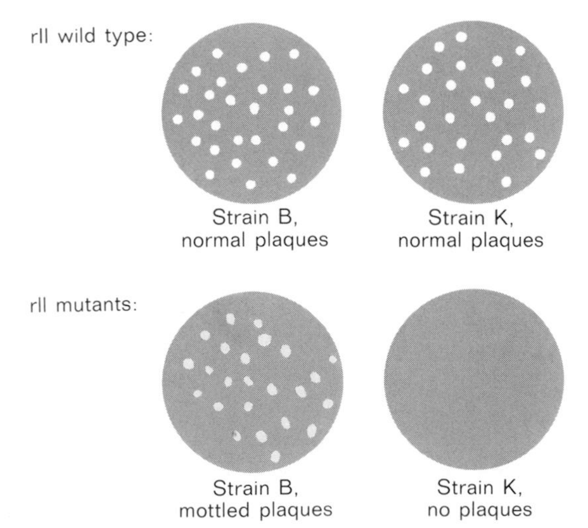

FIG. 5-1 *Plaque Assay for rII Mutants.* Mottled plaques arise when both *rII* wild type and *rII* mutant viruses are grown jointly. Plaques from *rII* mutants alone have a large, sharp halo. See also Chapter 5 frontispiece.

A second important finding to emerge from these studies led directly to the ability to perform the phase-shift experiments. The *rII* could be resolved into two distinct cistrons. Genes are normally assigned on the basis of altered phenotype, and the designation *rII* refers to the altered plaque-forming capacity. However, the fine-structure genetic analysis revealed that the *rII* gene actually behaved as two independent units. These units were called the A and B cistrons of the *rII* region, with the presumption that each cistron gives rise to one polypeptide chain. The A and B cistrons were discovered through a property known as *complementation*. When the large map of mutations of the *rII* region had been compiled and the various point mutations ordered, it was found that the various mutants fell into two categories, A and B. These categories divided the map into a left-hand and right-hand section according to the following property: Any mutant in A, when infected along with a mutant in B, would lead to wild-type behavior. In contrast, mutants in A, when mixed with other mutants in A, would lead to wild-type behavior only in the small number of cases that could be attributed to genetic recombination. The complementation phenomena, however, occurred in much greater numbers and could not be explained by recombination. Therefore two distinct cistrons were postulated. The arguments for two cistrons are summarized in Fig. 5-2.

The important phase-shift- (or frame-shift-) mutation experiments (to be described shortly) relied on two additional factors: a strain of T4 involving a small deletion just at the interface between the A and B cistrons, and the mutagen acridine, which acts through inserting or deleting bases in DNA. In the deletion mutant that eliminated a large part of A, but only a small region of B, the B activity was still present. However, acridine induced mutations in A on the same chromosome as the deletion and also exerted an effect on B, in that the B cistron was no longer active in complementation experiments of the type that demonstrated the presence of two cistrons. It was thought that these acridine mutants, by causing an insertion or a deletion, altered the phase of the genetic information, and by shifting the phase also confused the reading of the message in the B cistron. Without the deletion joining the A and B cistrons there would have been a normal interruption in the reading between the A and B regions and the B would have begun again in the proper phase, as indeed occurs in strains without the deletion. How-

Two mutants with alterations in different sections of rII (both form mottled plaques on B, no plaques on K):

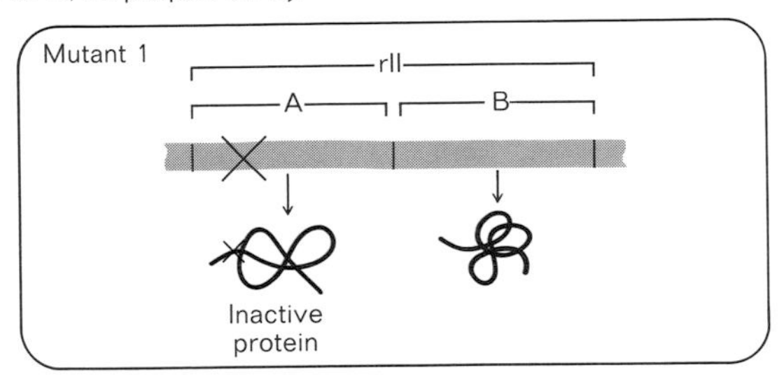

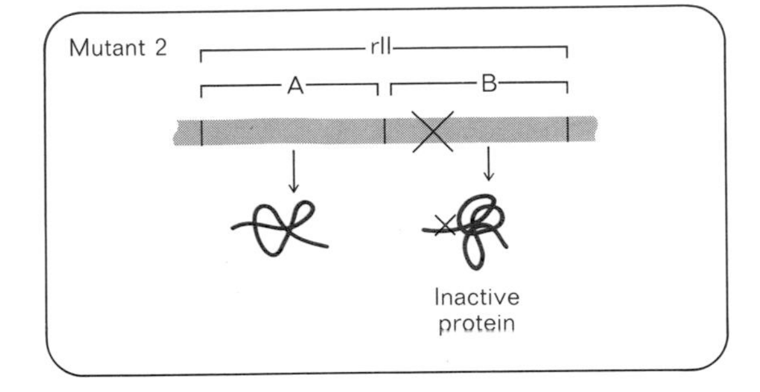

Infect same cells with both mutants:

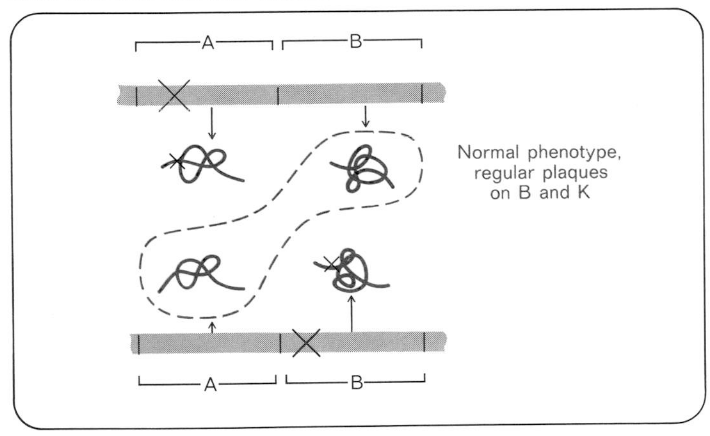

FIG. 5-2 *Complementation between Two Mutants.*

ever, by uniting the two cistrons through the deletion, the phase shift of the reading of A was propagated into the B region.

An important clue was also noted at that time. In certain cases the function of the B cistron could be restored by the presence of two mutations in the A cistron. Presumably one of these acridine-induced mutations was an *insertion* of one

1. Normal message:
   THE CAT ATE THE RAT AND WAS SAD ALL DAY
2. *Insert* one letter, entire message out of phase:
   THE ⊗CA TAT ETH ERA TAN DWA SSA DAL LDA
3. *Delete* one letter, entire message out of phase:
   THE CTA TET HER ATA NDW ASS ADA LLD AY
   ↓
   A
4. Combine *insertion and deletion*, message phase restored:
   THE ⊗CT ATE THE RAT AND WAS SAD ALL DAY
   ↓
   A
5. Combine *three insertions* or deletions, message phase restored:
   THE ⊗⊗⊗ CAT ATE THE RAT AND WAS SAD ALL DAY

FIG. 5-3 *Illustration of Phase Shift.* Message taken from a lecture by S. Brenner at the College de France, Paris, 1968.

base and the other a *deletion* of one base. The insertion mutation had the capacity to shift the phase one ahead; the deletion mutation would shift the phase one behind. Either alone would destroy the sense of the genetic message in B. However, when present together, one would compensate for the other, and apart from the region between them, the genetic message would be restored to the proper phase. Still, these experiments alone do not relate to the size of the unit of genetic information. But that information was easily obtained from the acridine experiments. The various mutations of *rII* induced by acridine could now be placed in two categories: plus mutations and minus mutations. Any mutation from A of the plus type when recombined with a minus mutation from A would restore the phase of B. Here, of course, the assignment of plus and minus is arbitrary and could be reversed. The important point for the size of genetic code came from the observation that if *three* plus mutations, or *three* minus mutations, were recombined in A on the chromosome, the phase of B was restored. As seen in Fig. 5-3, experiments of this type strongly suggest a triplet code.

Experiments with Randomly Ordered Synthetic Messenger RNA

The biological approach to the genetic code was illustrated in the discussion of phase-shift mutations. In principle this approach could reveal the exact genetic dictionary. If mutations of a precise chemical type, for example, modifying A to resemble G, could be induced in the genetic material by chemical mutagens and their effects on replacement in amino acids determined, all the assignments of the genetic code could be made. Experiments of this type were certainly in the minds of workers and would have been pursued. However, shortly after the development of the phase-shift experiments, a more biochemical approach to the coding problem was discovered that eventually superseded the genetic approach. A single experiment stands out as the critical observation that permitted a biochemical analysis of the genetic code. This experiment was performed by Nirenburg and Matthaei in 1961. In the course of their effort to reconstitute a protein-synthesizing system, these workers conducted an experiment in which poly U was added as the messenger RNA. In addition, ribosomes, transfer RNA, and other factors were included. Remarkably, poly U permitted protein synthesis to occur and only polypeptides of phenylalanine were formed. Thus these experiments provided a strong implication that some sequence of uridine represented the genetic code for phenylalanine. Subsequent experiments demonstrated that other RNA-like homopolymers would also direct specific incorporation of amino acids. Poly C was found to direct the incorporation of proline, and poly A, the incorporation of lysine. Thus by simple experiments of this type, three genetic-code assignments could be made. Given the likelihood of a triplet code on the basis of the phase-shift-mutation studies, the codons UUU for phenylalanine, CCC for proline, and AAA for lysine were indicated. Experiments with poly G were hampered by the strong tendency of this material to associate with itself, and the assignment of GGG for glycine was only made at a later time.

Experiments with synthetic homopolymers thus provided a number of key points in the genetic code, but clearly, to account for all 20 amino acids, codons containing combinations of nucleotides are involved. In order to investigate the other codons, experiments were initiated with synthetic messenger RNAs containing more than one type of nucleotide. The synthetic messenger RNA molecules are most conveniently synthesized with the enzyme polynucleotide phosphorylase. This enzyme catalyzes the reaction of nucleoside diphosphates to yield RNA polymers. Homopolymers such as poly U are formed simply by incubating UDP with the enzyme. Therefore incubations with two or more nucleoside diphosphates yield polymers of several nucleosides. In addition, by varying the ratio of the two nucleosides in the mixture, polymers of different nucleoside frequency could be

TABLE 5-1 Coding Properties of Polymers

	Observed Amino Acid Incorporation		Tentative Codon Assignments	
Amino Acid	Poly AC (5:1)	Poly AC (1:5)	Poly AC (5:1)	Poly AC (1:5)
Asparagine	24	5	2A, 1C	2A, 1C
Glutamine	24	5	2A, 1C	2A, 1C
Histidine	6	23	1A, 2C	1A, 2C
Lysine	100	1	3A	3A
Proline	7	100	1A, 2C; 3C	1A, 2C; 3C
Threonine	26	21	2A, 1C; 1A, 2C	2A, 1C; 1A, 2C

Source: J. F. Speyer, P. Lengyel, C. Basilio, A. J. Wahba, R. S. Gardner, and S. Ochoa, *Cold Spring Harbor Symposium of Quantitative Biology* **28:** 559 (1963).

synthesized and their properties tested as well in the cell-free system. Some results with mixtures of A and C are given in Table 5-1. It is clear from these results that varying the concentration of A:C in mixtures of polynucleotides will result in polymers with different coding properties. In addition to the proline synthesized by CCC and the lysine synthesized by AAA, the mixed copolymer permits synthesis of asparagine, glutamine, histidine, and threonine. Moreover, the ratio of A to C influences the occurrence of these additional amino acids. Experiments of this type provide highly suggestive patterns to the genetic code. For example, since AC copolymer with more A than C promotes the incorporation of more asparagine than histidine, one could reasonably expect that asparagine is coded by two A's and one C and histidine by two C's and one A. However, it is not possible from experiments of this type to determine the order of the bases in the codon and certain difficulties in assignment can be caused when incorporation is small.

Transfer RNA Binding

Experiments with homopolymers and copolymers thus began to reveal the makeup of the genetic dictionary but were hampered particularly by the inability to determine sequence. The elucidation of the genetic code was again at an impasse, and once more Nirenburg emerged as the man of the hour. Along with Leder, he discovered a system to reveal the precise chemical makeup of the codons in the genetic dictionary. Specific codon-directed binding of transfer RNA to ribosomes provided the assay needed. It was already known that the presence of messenger RNA on the ribosome stimulated the binding of transfer RNA, and Nirenburg and Leder then found that trinucleotides alone were sufficient to stimulate specific binding of transfer RNA. UUU stimulated the binding of transfer RNA charged with phenylalanine, and, interestingly, UUC also stimulated the binding of phenylalanine. Moreover, the experiments with all 64 possible codons they synthesized provided a nearly complete assignment of the genetic dictionary. The results of this approach and some later work are shown in Table 5-2.

Wobble and Evidence About the Genetic Code from the Structure of Transfer RNA

The role of transfer RNA in protein synthesis, involving binding to codons on the messenger RNA, implies a complementarity of structure in which an *anticodon* should be present in the transfer RNA molecule. In fact, as shown in Fig. 4-6, the second loop of the transfer RNA molecule has a 5′ ⟶ 3′ sequence, IGC. Recognizing that this is the transfer RNA for alanine, it can be seen that a possible codon is GCC, and the sequence IGC exhibits a certain complementarity. Notably, the unusual base inosine (I) is involved, rather than one of the four common bases. I in the anticodon pairs with the third base in the codon. Studying the genetic-code desig-

TABLE 5-2 The Genetic Code

First Base	Second Base: U	C	A	G	Third Base
U	UUU Phe	UCU Ser	UAU Tyr	UGU Cys	U
	UUC Phe	UCC Ser	UAC Tyr	UGC Cys	C
	UUA Leu	UCA Ser	UAA ***	UGA ***	A
	UUG Leu	UCG Ser	UAG ***	UGG Trp	G
C	CUU Leu	CCU Pro	CAU His	CGU Arg	U
	CUC Leu	CCC Pro	CAC His	CGC Arg	C
	CUA Leu	CCA Pro	CAA Gln	CGA Arg	A
	CUG Leu	CCG Pro	CAG Gln	CGG Arg	G
A	AUU Ile	ACU Thr	AAU Asn	AGU Ser	U
	AUC Ile	ACC Thr	AAC Asn	AGC Ser	C
	AUA Ile	ACA Thr	AAA Lys	AGA Arg	A
	AUG Met	ACG Thr	AAG Lys	AGG Arg	G
G	GUU Val	GGU Ala	GAU Asp	GGU Gly	U
	GUC Val	GCC Ala	GAC Asp	GGC Gly	C
	GUA Val	GCA Ala	GAA Glu	GGA Gly	A
	GUG Val	GCG Ala	GAG Glu	GGG Gly	G

*** = termination

The codons read in the 5′→3′ direction.

Source: F. Crick, "The Genetic Code III," *Scientific American*, October 1966.

nations in Table 5-2 reveals, for a great many assignments, that the third base is neutral, or can be either a purine or a pyrimidine. Crick suggested that I can pair with uracil, adenine, or cytosine. Thus in the actual codon-anticodon bonding in protein synthesis there is a greater degree of freedom than present in the original base pairing studied in the structure of DNA. This greater degree of freedom was called *wobble* by Crick because of the ability for these anticodons containing unusual bases to form hydrogen bonds if the structures are wobbled slightly from their usual configuration for hydrogen-bond formation in the strict A-T and G-C pattern. Some examples for wobble are shown in Fig. 5-4 and

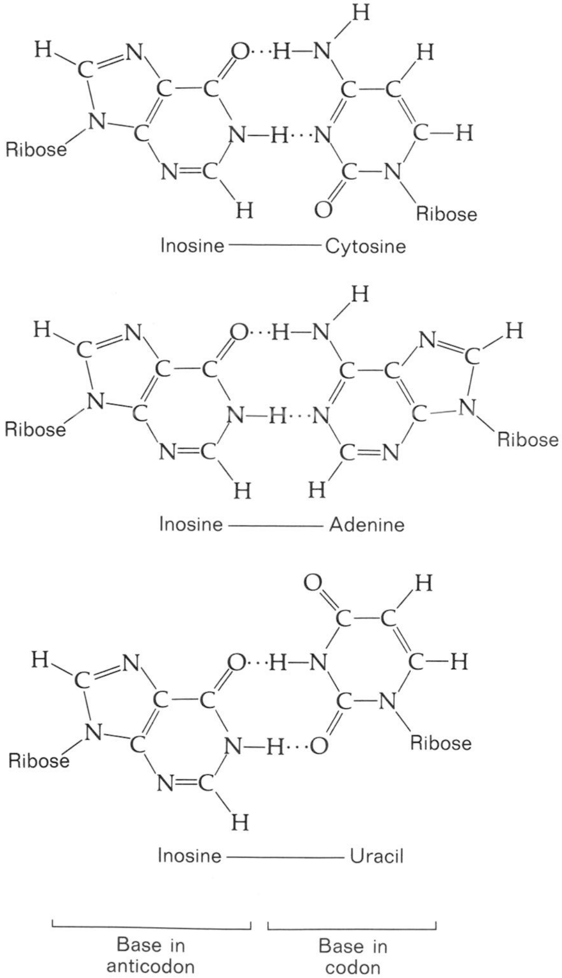

Inosine —— Cytosine

Inosine —— Adenine

Inosine —— Uracil

Base in anticodon | Base in codon

FIG. 5-4 *Examples of Wobble Pairing.* From F. Crick, *Journal of Molecular Biology,* **19:** 584 (1966).

provide a new pairing pattern as summarized in Table 5-3. A substantial number of anticodon sequences have been established from the chemical studies of transfer RNA molecules, and in all cases they correspond to the respective codon assignments already described. Thus the chemical sequencing of transfer RNA provides a strong line of evidence for the assignment of the genetic code since in this case a "synthetic" experiment is not involved, but rather the code is deduced directly from the analysis of natural products—the transfer RNA molecules as they are isolated from cells.

TABLE 5-3 Pairing Combinations with the Wobble Concept

Base in Anticodon	Base in Codon
G	U or C
C	G
A	U
U	A or G
I	A, U, or C

Source: F. Crick, *Journal of Molecular Biology,* **19:** 584 (1966).

Experiments with Ordered Synthetic Messenger RNA Molecules

The transfer RNA binding studies led to a nearly complete assignment of the genetic dictionary, but several reservations remained. First, a number of the codons bound transfer RNA poorly, making quantitative studies of their coding properties difficult. In addition the system did not actually involve protein synthesis, and it could be argued that some differences in properties arise in the actual biosynthetic process for protein formation. All of these reservations were overcome and final assignments of the genetic code made with confidence as a result of studies by Khorana and his associates with synthetic messenger RNA of known sequence. On the basis of their elegant chemical and enzymatic methods, these workers were able to synthesize several RNA polymers with known sequences. For example, the material with the sequence $(CU)_n$ was synthesized and found to produce a polypeptide in which leucine and serine alternate since the message could be read as CUC or UCU (see Fig. 5-5). When three bases are used in the ordered sequence, such as UUC in repeating blocks, three amino acids are incorporated into polypeptides since a different amino acid is specified depending on the phase in which the ordered messenger RNA is read. Some assignments of codons from this type of study are shown in Table 5-4. Finally, a number of polymers in which four nucleotides are present were synthesized. The incorporation of protein stimulated by these polymers permitted further confirmation of the assignments of the genetic code (see Table 5-4). Studies with synthetic ordered messenger RNA molecules also revealed a number of interesting properties concerning initiation and termination of protein synthesis. These properties will be considered in Chapter 6.

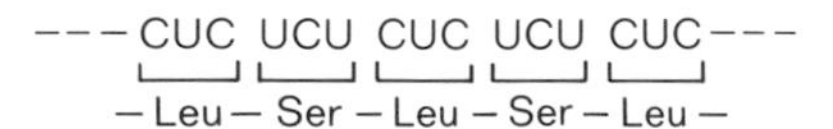

FIG. 5-5 *Coding Properties of Regular Copolymers.* From D. S. Jones, S. Nishimura, and H. G. Khorana, *Journal of Molecular Biology,* **16:** 454 (1966).

Mutational Studies and the Genetic Code

Two experimental approaches involving mutations have also been useful in establishing the validity of the genetic code. In a great many proteins studied in recent years, changes in one amino acid have been noted, frequently with an alteration of properties. Examples can be taken from hemoglobin, the mutant forms of viruses, and studies used to illustrate colinearity with the tryptophan synthetase gene. In almost all cases, the change in the amino acid can be shown to correspond to a single base change in the genetic code. Interestingly, in one exception, a change from glutamic acid to methionine in tryptophan synthetase, two base changes must be invoked. And in this case the back-mutation rate, or reversion to wild type, is much, much less frequent than the usual rate found with all of the other mutations studied.

Phase-shift mutations of the type used to establish the triplet nature of the genetic code have also provided systems to confirm the basic assignment of the genetic dictionary. Working again with the bacteriophage T4, but concerned with the gene and protein for lysozyme rather than *rII*, George

TABLE 5-4 Coding Properties of Regular Polymers

Three Bases

Polymer	Codons Recognized			Polypeptide Produced	Codon Assignments
$(AAG)_n$	AAG	AAG	AAG	Polylysine	AAG = Lys
	AGA	AGA	AGA	Polyarginine	AGA = Arg
	GAA	GAA	GAA	Polyglutamate	GAA = Glu

Four Bases

Polymer	Codons Recognized				Amino Acids Incorporated	Codon Assignments
$(UAUC)_n$	UAU	CUA	UCU	AUC	Tyrosine	UAU = Tyr
	1	2	3	4	Leucine	CUA = Leu
					Serine	UCU = Ser
					Isoleucine	AUC = Ile

Source: H. G. Khorana et al., *Cold Spring Harbor Symposium of Quantitative Biology*, **31:** 39 (1966).

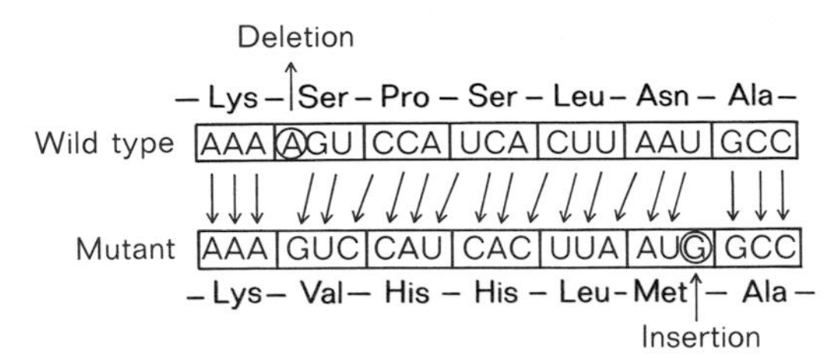

FIG. 5-6 *Phase-shift Mutations.* Based on E. Terzaghi, Y. Okada, G. Streisinger, J. Emrich, H. Inouye, and A. Tsugita, *Proceedings of the National Academy of Sciences,* **56:** 500 (1966).

Streisinger and his colleagues have characterized a double mutant in this gene. The double mutant involves the deletion of a single base, followed by the insertion of a base 15 nucleotides away. As summarized in Fig. 5-6, the amino acid sequence of the altered mutant derives from the sequence of the wild type just as predicted by the genetic code. These latter experiments thus provide an in vivo confirmation of the various experiments that led to the codon assignment, further assuring their overall validity.

General Aspects

A striking characteristic of the genetic dictionary is its *degeneracy*. It has been found that, in many cases, several codons code for the same amino acid. Yet, for methionine, AUG is the only codon. Similarly, tryptophan is coded for only by UGG. In addition, the functions of UAA, UAG, and UGA have not yet been discussed. The significance of the role of AUG and the unspecified codons will be apparent in the following chapter, a discussion of the integration of the machinery of protein synthesis to initiate, extend, and terminate biosynthesis of proteins. The degeneracy also provides a level of security, particularly in relation to *suppression* which also will be considered in the following chapter.

The most mysterious aspect of the genetic code is the particular arrangement of amino acids and codons that was selected several billions of years ago when the genetic code first arose, during the origin of life. Thinking on the subject falls into two categories. On one hand, the assignment of amino acids and codons may be completely random, with the codon bearing no stereochemical relationship to its amino acid, and the amino acid coded by similar codons, selected only by chance. On the other hand, one could imagine a specific stereochemical relationship between, for example, valine and its codon GUU. At the present time there is no evidence for such stereochemical relationships. Only certain curiosities exist, such as the fact that glycine, the least interactive amino acid by virtue of its lack of a side chain, is coded for by G's, the most interactive nucleotides. Guanine nucleotides are very "sticky" and readily tend to form aggregates. Could it be that the high interactivity of G is most compatible with the lack of side chains of glycine in some direct, stereochemical sense? This question cannot really be answered at the present time. Nor is there an explanation for the fact that the codons for five hydrophobic amino acids—phenylalanine, leucine, isoleucine, valine, and methionine—contain U in the second position. In considering a stereochemical basis for the origin of the genetic code, it should be kept in mind that not every amino acid must be stereochemically related to its codon. Conceivably, if only one amino acid and one codon bore some special affinity, such as glycine and the GGG codon, this would be enough to orient the genetic code. Subsequent assignments would be made in a "conservative" way, so as to minimize deleterious effects of single base changes in the codon. Thus we might expect alanine to be related to glycine by a single base change, as it is. Moreover, similar amino acids such as leucine and isoleucine and valine are related by single base changes. But the wisdom of hindsight may be misleading. Without a great deal more knowledge it is unlikely that a hypothesis of the origin of the code can be enunciated.

In fact it is quite possible that these questions will never be answered. To look seriously at the origin of the genetic code, one would have to investigate the origin of the entire mechanism of protein synthesis. To reconstruct the factors that led to the particular assignments of the genetic code of today, it would be necessary to understand the evolution of the machinery of protein synthesis itself. One would need knowledge of the chemistry of "primitive" proteins, and the properties of "primitive" nucleic acids. Thus the questions involved get extremely complicated, and, quite possibly, answers are out of reach.

PROBLEMS

1 What number of insertion mutations in the A cistron of *rII* other than three would in principle give in-phase reading of the B cistron?

2 (a) Which amino acids would be incorporated in a protein-synthesizing system primed with a random polymer of A and U?
   (b) Which amino acids would be incorporated with a regular polymer of A and U alternating?

3 (a) Which amino acid(s) has the largest number of codons coding for it?
   (b) Which amino acid has the smallest number of codons coding for it?

4 (a) Which codons could be recognized by a transfer RNA molecule with $^{3'}$AAG$^{3'}$ in the anticodon?
   (b) With which amino acid would you expect this transfer RNA to be charged?

Chapter 6 Frontispiece *Electron micrograph of the endoplasmic reticulum of a rat pancreas cell. The small black dots along the membrane of the endoplasmic reticulum are individual ribosomes. These structures are the sites of protein synthesis in eucaryotic cells. Magnification 62,000×. Courtesy G. E. Palade.*

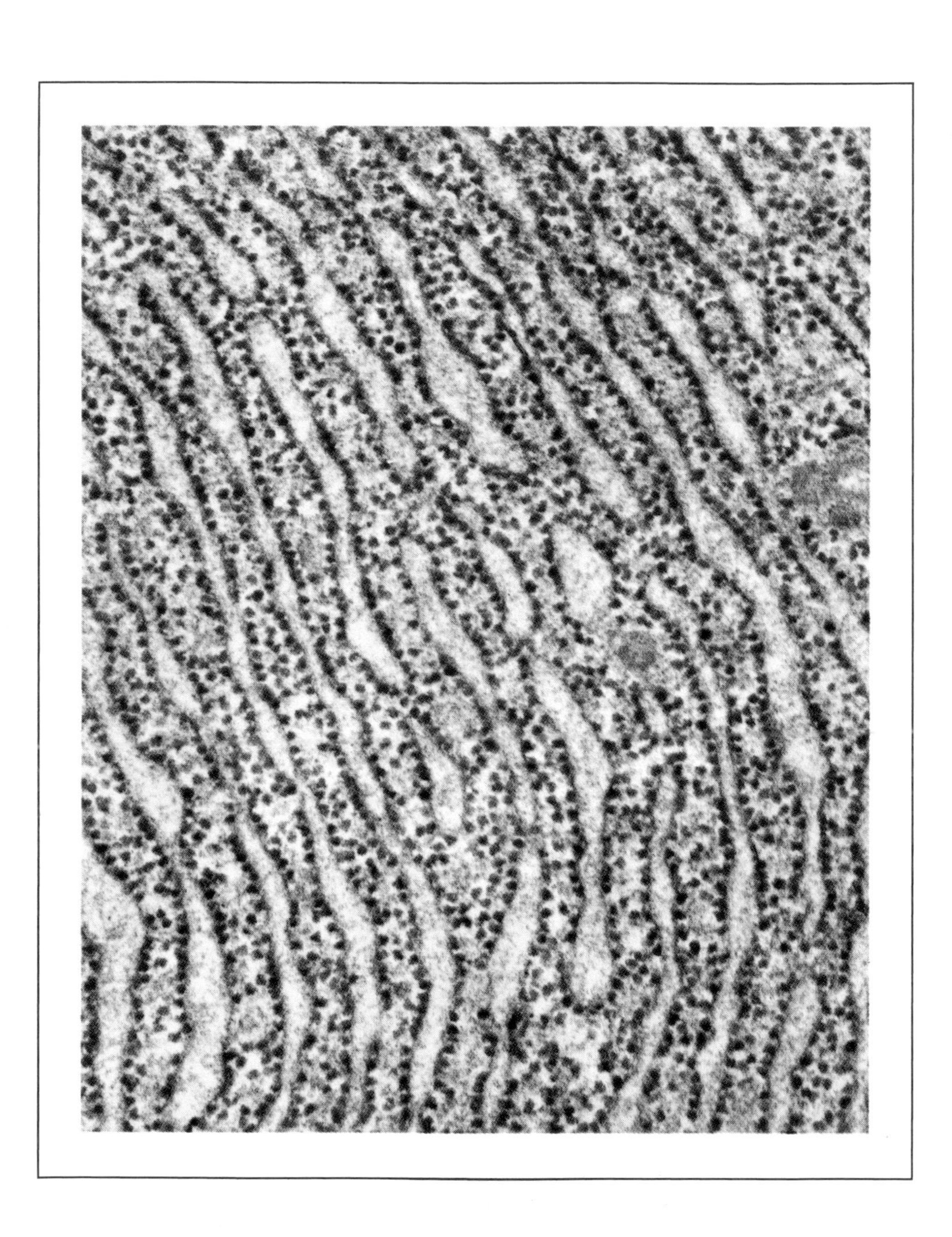

COORDINATION OF PROTEIN SYNTHESIS

6

The previous chapters hinted at the enormous complexity of the ribosome and its complicated function in the synthesis of proteins. This chapter will consider in greater depth the detailed steps involved in the synthesis of protein: the *initiation* of the synthesis on the ribosome, the *elongation* or continued synthesis for completion of the protein structure, and the *termination* of the synthesis to release a completed protein. This section will also describe the *direction* of synthesis, the number of ribosomes that cooperate on structures called polysomes, and the very interesting subject related to protein synthesis known as *suppression*.

Direction of Protein Synthesis

As noted in Chapter 4, the messenger RNA molecules, which transfer the genetic information from DNA to the ribosomes for protein synthesis, are generated from DNA by the enzyme RNA polymerase in the 5′ ⟶ 3′ sense. This direction was established in experiments with the inhibitor of messenger RNA synthesis, 3′-deoxyadenosine (abbreviated 3′-dA), an analog with the deoxy at carbon 3 instead of the usual carbon 2. The inhibitor is phosphorylated to 3′-dATP by cells and incorporated into growing messenger RNA polymers. But once incorporated, continued reaction is inhibited since the RNA chain no longer possesses a free 3′-hydroxyl to attach to the phosphate of the incoming nucleotide triphosphate. This finding implies that synthesis must occur in the 5′ ⟶ 3′ direction.

The next topic to consider is the direction of synthesis of the protein itself. Proteins are composed of a linear sequence of amino acids joined by peptide bonds, and the sequence can be expressed either beginning from the end with a free amino group or the end with a free carboxyl group. Quite early in the considerations of protein synthesis, it was demonstrated that the addition of amino acids to form a complete protein occurs by starting with the amino-terminal amino acid and adding the next residue to its carboxyl side. Each additional amino acid is added to the existing carboxyl-terminal position. Proof for this direction of synthesis came from experiments on hemoglobin synthesis in immature rabbit blood cells known as *reticulocytes*. Reticulocytes present a very convenient system for studying protein synthesis since hemoglobin is virtually the only protein formed in these cells. Reticulocytes were labeled by a brief exposure to a radioactive amino acid and the hemoglobin was isolated and digested into fragments. When the fragments were arranged according to their position in the sequence of hemoglobin, which had already been established, increasingly large amounts of radioactivity were found in fragments closer to the carboxyl-terminal end of the polypeptide chain. This observation, summarized in Fig. 6-1, strongly implies that amino acids are added onto an existing portion of the protein beginning at the amino-terminal end, and that the last amino acid added is the carboxyl-terminal residue.

Several of these points on "direction" are related to Streisinger's observation, summarized in the previous chapter. He reported that phase-shift mutations can be related to the amino acid sequence in mutant lysozyme from the bacteriophage T4. The comparison of altered codons and altered amino acids implies that the sense or direction of both languages can be correlated. The data indicate that the direction (amino end to carboxyl end), N $\longrightarrow$ C, of the protein, and

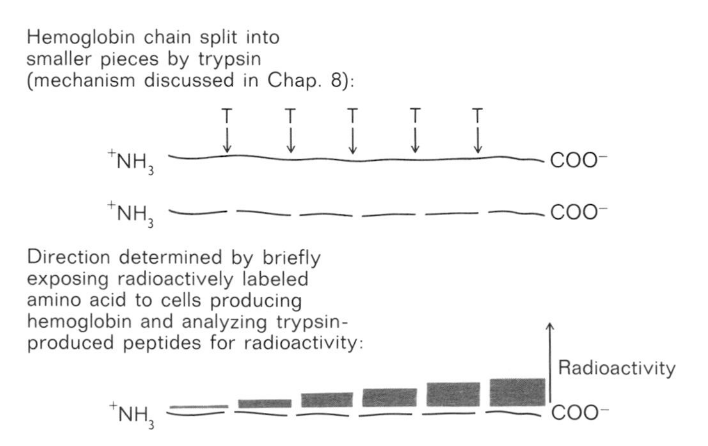

FIG. 6-1 *Hemoglobin Synthesis*. Experiment showing that a pulse of radioactive amino acids leads to preferential incorporation of radioactivity toward the carboxyl-terminal end of the protein. These results indicate that the protein is synthesized beginning at the amino-terminal end.

$5' \longrightarrow 3'$ for the nucleic acid sequence, are colinear. These directions have been employed in most of the previous considerations of proteins and nucleic acids.

Initiation of Protein Synthesis

To begin protein synthesis, the reading of the messenger RNA molecule must start in such a way as to maintain the correct phase of the message. Khorana's experiments showed that molecules initiated randomly will produce proteins with a variety of amino acids depending on the phase in which the message is read. This same phase problem is what enabled Crick and his colleagues to establish the triplet nature of the genetic code. Also, if a protein lacks a certain fraction of its sequence due to improper initiation, it may be of little value to the cell. Therefore not only must the message be started in phase, but it must be started at the point on the messenger RNA that corresponds to the first amino acid residue to be incorporated. Thus a precise initiation system for the cell is necessary, and a very accurate system for initiation has evolved.

Two important observations permitted the elucidation of the mechanism of initiation. The first was the discovery that proteins in *E. coli* are all synthesized with the amino acid N-formylmethionine as the first residue. At some stage in the formation of the newly synthesized protein, the formyl group is cleaved from the methionine, and then the methionine is also released from the protein by a specific peptidase. But protein synthesis will not begin unless the first residue is formylmethionine.

The second important discovery was the presence of a special transfer RNA molecule specific for methionine that allowed formylation of the methionine. Other methionine-specific transfer RNA molecules were known, but only the special one would permit the bound methionine residue to be formylated by the enzyme known as transformylase. These two observations provided a strong indication for the role of N-formylmethionine in initiation. However, the exact relation of formylmethionine to the genetic code was unclear. It was noted that the codon AUG is specific for methionine. Studies both with the formyl-accepting and non-formyl-accepting transfer RNAs specific for methionine showed that both were bound to ribosomes in the presence of AUG. Yet studies on protein synthesis strongly suggested that only formylmethionine would be incorporated at the initiation point on messenger RNA.

The actual initiation point of protein synthesis is governed by three proteins known as initiation factors. The factors couple the binding of formylmethionyl–transfer RNA, messenger RNA, and ribosomes to form a complex with the transfer RNA bound to AUG at the initiation site on the messenger RNA. In contrast, when AUG is encountered in the middle of a message, only the nonformylated transfer RNA is capable of reacting at the ribosome to incorporate methionine. Both formyl-accepting and non-formyl-accepting methionine transfer RNAs have now been completely sequenced and, remarkably, they differ only slightly in their structure. Therefore the discrimination between these two transfer RNAs in terms of the formylation reaction and the initiation steps must be very subtle.

The three factors that accomplish the events of initiation are known as F1, F2, and F3. The F1 participates in the attachment of the formylmethionyl–transfer RNA to the 30S subunit and in the binding of the 50S ribosomal subunit to the initiation complex. F2 is involved in messenger RNA binding to ribosomes, attachment of formylmethionyl–transfer RNA (along with F1), and the binding of GTP. The hydrolysis of GTP, a high-energy phosphate (see Chapter 11) takes place during formation of the initiation complex of protein synthesis. F3 is responsible for the binding of the messenger RNA to the 30S ribosomal subunit. These various steps are summarized in Fig. 6-2.

Elongation

Various stages in initiation, summarized above, lead to a 70S ribosome bound to the messenger RNA, with the initiation codon AUG in the proper position for codon-anticodon binding. Since two amino acids must be joined at the ribosome, a minimum of two sites for amino acids must be present on the ribosome. Evidence now suggests that there are two sites, known as D and A for donor and acceptor. Amino acid–bearing transfer RNA molecules enter at site A. This site would be the likely point of entry of formylmethionyl–transfer RNA. In a second step the aminoacyl–transfer RNA is "translocated" to site D along with an advancement of the messenger RNA by three nucleotides and a second charged transfer RNA enters at site A. Then, with formylmethionine

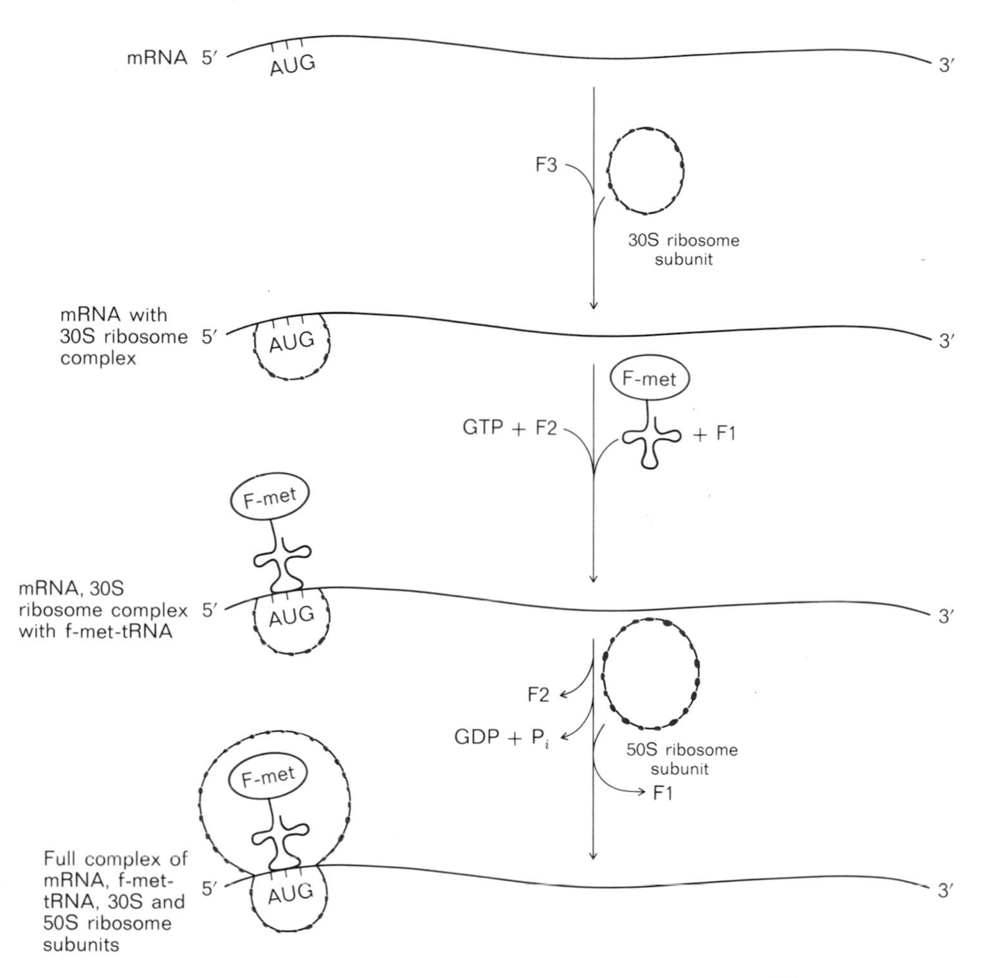

FIG. 6-2 *Initiation of Protein Synthesis.* Adapted from J. D. Watson, *Molecular Biology of the Gene,* 2nd ed., copyright © 1970 by J. D. Watson, W. A. Benjamin, New York, 1970.

at site D, the amino acid that enters at A would be the normal terminal amino acid for the protein under synthesis, for example, alanine. With the two amino acids in place, a peptide bond is formed by attack of the amino group of alanine at the acyl bond of the methionyl–transfer RNA. The transfer RNA in site D "donates" its amino acid or peptide to the aminoacyl–transfer RNA in site A, the "acceptor" site. The transfer RNA originally attached to the formylmethionine is then liberated and synthesis can continue with the translocation of the formylmethionyl-alanyl–transfer RNA molecule at site A to site D. Now another charged transfer RNA can enter site A, and the cycle continues. Therefore, in the normal process of elongation, site D will contain the already synthesized peptide chain attached to the transfer RNA molecule of the carboxyl-terminal amino acid of the peptide, and site A will be the point of entry of the next amino acid (on its transfer RNA) to be added to the growing chain. The existing peptide chain in site D is donated, the newly formed peptidyl–transfer RNA in site A is translocated to site D, and the next amino acid is awaited. This elongation cycle is summarized in Fig. 6-3 and is believed to involve at least two additional factors: T and G (T has two parts, Tu and Ts). T facilitates binding at the A site, and G catalyzes the translocation step to move the peptidyl–transfer RNA from site A to site D. This step is accompanied by the hydrolysis of the energy-rich molecule GTP to drive the reaction. A third activity that catalyzes the peptide-bond formation is localized on the 50S ribosomal subunit itself.

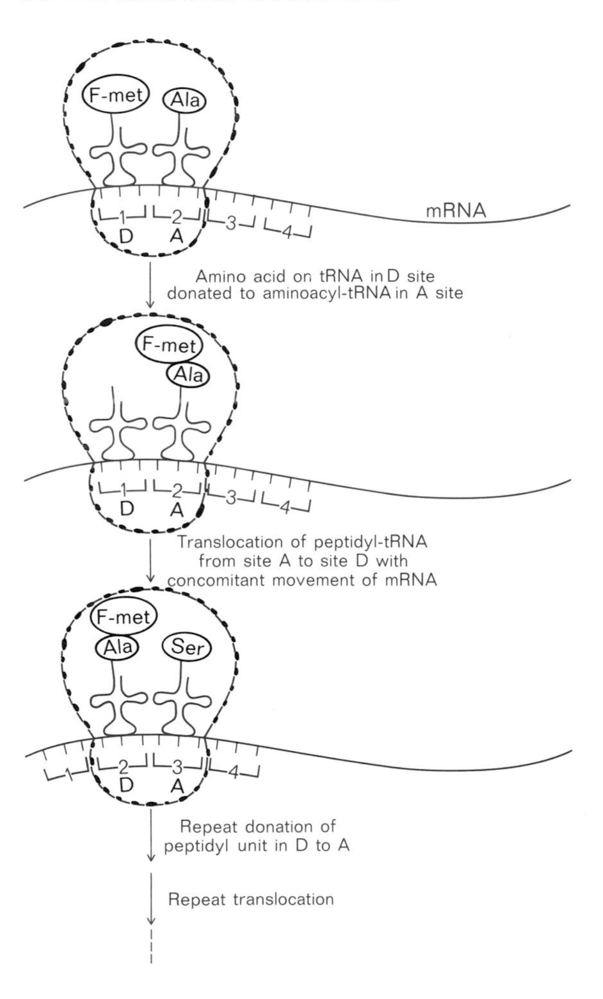

FIG. 6-3 *Elongation Cycle in Protein Synthesis.* The formylmethionyl residue may be cleaved from the polypeptide at a later stage of protein synthesis. For additional details, see J. Lucas-Lenard and F. Lipmann, *Annual Review of Biochemistry,* **40:** 409 (1971).

The elongation process, with its movement of bulky peptidyl–transfer RNA complexes from one site on the ribosome to another, represents one of the most complicated events that has been studied in molecular terms by biochemists. In summary, the process includes the following: the ribosome must recognize the messenger RNA and form a complex, the ribosome-messenger RNA complex must shift so that different regions of the messenger RNA are aligned for "reading" by the transfer RNA molecules, the proper transfer RNA molecules must be introduced in a position corresponding to the codons of the messenger RNA, the whole complex must be properly maneuvered through the peptide-synthesis reaction and, finally, the transfer RNA must be eliminated from the surface of the ribosome while the peptide chain grows. Not only is it complicated, but it is rapid—evidence suggests that a peptide chain of 100 amino acids can be synthesized in about 5 seconds. The linear chain then folds into a precise three-dimensional structure, probably while it is being synthesized, and eventually is released from the ribosome. The rules and forces governing the folding will be discussed in subsequent chapters.

The Termination of Protein Synthesis

Almost as important as the proper initiation of synthesis is the correct termination of synthesis. Failure to terminate could produce a protein with extra, perhaps structurally deleterious, amino acids. Moreover, failure to terminate could lead to possibly out-of-phase reading of another cistron, especially with multicistronic messenger RNA. Chapter 5 noted that in the genetic code 3 of the 64 codons do not specify any known amino acids. Careful studies with *E. coli* and its bacteriophages have revealed that these codons are signals for chain termination. When they arise by mutation in the middle of the cistron, premature termination occurs with the release of an incomplete peptide chain. The termination involves the release of the last transfer RNA molecule involved in the synthesis as well. It now appears likely that there are specific release factors that recognize the termination codons and hydrolyze the acyl bond to the transfer RNA to release the peptide chain. Why there are three termination codons in the genetic code is a very interesting point that is intimately related to the problem of suppression.

Suppression

The concept of suppression arose from the observation that certain mutations causing chain termination of a protein in some strains of *E. coli* would permit normal synthesis when transferred to other "suppressor" strains of *E. coli* by genetic processes. Frequently, however, a "new" amino acid was placed at the position of premature termination in the suppressor strain. Moreover, the amino acid inserted would depend on the suppressor strain used. These observations led to the view that there are genes, known as *suppressor genes,* which alter the ability of a cell to read the genetic code. It is now clear that these suppressors arise from mutations in the transfer RNA molecules themselves. For example, the transfer RNA molecule for tyrosine, which recognizes the codon UAC, would bear in its structure the anticodon $^{3'}$AUG$^{5'}$ (reading from 3′ $\longrightarrow$ 5′). Should a mutational event alter the base in the codon in such a way as to transform the anticodon to $^{3'}$AUC$^{5'}$, then it would recognize UAG. If this termination codon $^{3'}$UAG$^{5'}$ were encountered, the altered tyrosine transfer RNA would recognize it and insert tyrosine at that point. As might be surmised, the various suppressors that insert different amino acids in response to the termination codon all insert amino acids that are related by their codon to the termination codon with one base change. Modified transfer RNAs with changes involving two bases are much less likely to occur.

Two other important points arise from the consideration of suppression. One is the value now seen for cells to possess several distinct transfer RNA molecules for each amino acid. Without additional copies, the change in a tyrosine transfer RNA molecule, for example, would render it incapable of reading the normal codon from tyrosine and completely interrupt protein synthesis. However, the presence of additional kinds of tyrosine transfer RNA molecules ensures that tyrosine will still be incorporated in the proper places in the sequences of all proteins being synthesized, in addition to the occasional occurrence in response to a termination codon. The second point of interest concerns the recent discovery that termination codons, as they occur at the end of natural cistrons, frequently come in groups. For example, the messenger RNA of the bacteriophage R17 contains all three termination codons—UAA, UAG, and UGA—at the end of the gene for the coat protein of the phage (Fig. 6-4). Thus, if the cell should develop a suppressor transfer RNA by some mutational event,

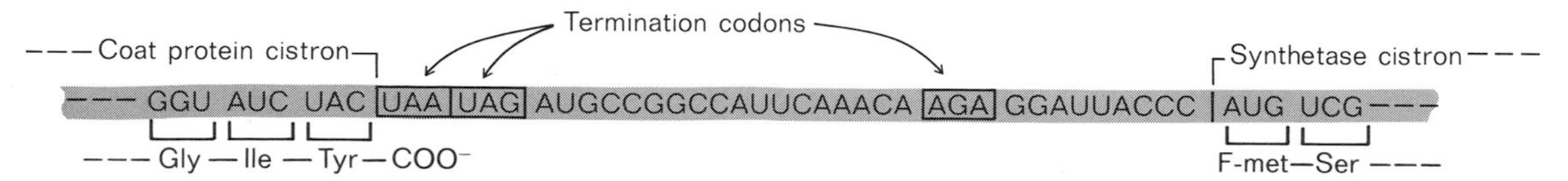

FIG. 6-4 *R17 RNA.* A part of the RNA of R17, containing the region between the coat-protein cistron and the synthetase cistron. Three termination codons are found in this region. An initiation codon is also present, but whether a short polypeptide is made before the third termination codon is reached is not known. (Other aspects of R17 structure and function are presented in Chapter 17.) Data from J. L. Nichols, *Nature,* **225:** 147 (1970).

it would only insert an amino acid at one of the termination codons. It is unlikely, however, that one cell would possess suppressors for all three termination codons. Therefore the presence of all three termination signals at the end of the cistron ensures that by one means or another termination is very likely to occur.

Polysomes

The previous sections considered individual ribosomal units in relation to the messenger RNA molecules. However, in principle the geometry of the ribosome, coupled with the size of the messenger RNA molecule, would permit several ribosomes to function actively on a single molecule of messenger RNA corresponding to one cistron. Each ribosome would be operating at a different point on the sequence, each having started at the same initiation point. The ribosomes would carry polypeptide chains at different stages of completion, the ones farther along the sequence carrying those chains closer to completion. Electron micrographs, of the type shown in Fig. 6-5, clearly reveal the presence of clusters of ribosomes—called *polysomes*—in just such a manner. For a protein the size of hemoglobin, with about 150 amino acids, it is quite common to observe 4 to 6 polysomes on a single messenger RNA molecule. With larger proteins and their correspondingly larger messenger RNA molecules, even much larger clusters of ribosomes on a single messenger molecule can be observed.

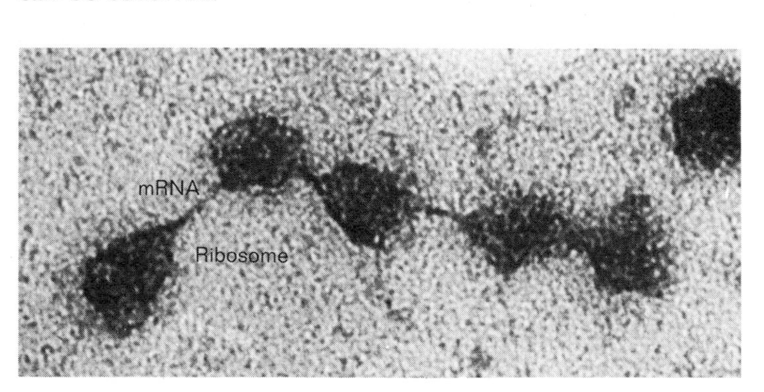

FIG. 6-5 *Electron Micrograph of Polysomes.* Courtesy of A. Rich. Prior publication in *Scientific American,* December 1963.

Variation of Synthesis Among Species

Most of the studies described so far concerning the genetic code have been derived from molecules found in *E. coli*. One exception is the alanine transfer RNA from yeast, the first RNA molecule sequenced, which provided some important evidence about the nature of the genetic code. It is reasonable therefore to ask whether the machinery and details of protein synthesis, including the genetic code, are identical for all other living species. In terms of the genetic code itself, studies with the binding of transfer RNA from both higher plants and higher animals to ribosomes in the presence of nucleotide triplets involving transfer RNA molecules give results parallel to those found with *E. coli* transfer RNA and strongly suggest that the genetic code is universal, or at least "global." While the genetic dictionary is thus the same from species to species, it is also clear that many other details of protein synthesis differ among species. The ribosomes of eucaryotic organisms are all considerably larger than the ribosomes found in *E. coli,* with 40S and 60S subunits contributing to an 80S ribosome. In addition, it is now known that initiation of protein synthesis in plant and animal cells occurs not with formylmethionine, as in *E. coli,* but simply with methionine. However, even in eucaryotic cells, the organelles such as chloroplasts and mitochondria retain an initiation system using formylmethionine. Furthermore, the accessory enzymes of protein synthesis are likely to differ widely among species. In certain instances a transfer RNA from one species can be activated by a synthetase from another species, but in many cases homologous interactions do not occur. For example, the phenylalanine synthetase enzyme of the mold *Neurospora,* when reacted with transfer RNA molecules from *E. coli,* will charge the alanine transfer RNA with phenylalanine, and the valine transfer RNA of *E. coli* with phenylalanine. Thus the incorrect aminoacyl-transfer RNA is produced by these components. Moreover, the same phenylalanine synthetase of yeast, when confronted with phenylalanine transfer RNA from *E. coli,* does not charge the transfer RNA. Therefore, while the genetic code is virtually identical among species, marked differences in the accessory enzymes may also be expected. Probably the mechanism of control of protein synthesis, which regulates the rate and extent of formation of individual proteins, also varies considerably among species, particularly between lower and higher organisms. We will return to this question after taking a closer look at proteins and their role in metabolism.

PROBLEMS

1 Is formylmethionine or methionine the initiating amino acid of protein synthesis for
(a) Bacteria
(b) Mitochondria
(c) Chloroplasts
(d) Cytoplasm of eucaryotic cells

2 The inhibitor of messenger RNA synthesis, 3′-deoxyadenosine is incorporated but interferes with further extension of RNA polymers. What might reasonably be expected for the fate of 5′-deoxyadenosine?

3 Detailed studies of protein synthesis revealed the specific directions of the colinearity described in Chapter 1. Which of the following directions are correct?
(a) N $\longrightarrow$ C corresponds to 5′ $\longrightarrow$ 3′
(b) N $\longrightarrow$ C corresponds to 3′ $\longrightarrow$ 5′

4 The D and A sites are named for the functions "donor" and "acceptor." What do they donate and what do they accept?

5 A possible anticodon for Trp transfer RNA is $^{3'}$ACC$^{5'}$. If a double mutation occurred in the transfer RNA gene to change the codon to $^{3'}$AUU$^{5'}$, would the new transfer RNA provide a viable suppressor?

6 The "genetic code" is the same in all species and extends to the specificity of synthetases for amino acids and transfer RNA molecules. Is this statement correct?

Chapter 7 Frontispiece *Collage of crystals of the enzymes of glycolysis. First row: phosphorylase, phosphoglucomutase (actual crystals unavailable), hexokinase; second row: phosphoglucoisomerase, phosphofructokinase (actual crystals unavailable), aldolase; third row: triosphosphate isomerase, glyceraldehyde-3-phosphate dehydrogenase, phosphoglycerate kinase; fourth row: phosphoglycerate mutase, enolase, pyruvate kinase. Courtesy H. C. Watson. Prior publication in Cold Spring Harbor Symposium of Quantitative Biology,* **36:** *iv (1971).*

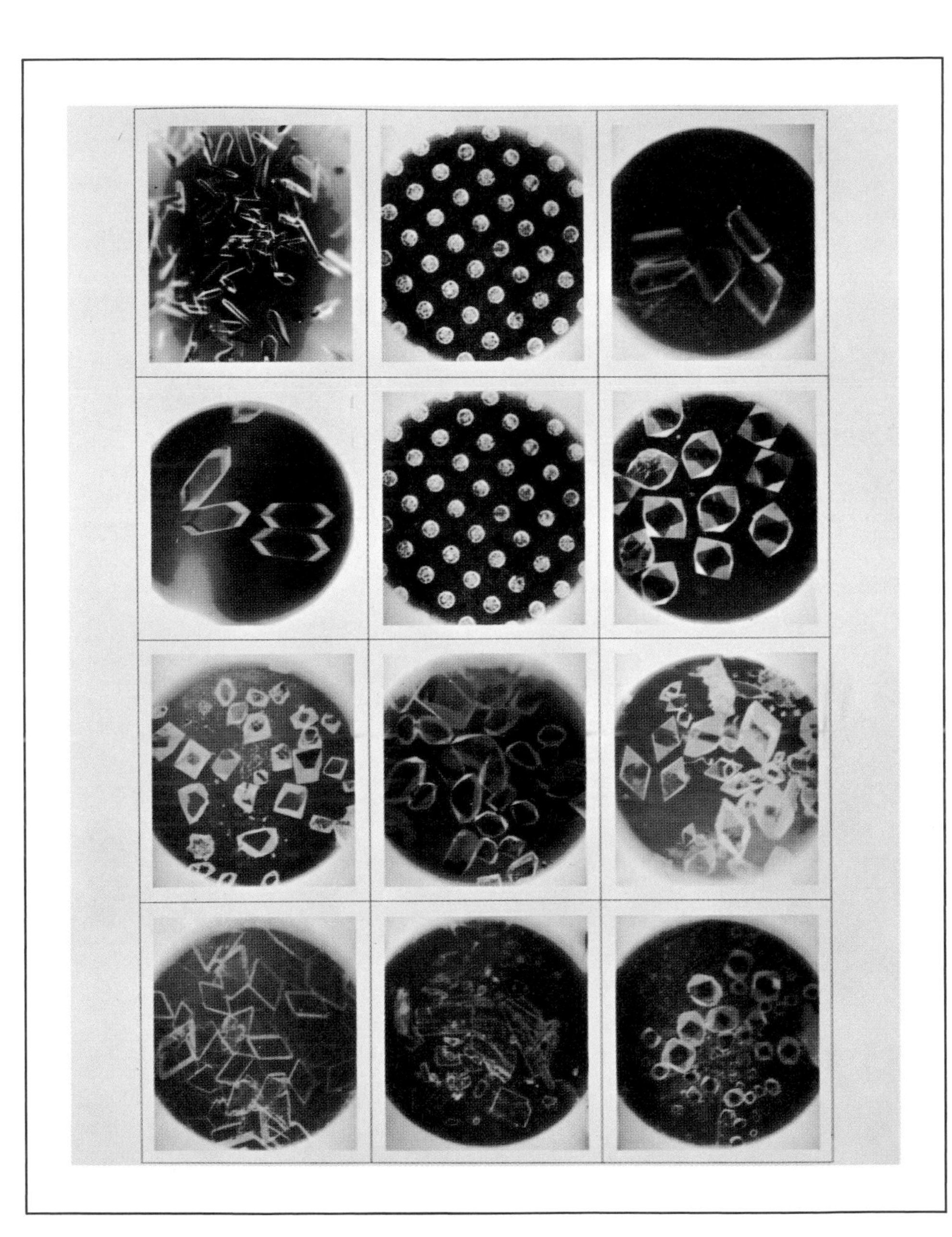

AN INTRODUCTION TO PROTEINS

7

The power of biology lies in the proteins. They are responsible for the many forms and functions of living things. Every color of life, every taste and smell, relies on proteins to synthesize the pigments, metabolize the precursors, generate the chemical agents, transform the energy sources, or transmit the sensory signals. Each of the many reactions described here and in the next three chapters is catalyzed by a distinct protein. While we have emphasized the cool, simple logic of DNA as the fundamental vehicle of evolution, here we must note the encompassing role of proteins. Evolution itself is measured in terms of the development of new proteins with improved capabilities and innovative functions.

The wide range of structural patterns found in proteins is staggering and seemingly endless. Just when the role of sulfhydryl groups or of histidine residues in proteins seems to be understood, a new enzyme appears with a totally unexpected involvement. Or proteins with a totally unexpected catalytic mechanism or regulatory role may be revealed. Thus virtually each protein studied in detail emerges with a unique character. Since the number of proteins biochemists have examined is considerable, it would be extremely difficult to describe all of these characters. Instead we will examine a few selected cases in detail, on the assumption that more is learned by studying a small number in depth than many superficially. We will attempt to explain just what proteins are, what they do, and how they do it. This task is not an easy one, for proteins as the catalysts and regulators of cellular life have evolved exceedingly intricate mechanisms of action. Other proteins in purely structural roles have evolved with special properties to provide a three-dimensional matrix. To provide some orientation to proteins, five topics will be briefly raised to illustrate the complexity of the problems that must be faced in attempting to understand the structure and action of proteins.

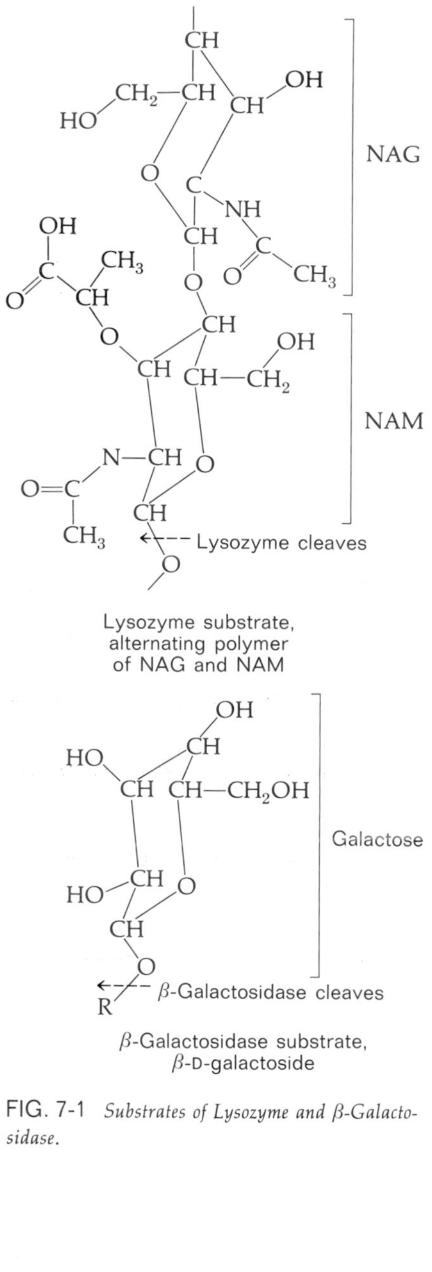

NAG

NAM

Lysozyme cleaves

Lysozyme substrate, alternating polymer of NAG and NAM

Galactose

β-Galactosidase cleaves

β-Galactosidase substrate, β-D-galactoside

FIG. 7-1 *Substrates of Lysozyme and β-Galactosidase.*

Correlation of Structure and Function

An eventual goal of protein chemistry is to consider the function of an enzyme and be able to surmise roughly what that structure would be in terms of the requirements of the reactions the enzyme catalyzes. The ability to make such a prediction would require a detailed understanding of the architectural principles of protein structure, as well as the role of the various amino acids in catalytic properties. Evolution would also have to be taken into account. Although biochemistry has made tremendous progress in protein research, the current understanding of proteins is still extremely far from this goal. To illustrate the nature of the problem and our perspective of the subject at this stage, let us consider two enzymes with related functions, lysozyme and β-galactosidase.

The two enzymes catalyze similar reactions. Lysozyme hydrolyzes polysaccharides of the type found in cell walls (see Fig. 7-1). Cell walls contain hexose polymers with an alternating sequence of NAG (N-acetylglucosamine) and NAM (N-acetylmuramic acid) residues. The enzyme β-galactosidase hydrolyzes galactosides, also hexose derivatives (Fig. 7-1).

In the case of lysozyme, the entire structure is known (Fig. 7-2). The structure was deduced from the complex x-ray diffraction pattern of crystals of the enzyme by David Phillips and his colleagues who used techniques of Fourier analysis and heavy-atom substitutions. The methods of x-ray crystallography are very complicated and will not be discussed here, although we will rely on the results of x-ray crystallography in the description of protein structures. The details of lysozyme can be summarized by noting that the protein contains 129 amino acids to give a molecular weight of 14,600. Four disulfide bonds are present and cross-link the polypeptide chain. In addition the various catalytic and binding groups for holding and hydrolyzing the substrate can be specified (see Fig. 7-3).

A striking contrast in structure arises when we consider β-galactosidase. The galactosides hydrolyzed by this enzyme are not very different from the substrates of lysozyme. However, in contrast to lysozyme, β-galactosidase contains four polypeptide chains, each identical to the others, and each with a molecular weight of 135,000. Thus the total molecular weight of the enzyme is 540,000, or 37 times larger than lysozyme. We will study more about β-galactosidase, particularly in the area of control of protein synthesis, since this

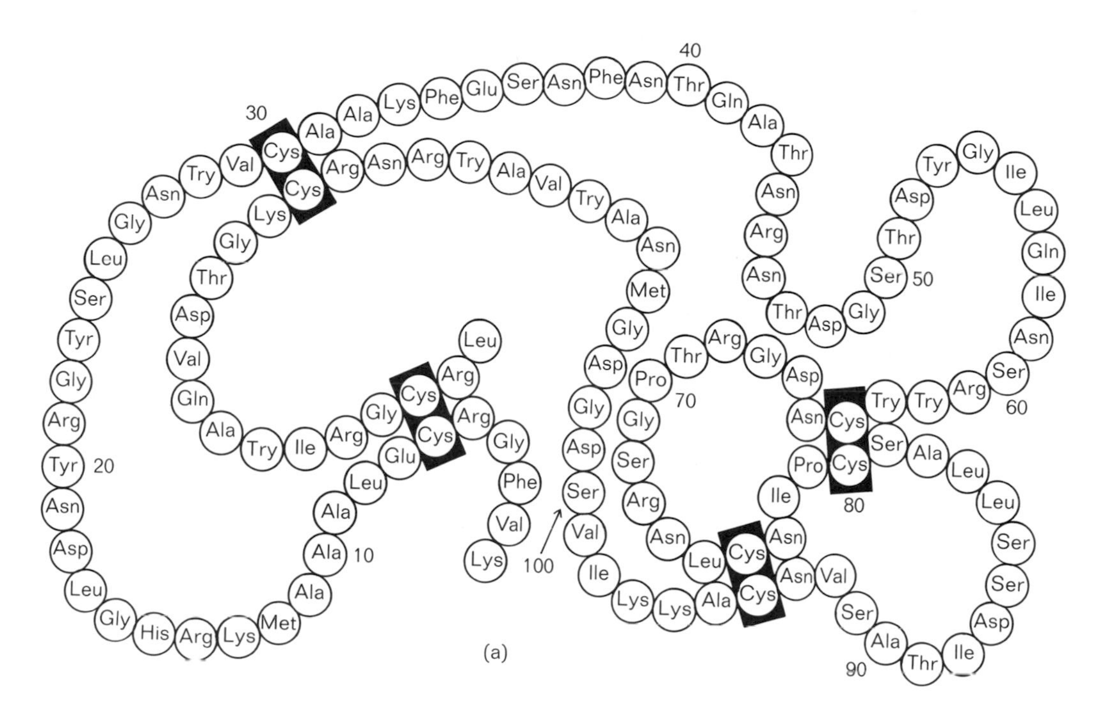

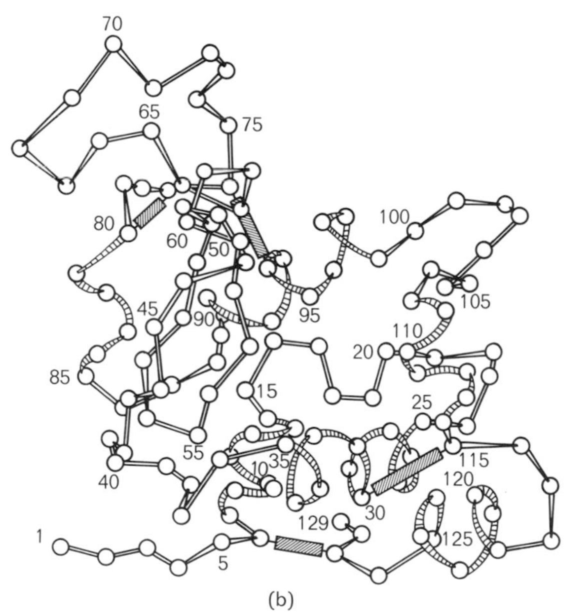

FIG. 7-2 *Structure of Egg White Lysozyme.* (a) Primary structure including positions of the four disulfide bonds. Tryptophan is abbreviated as Try in this drawing. From R. E. Canfield and A. K. Liu, *Journal of Biological Chemistry,* **240:** 1997 (1965). (b) Schematic drawing of the main chain conformation of lysozyme. By W. L. Bragg, from C. C. Blake et al., *Nature,* **206:** 757 (1965).

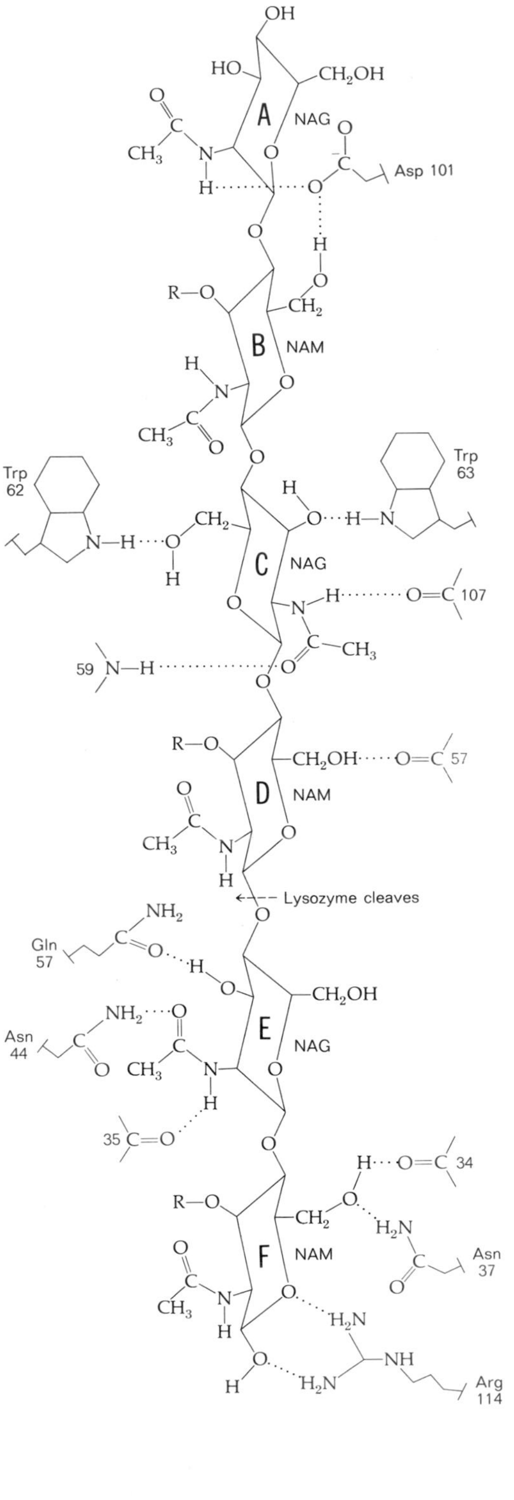

FIG. 7-3 *Substrate Interactions with Lysozyme.* From R. E. Dickerson and I. Geis, *The Structure and Action of Proteins,* Harper & Row, New York, 1969.

enzyme has served as the case study for many experiments in control systems. However, it is impossible at the current stage in our understanding of proteins to offer any reason why β-galactosidase should be 37 times larger than lysozyme in order to carry out a reaction that is relatively similar to the reaction carried out by lysozyme. All we can say is that evolution must play a large role.

Divergent and Convergent Evolution

In a few cases more precise statements can be made on the nature of evolution as it applies to the development of protein structure. The serine proteases, in particular, have been studied in detail and serve to illustrate both divergent and convergent evolution. For example, the mammalian digestive enzyme chymotrypsin, for which the three-dimensional structure has been solved at the atomic level (see Fig. 7-4), cleaves peptide bonds in proteins on the carbonyl side of aromatic acids, especially tyrosine and phenylalanine. The active site can be related to the specificity of the enzyme, and the details are summarized in Fig. 7-5. We see that an aspartyl residue, a histidyl residue, and a seryl residue, at positions 102, 57, and 195, respectively, interact to form a charge relay system that triggers the reaction by abstracting protons from the active serine (195). In addition the serine at position 189, which binds the amino acid residue to be cleaved, lends specificity to the enzyme by favoring noncharged groups.

When we examine trypsin, which has a structure very similar to chymotrypsin, we notice that most of the changes are only at the surface of the molecule. Even with these changes and the addition of two more disulfide bonds, there are no major alterations in the overall conformation. However, one important difference does emerge. We can relate this change very well to the specificity of trypsin. Trypsin hydrolyzes peptide bonds at the carbonyl side of lysine and arginine residues. These residues have a positive charge. In this case the negatively charged aspartyl group in the specificity site of trypsin directs the protein to the positively charged residues. Recall that a serine occupies the corresponding position in chymotrypsin. Here then is a case in which a similar sequence of amino acids has diverged to yield two proteins, chymotrypsin and trypsin, with very different specificities, simply by changing a noncharged to a negatively charged residue at the substrate-recognition site.

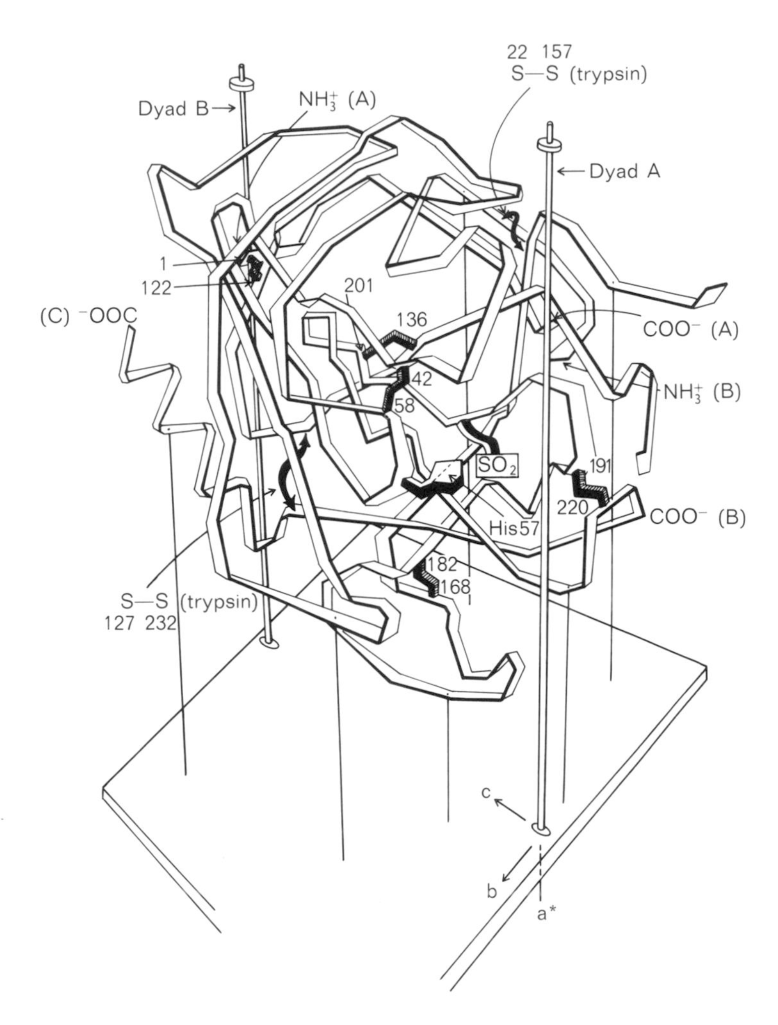

FIG. 7-4 *Structure of Chymotrypsin.* The molecule contains three polypeptide chains (A, B, and C) formed by cleavage of the chymotrypsinogen molecule during activation (see Fig. 13-5). The five disulfide bonds are shown by cross-hatching. The positions of two additional disulfide bonds present in trypsin at homologous positions are indicated by the thick arrows. The structure is from P. B. Sigler, D. M. Blow, B. W. Matthews, and R. Henderson, *Journal of Molecular Biology,* **35:** 143 (1968).

The third protease we will consider is subtilisin. This enzyme bears no resemblance to chymotrypsin or trypsin in sequence or structure. Therefore it seems to have had a totally independent evolutionary origin from the trypsin and chymotrypsin enzymes. Yet, the active site contains the same charge relay system found in the first two proteases, an aspartyl, a histidyl, and a seryl residue, numbers 32, 64, and 221 in subtilisin. All three residues are located in space in such a way as to provide exactly the same catalytic mechanism as found in the trypsin and chymotrypsin. However, even the order of the three residues in the primary structure differs from the order found in trypsin and chymotrypsin. Here we have a clear example of convergent evolution, where an inde-

FIG. 7-5 *Mechanism of Action of Chymotrypsin.* The three amino acids that form the charge relay system are shown at the top with α-carbons shaded. Modified from R. E. Dickerson and I. Geis, *The Structure and Action of Proteins,* Harper & Row, New York, 1969.

Charge relay system:

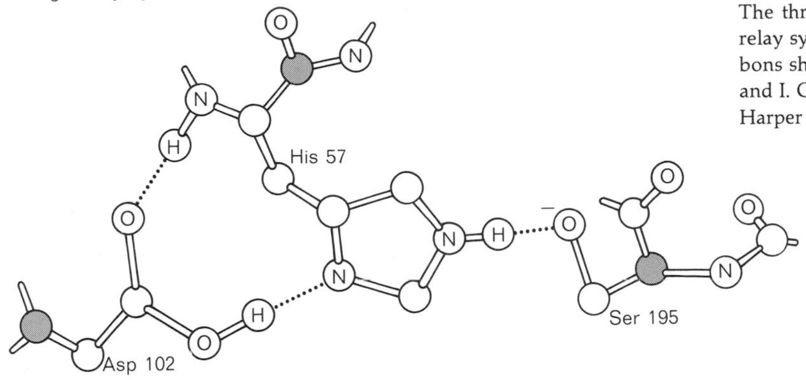

His 57

Asp 102

Ser 195

Cleavage mechanism:

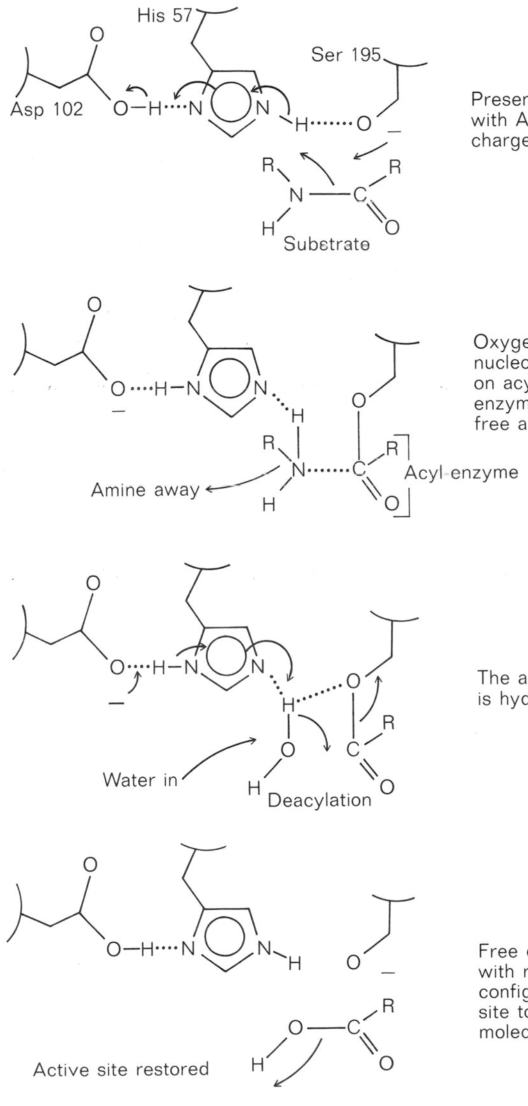

Presence of His 57 in conjunction with Asp 102 stabilizes negative charge on oxygen of Ser 195

Substrate

Oxygen of Ser 195 makes nucleophilic attack (see Appendix III) on acyl carbon, forming acyl-enzyme intermediate and releasing free amine

Amine away

Acyl-enzyme

The acyl-enzyme intermediate is hydrolyzed by water

Water in

Deacylation

Free carboxylic acid is formed with restoration of the initial configuration of the enzyme active site to await next substrate molecule

Active site restored

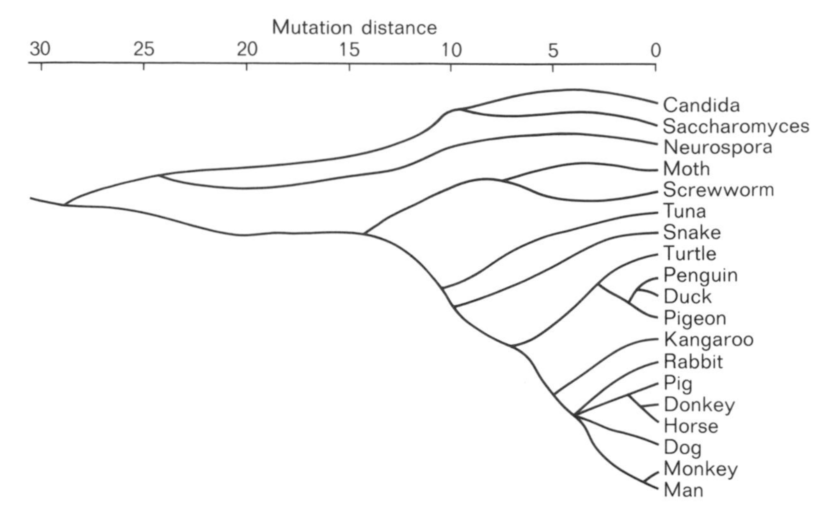

FIG. 7-6 *Species Variations in Cytochrome c.* The mutation distance is defined as the number of base changes in cytochrome c genes needed to account for the difference in amino acid sequence for any two species. The actual mutation distance for any two species is given by the distance from one to the point of divergence of the second and back to the other. From W. M. Fitch and E. Margoliash, *Science,* **155:** 279 (1967).

pendent line of evolution has converged on a mechanism that had also been selected by another enzyme. Thus, while we are very far from being able to explain differences such as those found in lysozyme and β-galactosidase, we can discern some factors, such as divergent and convergent evolution, that play an important role in shaping protein structure.

The role of evolution in proteins can be further related to changes in a protein from species to species. For example, the protein cytochrome c, an important factor in metabolic respiration, has been examined in a wide range of species. Two major points emerge. First, there is a considerable variety among species, even though in every case the function of the protein is very similar. A "family tree" is shown in Fig. 7-6 and illustrates just how "tentative" proteins are. Alteration through mutation and testing in the competition of selective pressure are occurring at all times in all living systems.

The second point to be noted is the invariability of certain aspects of the structure, even in the midst of this continuous evolution. Examination of the detailed sequences (Fig. 7-7) reveals that about one-third of the amino acid residues are identical in all species examined. Presumably each serves some essential demand with such strong requirements that any change at these positions weakens performance. Eventually protein chemists hope to understand just what role each amino acid plays and why certain ones can be replaced while others are uniquely able to fulfill a task. Quite obviously, much is left to be discovered.

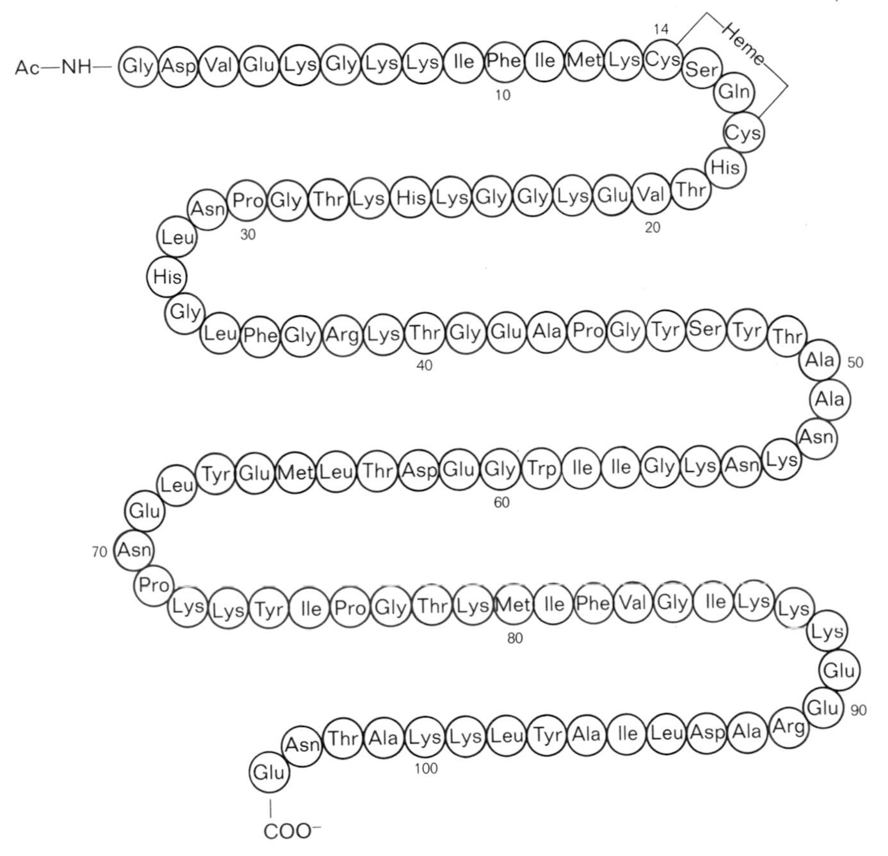

FIG. 7-7 *Amino Acid Sequence of Human Heart Cytochrome c.* The amino acids indicated in color are invariant positions; that is, they are the same in all species studied so far. From H. Matsubara and E. L. Smith, *Journal of Biological Chemistry*, **237:** 3575 (1962).

The Basis of Catalytic Power

In the illustrations above we have discussed some aspects of specificity, particularly for the differences in chymotrypsin and trypsin. Earlier (Chapter 4) some familiarity was gained with the limits of specificity in proteins from the consideration of the amino acid synthetase enzymes. However, enzyme action has a second important aspect, and this is the concept of *catalysis*. The reactions enzymes catalyze have a certain equilibrium that is maintained in the presence and absence of the enzyme. Thus, if the equilibrium lies in favor of the products over reactants by a factor of 100, even in the presence of the enzyme the equilibrium will lie in the favor of products over reactants (or substrates when considering enzymes) by the same factor of 100. The big advantage of enzymes, however, is that they will alter the *rate* at which the equilibrium is reached. Since they are not consumed in the reaction, enzymes are *catalysts*. They can be enormously powerful, speeding the rate of attainment of equilibrium by, in some cases, as much as 10^{18}. The catalytic properties of proteins are one of the most fundamental qualities of biochemical life. Enzymes permit replication of genes to occur quickly, and the metabolic advantages of favorable changes in protein sequence can be expressed rapidly.

A number of factors can be cited by analogy with chemistry to account for the enormous rate enhancement of catalytic power. Among the factors likely to be involved are:

1 *Proximity of substrates,* which is the close contact of reacting groups brought together on the surface of the enzyme molecule. Substrates are rarely present in cells at above millimolar concentrations. However, on the surface of the enzyme they may be effectively "concentrated" to levels corresponding to solutions in the molar range. An approximate rate enhancement of 10^4 may be assigned for proximity of substrates.

2 *General acid-base catalysis,* which is the ability of amino acids of the protein to act as acids or bases in the catalysis of the reactions considered. Many biochemical reactions are catalyzed by amino acid residues acting as acid or base (see Chapter 9, the section on the mechanism of ribonuclease, for example). The presence of the amino acid would be expected to speed reactions by as much as several orders of magnitude on the basis of simple chemical models. An approximate rate enhancement of 10^2 may be assigned for general acid-base catalysis.

3 *Proximity of catalytic groups,* which is the close contact on the surface of the enzyme by the substrate and catalyst. While acid-base catalysis has a marked effect on model reactions, the catalytic group is rarely present above millimolar concentrations. By the same argument applied to substrates, the effective concentration on the surface of the enzyme may be thousands of times higher. An approximate rate enhancement of 10^3 may be assigned for proximity of catalytic groups.

The total enhancement from factors known in model reactions is thus approximately 10^9. In terms of the overall factor of approximately 10^{18}, the known factors still leave as unaccounted a factor of about 10^9 in catalytic power. Determining the origin of this unknown factor of about 10^9 in catalysis remains one of the main problems in understanding enzymes. Special strains, distortions, and orientations forced on the substrate by the conformation of the enzyme have been proposed, but little of a conclusive nature can be stated at this time.

Cooperativity in Ligand Binding

While the extension of physical organic chemistry leaves a big gap in the quantitative explanation of catalytic power, it is a matter of degrees. Catalysis in simple chemical systems has been widely studied. When we come to consider cooperativity, however, it is more than a matter of degrees, for cooperative ligand binding is almost unique to proteins. Cooperative ligand binding occurs in proteins with more than one active site. The classical example of a cooperative system is the protein hemoglobin. Hemoglobin possesses four iron-containing hemes (see Fig. 7-11 for the structure) and the ligand oxygen binds to each of these at the iron. The hemes are located at some distance from one another on four individual polypeptide chains, two of type α and two of type β (Fig. 7-8). Hemoglobin, along with myoglobin, was one of the

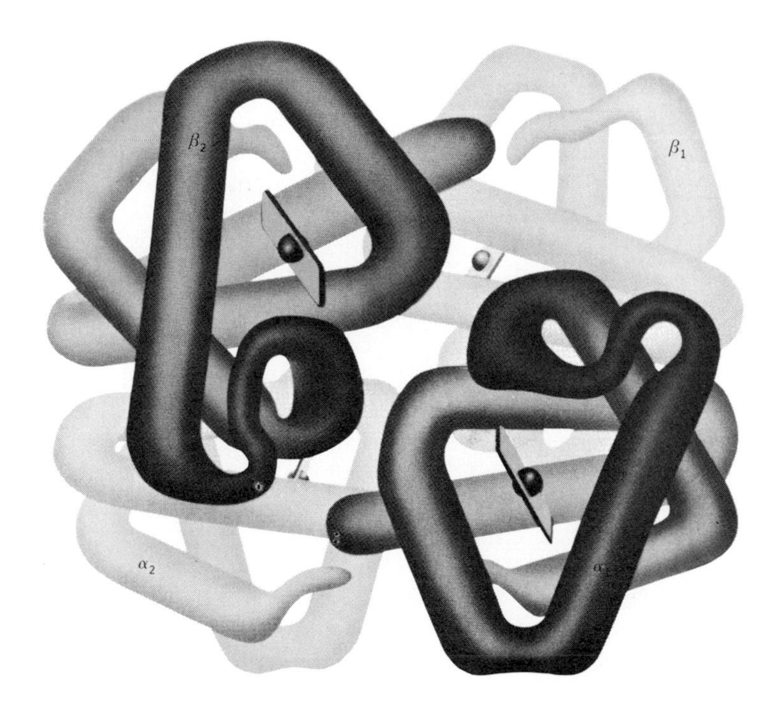

FIG. 7-8 *Structure of Hemoglobin.* Two α-chains and two β-chains are present in the arrangement indicated. Each chain contains a heme (indicated by the planes) with an iron atom (represented by a ball) at the center. The structure of oxyhemoglobin is shown. Upon deoxygenation, the left and right halves rotate slightly to bring the β-chains closer together and cause a new crystal type. See also Chapter 10 frontispiece and Fig. 10-10 to 10-18. From R. E. Dickerson and I. Geis, *The Structure and Action of Proteins,* Harper & Row, New York, 1969.

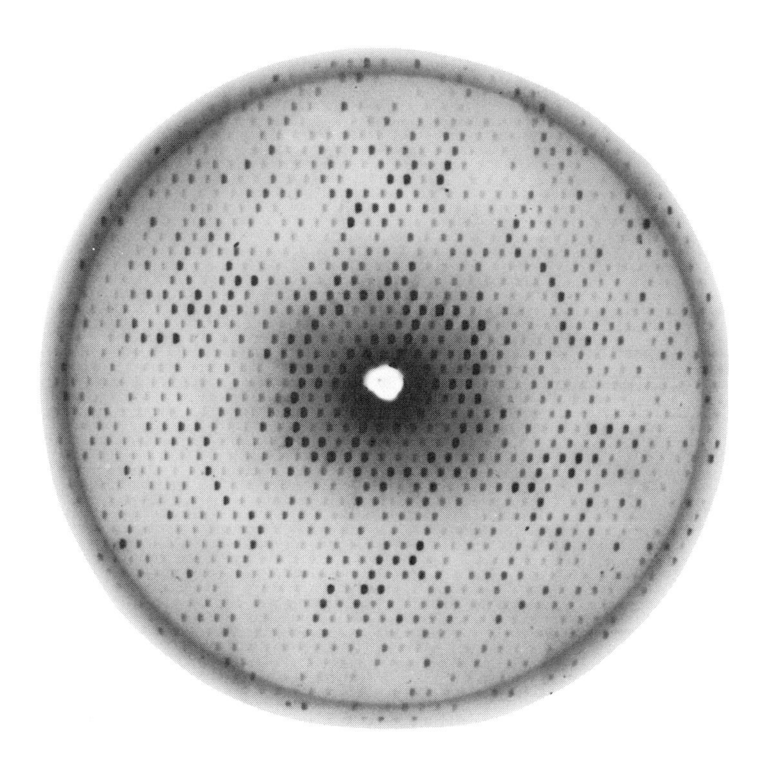

FIG. 7-9 *X-ray Diffraction Pattern.* The pattern was produced from a single crystal of hemoglobin exposed to an x-ray source. Courtesy J. K. Moffat.

first proteins studied extensively by x-ray crystallography. The structures of these two proteins were determined by Max Perutz and John Kendrew. Their work also included the development of many of the ingenious methods for extracting the wealth of structural information from the myriad of spots of an x-ray diffraction pattern (Fig. 7-9).

The cooperativity of hemoglobin is expressed by the change in affinity for oxygen during the course of the oxygen-binding reaction. This cooperative property can be most easily illustrated by contrasting hemoglobin with myoglobin. Myoglobin has a single polypeptide chain (closely related to a hemoglobin chain as shown in Fig. 7-10) and a single oxygen-binding site (Fig. 7-11). Therefore it can only bind oxygen in a single step, without the possibility of cooperativity. In this case the ligand-binding reaction can be described by a single constant-equilibrium reaction:

$$\text{Mb} + \text{O}_2 \rightleftharpoons \text{MbO}_2$$

However, in the case of hemoglobin, the binding of one molecule of oxygen at one site influences the affinity for the

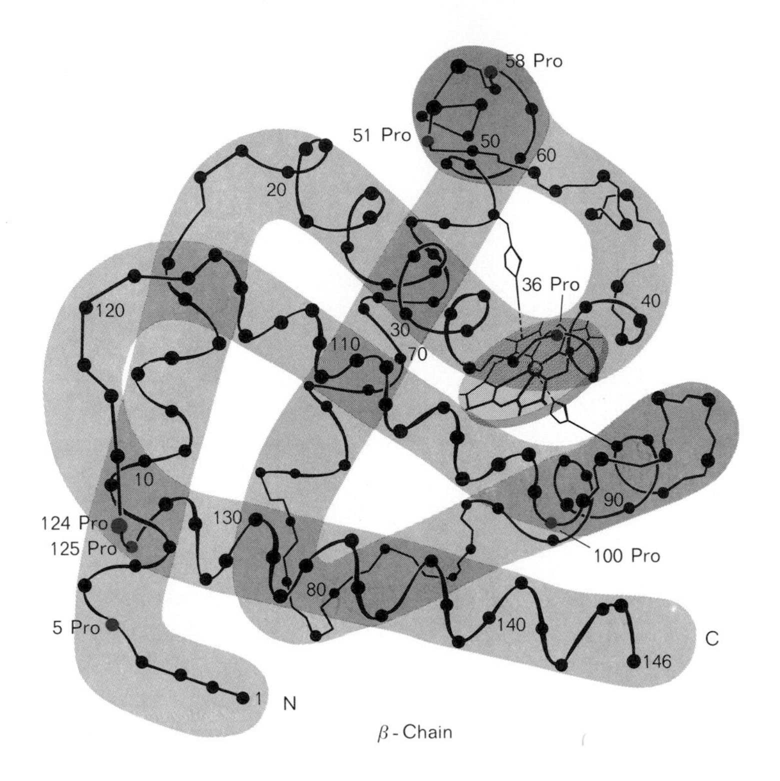

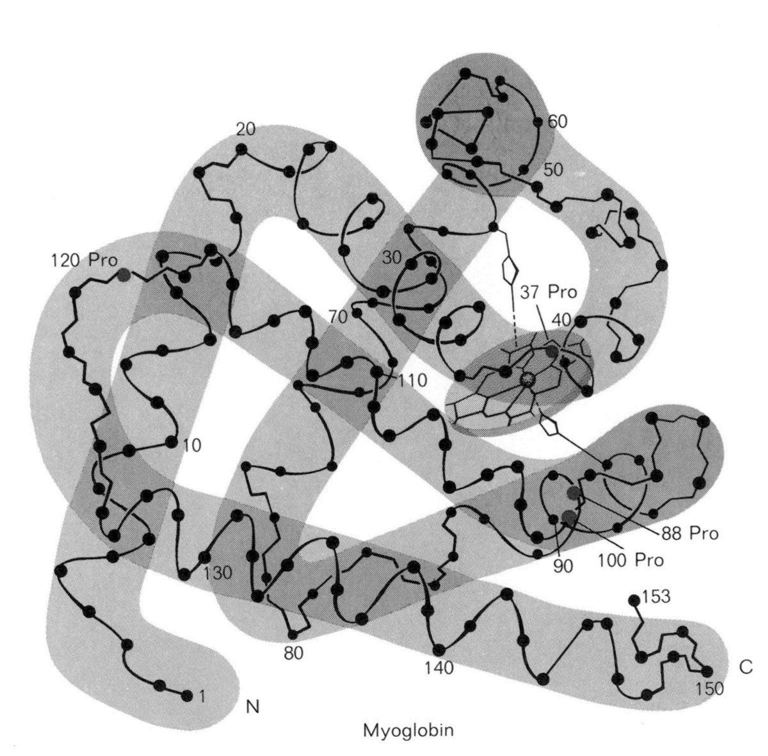

FIG. 7-10 *Folding of Myoglobin and Hemoglobin Chains*. Hemoglobin contains two α-chains and two β-chains resembling each other (the same amino acid is in about half of the positions) as well as the single-chain myoglobin. Note the occurrence of proline residues at the ends of the helical regions. From M. Perutz, "The Hemoglobin Molecule," copyright © *Scientific American, Inc.*, November 1966. All rights reserved.

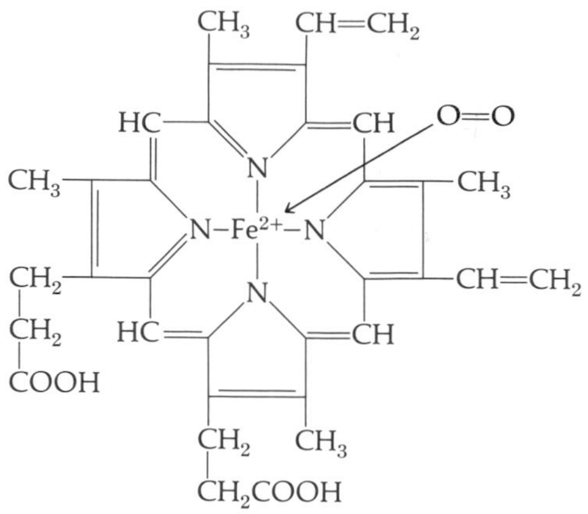

FIG. 7-11 *Heme Structure.* The structure includes iron and a porphyrin core with the various side chains included to give a compound known as protoporphyrin IX. The iron is ferrous (Fe^{2+}) and binds with oxygen with no change in valence. An additional bond to iron perpendicular to the heme ring is provided by an imidazole group of a histidine residue.

binding of additional molecules at other sites. A hemoglobin molecule with two oxygens bound will have a greater affinity for the third than a molecule with no oxygen bound. Perutz has summarized this situation with a biblical analogy, "To him who hath shall be given." The contrast between cooperative and noncooperative ligand binding is normally expressed in terms of a saturation curve of the type shown in Fig. 7-12. The noncooperative binding of myoglobin, as an example, gives a hyperbolic shape to the oxygen equilibrium curve,

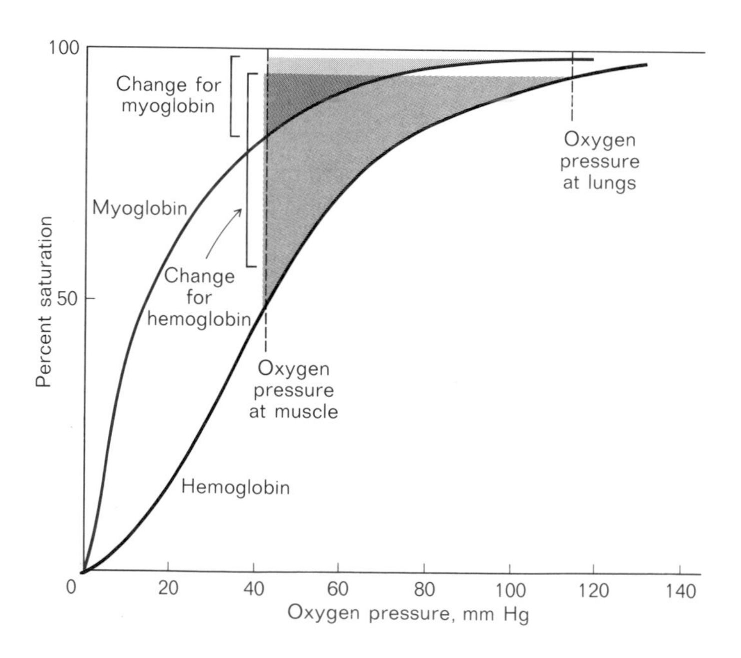

FIG. 7-12 *Oxygen Binding to Hemoglobin and Myoglobin.* The S-shaped oxygen-binding curve for hemoglobin permits a large fraction of the oxygen bound at the lungs to be released at the muscles and other tissues.

whereas the binding to oxygen by hemoglobin gives a sigmoidal, or S-shaped, equilibrium curve. This behavior is advantageous to animals since the efficiency of oxygen transport is greatly enhanced—red cells load up at the lungs and can unload at the tissues much more completely because of the sigmoidal curve. Just how the binding of oxygen to one heme site can influence the affinity of the other hemes for oxygen is a very fascinating question. It will occupy a large part of our consideration of proteins since cooperativity of this type, which is found in numerous enzymes with regulatory functions, plays a central role in metabolic regulation. It also illustrates one of the more sophisticated accomplishments of biochemical evolution.

Spontaneous Folding

We have considered structure-function relationships, evolution, catalysis, and cooperativity—all fundamental subjects as applied to proteins. However, in some respects we have saved the best for last. An even more fundamental aspect of proteins is the assembly of the newly synthesized sequence of amino acids (*primary structure*) into an active, compact, intricate, three-dimensional structure (*tertiary structure*). This protein folding is the last stage of protein synthesis and one that is usually taken for granted. In the earlier chapters we followed the genetic information through the mechanism of protein synthesis to the completion of the protein sequence according to the genetic code. Presumably, the protein chain as it is being synthesized, or as soon as it is completed, quickly and spontaneously folds into the precise three-dimensional structure maintained in the cell. Therefore we can say that primary structure determines tertiary structure. A secondary structure is also recognized as the short-range interaction that governs, for example, α-helix formation (see Chapter 8). Were our understanding of this folding complete, one could predict a precise three-dimensional conformation of the protein simply by knowing the amino acid sequence. In fact, a number of protein chemists are attempting very actively to predict conformation, but it is an extremely complicated problem. There is an enormous number of possible conformations for any protein, and it is a difficult task to attempt to decide which conformation will be spontaneously favored. Moreover, there may be precise sequences of steps that further influence the folding. Yet the fascination of the problem holds many minds' attention. The problems of assembly will be further considered in the next chapter, devoted to the structural aspects of proteins per se, including the various considerations that influence their conformation. Then in Chapter 9 we will return to the more functional considerations of enzyme catalysis, and, in Chapter 10, to control through cooperativity.

PROBLEMS

1 The charge relay system of chymotrypsin involves a serine, a histidine, and an aspartate, contributing a hydroxyl group, a basic imidazol group, and a carboxyl group, respectively. Which other amino acids have
   (a) A hydroxyl group
   (b) A basic group
   (c) A carboxyl group

2 The α- and β-chains of hemoglobin have similarities in structure. Are these similarities likely to reflect divergent or convergent evolution?

3 What would be the physiological consequences if the oxygen binding curve of hemoglobin were identical to the curve for myoglobin?

4 If a reaction catalyzed by an enzyme has a half-time of 1 msec and the rate is enhanced by a factor of 10^{18} over the value that prevails in the absence of the enzyme, what would be the half-time of the reaction in the absence of the enzyme?

Chapter 8 Frontispiece *Electron micrographs of intact and fragmented myosin molecules. Myosin has a long rodlike structure (composed of two α-helical polypeptide chains wound into a larger double-stranded helix) with each chain terminating in a large globular head. Molecules shown are (from top to bottom) complete myosin (double-headed), single-headed myosin, rods alone, heads alone, and fragments of rods. Magnification 155,000×. Courtesy S. Lowy. For additional details see Chapter 12 and Journal of Molecular Biology,* **42:** 1 *(1969).*

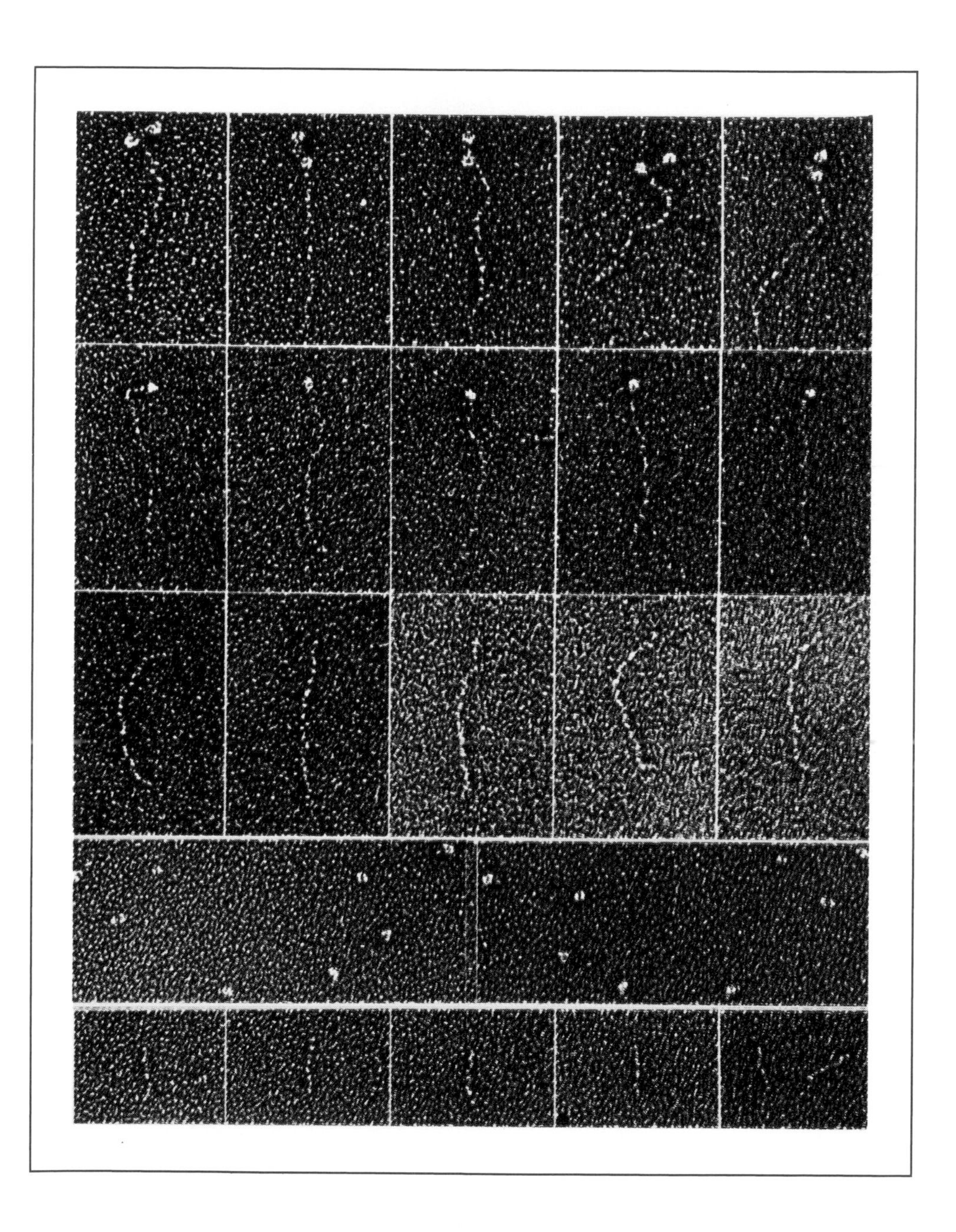

STRUCTURAL ASPECTS OF PROTEINS

8

Protein synthesis is a very intricate process. We have already studied the complex machinery necessary for the transcription and translation of genetic information through the nucleic acids to provide the formation of peptide bonds on the ribosome surface. Yet, as complicated and delicate as these steps are, they are simple in some respects compared to the last stage of protein synthesis, the acquisition of tertiary structure from primary structure. It is now fully established that the amino acid sequence of the protein alone is enough to determine its unique tertiary folding. We will review typical experiments that prove this point—that the linear sequence of amino acids folds spontaneously, even in isolated proteins, into the precise three-dimensional architecture of the active protein. However, it is still not fully clear whether the folding process for isolated proteins in a test tube is identical to the process that takes place on the ribosome surface. Experiments related to this question will be discussed with reference to the enzyme ribonuclease. We will mention the way in which an enzyme such as ribonuclease would be isolated and characterized and go on to discuss how we could then disrupt the structure to determine if it will spontaneously refold when the disrupting agent is removed. However, 100% disruption is difficult to establish, and the only fully rigorous way of establishing that the primary structure determines the tertiary is by complete synthesis of the enzyme. In the case of ribonuclease, such a synthesis can be described. We will follow the steps in first determining the amino acid sequence of the protein and then reconstructing it one amino acid at a time. The last topics we will consider in this chapter will relate to the precise structural principles governing the folding of amino acid sequences into tertiary structures. We will consider the helices and sheets arising from nearest-neighbor interactions of amino acids

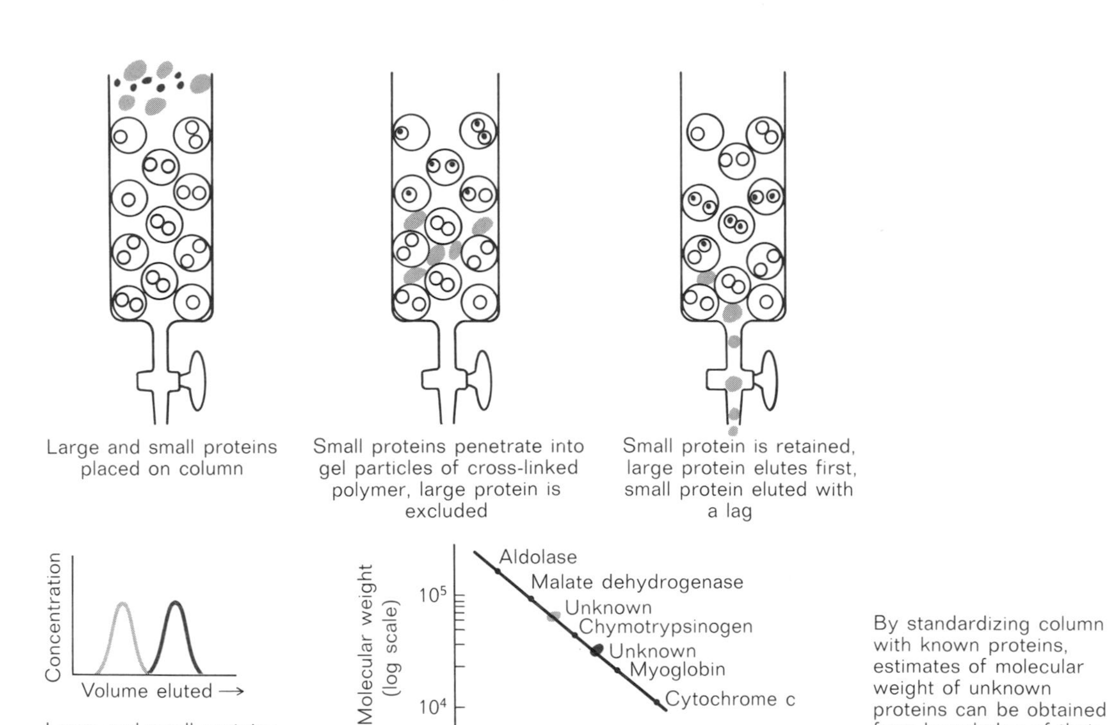

FIG. 8-1 *Gel Chromatography.*

in the primary sequence. These interactions are known as *secondary structure.* Finally, we will return again to ribonuclease to examine some detailed examples of tertiary structure as found by crystallographic analysis. From this point of view we will consider the efforts to establish a stereochemical code of protein folding to complement the genetic code for information storage in DNA and thus link the genetic information to its final expression in the three-dimensional structure of proteins.

The Enzyme Ribonuclease

Ribonuclease, secreted by the pancreas, is a relatively small protein that cleaves RNAs. The enzyme cleaves the RNA at 5′ linkages where the 3′ linkage on the same phosphate is attached to a pyrimidine as was summarized in Chapter 4. Products of the ribonuclease reaction terminate in pyrimidine nucleotides with a free 3′-phosphate group.

To obtain an enzyme such as ribonuclease in a purified form, procedures based on separation from other proteins

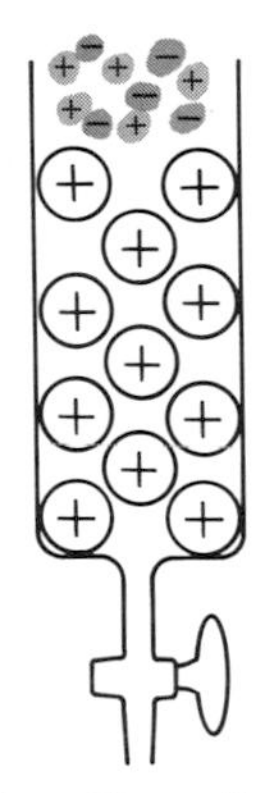

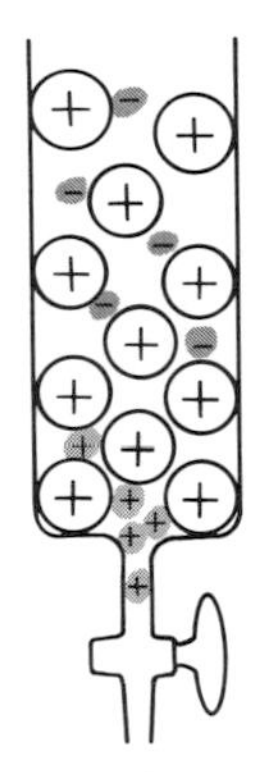

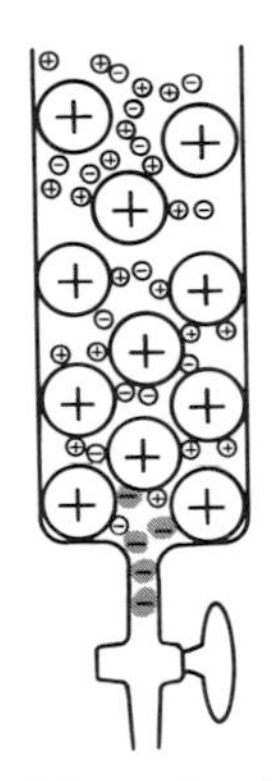

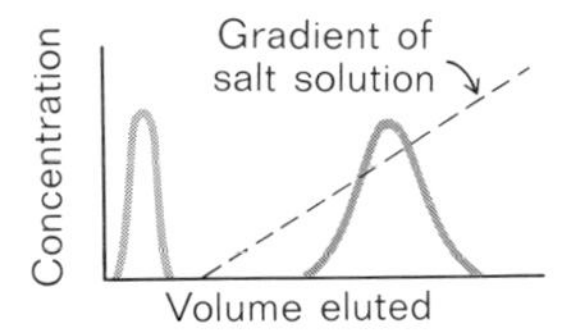

FIG. 8-2 *Ion-exchange Chromatography.* Charged molecules are released from the resin by displacement with salt. The molecules could also be released by deprotonating the resin with increased pH.

present in cell extracts must be employed. Two approaches are generally followed, fractionation based on charge and fractionation based on size. Proteins have a wide range of charge, solubility, and size properties that can be used to separate one from another. Fractionation based on size can involve some transport process (such as sedimentation) or chromatography on a gel, which may exclude particles according to size. In gel chromatography, the larger the particle, the sooner it would appear in the eluate from a column (see Fig. 8-1). Ion-exchange methods (Fig. 8-2) that separate according to charge also involve chromatography, but with resins that bear a net negative or positive charge. The enzyme can be made to stick to a column that has an opposite charge. The enzyme may then be eluted by buffers that change the pH and hence change the charge on the protein or the column. Elution can also be achieved by buffers that raise the salt concentration and hence lower the strength of bonding between charged groups. When ribonuclease was isolated by procedures of this type, it was possible to establish its specificity rigorously.

Studies on the isolated preparation of ribonuclease established a molecular weight of 14,000 for the protein. Further

FIG. 8-3 *Amino Acid Sequence of Ribonuclease.* Disulfide bonds are indicated in color. From D. G. Smyth, W. H. Stein, and S. Moore, *Journal of Biological Chemistry,* **238:** 227 (1963).

structural work ultimately led to the determination of the entire sequence of amino acids present in the molecule (Fig. 8-3). The sequence and its determination will be discussed in greater detail in the section on synthesis of ribonuclease later in this chapter. It is presented here to note that the enzyme contains four disulfide bonds that will require special consideration in the discussion of the disruption of structure that follows.

Disruption of Tertiary Structure

With the purified enzyme available, it should be possible to disrupt the tertiary structure with some treatment, relieve the treatment, and determine whether the enzymatic activity and native structure return to the molecule. This process would establish the adequacy of the primary structure in the determination of the tertiary structure. To carry out such a study we need, first of all, the competent agents for disrupting structure, and, secondly, criteria that the structure has been disrupted. A popular agent for disrupting structure is urea; a related compound, guanidine hydrochloride, is also widely used. Structures of these compounds are shown in Fig. 8-4.

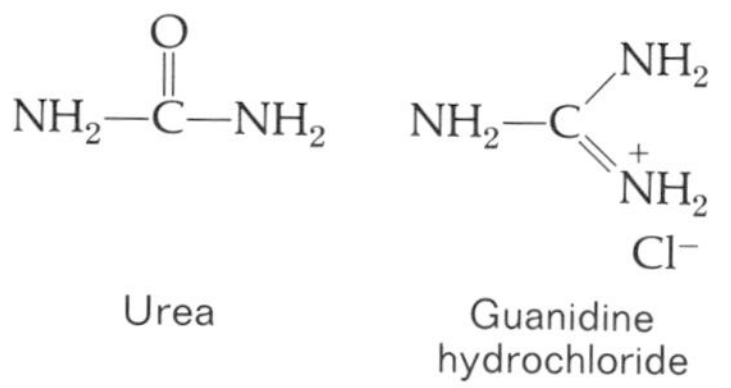

$NH_2-C(=O)-NH_2$ — Urea

$NH_2-C(-NH_2)=\overset{+}{N}H_2\ Cl^-$ — Guanidine hydrochloride

FIG. 8-4 *Structures of Protein-unfolding Agents.*

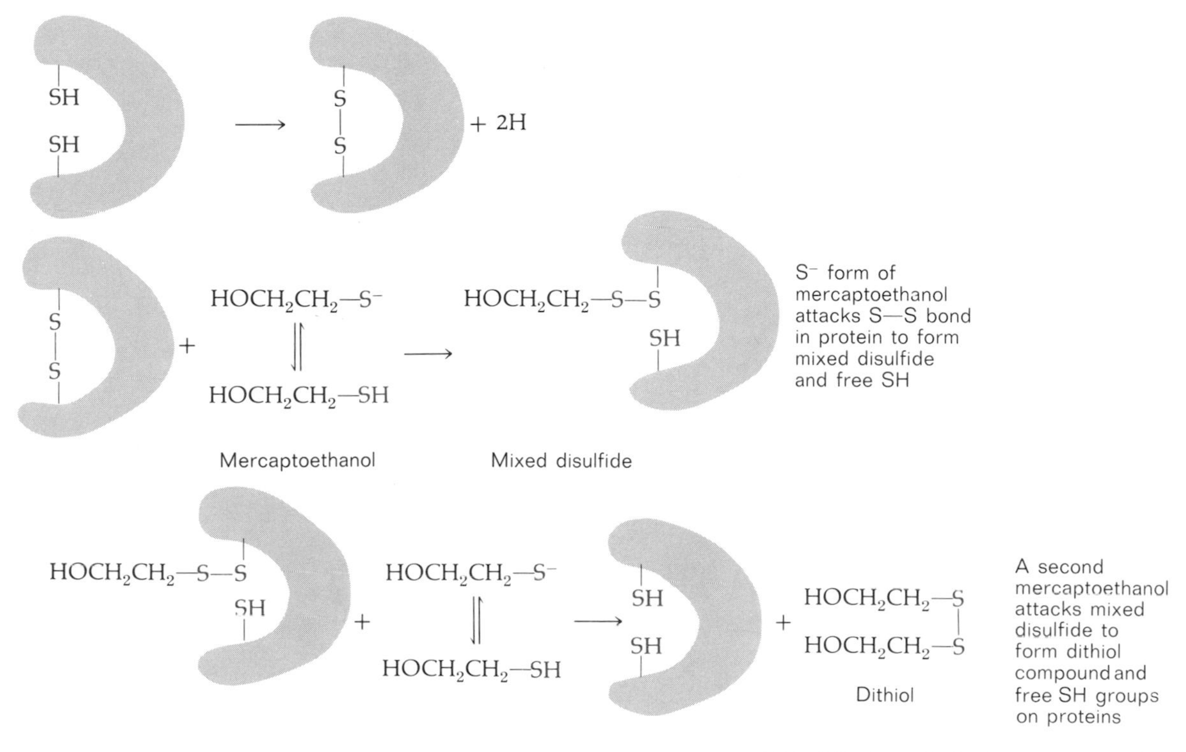

FIG. 8-5 *Disulfide Bonds.* Cysteine residues can combine in disulfide bonds. The free SH groups can be regenerated by excess mercaptoethanol due to disulfide interchange.

They are employed in the range of 6 to 8 *M* to obtain full disruption of structure. With a protein containing disulfide bonds, for example, ribonuclease with four such bonds, mercaptoethanol or another reducing agent must also be added to break the disulfide bonds for full disruption of structure (see Fig. 8-5).

A very useful criterion for the disruption of structure is the viscosity of the molecule. Normally, enzymes, as compact globular structures, have a low viscosity. We can describe their viscosity in terms of the intrinsic viscosity $[\eta]$. This parameter eliminates from consideration the effects of concentration by extrapolation of measured viscosities to infinite dilution. Intrinsic viscosity is thus a property of the individual molecules. Interestingly, a value for the intrinsic viscosity of 2.5 in units of cubic centimeters per gram is predicted on the basis of deductions first proposed by Einstein for any spherical particle regardless of its size. Thus Table 8-1 shows that a molecule such as ribonuclease has an intrinsic viscosity of about 3.3 cc/g, slightly more than the minimum 2.5 cc/g since it is not a perfect sphere. Molecules of even greater size, such as serum albumin, retain intrinsic viscosities in the

TABLE 8-1 Intrinsic Viscosity [η]

	Molecular Weight	Native Structure, cc/g	Random Coil, cc/g
Ribonuclease	13,700	3.3	13.9*
Serum Albumin	65,000	3.7	22
Bush Stunt Virus	8.9×10^6	4.0	—†
DNA	5×10^6	5000	150‡

* Urea plus reduction of S—S bonds.

† No value available since structure dissociates into small units.

‡ Heat denatured.

Source: After H. K. Schachman, *Cold Spring Harbor Symposium on Quantitative Biology*, **28:** 409 (1963).

same range. However, as shown in Table 8-1, when proteins are placed in urea the intrinsic viscosity rises markedly, to values that would be expected for a simple random coil—a polymer of amino acids with no fixed structure. Moreover, in this case the intrinsic viscosity is no longer independent of molecular weight but rises with the size of the molecule. (We can contrast the behavior of globular proteins with that of DNA, which has a very high intrinsic viscosity due to its rodlike shape. When it is brought to a random-coil configuration by heating, a marked drop in viscosity occurs.) Thus ribonuclease in urea with disulfide bonds reduced shows all the evidence of being a random coil. It is not possible to exclude completely the possibility that small pockets of organized structure remain, but there are no indications of such residual organization.

Loss of structure can also be detected by simple spectral measurements. As seen in Fig. 8-6, ribonuclease absorbs strongly in the ultraviolet region of the spectrum, as is typical of most proteins. The absorption in the 280 nm region is due to contributions from tyrosine, tryptophan, and phenylalanine residues—the aromatic amino acids. The exact shape of the spectrum in this region depends on the portions of the three amino acids as well as the environment of the amino acids. Tyrosines in a buried position deep in the molecule often have a different spectrum from other tyrosine molecules on the surface of the molecule in contact with water. As indicated in the figure, when ribonuclease structure is disrupted, the spectrum shifts to the left. This provides another indication that an alteration of environment has taken place, as would be expected to occur with the formation of a random-coil structure.

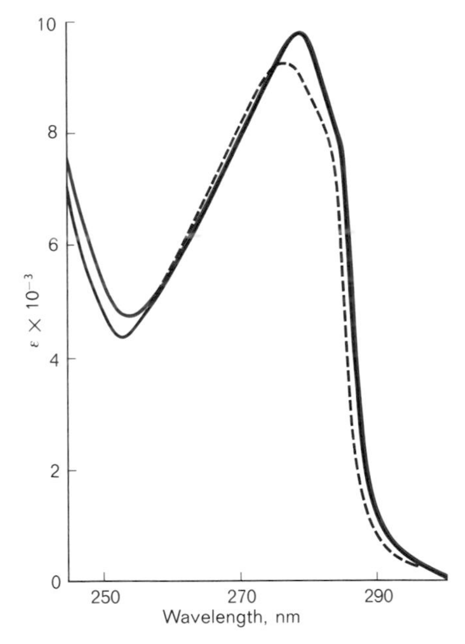

FIG. 8-6 *UV Spectra of Ribonuclease.* Native ribonuclease (solid curve), reduced ribonuclease (dashed curve), and reduced reoxidized ribonuclease (colored curve). From F. H. White, *Journal of Biological Chemistry,* **236:** 1353 (1961).

Refolding of Ribonuclease

With the structure fully disrupted in 8 *M* urea plus mercaptoethanol to break the disulfide bonds, it should now be possible to remove the disrupting agents and determine whether the native structure reforms spontaneously. The urea and mercaptoethanol can be removed by passage through a Sephadex column that excludes the protein first while permitting the smaller urea and mercaptoethanol to diffuse into the matrix of the gel (see Fig. 8-1). In this way the protein can rapidly be obtained free of urea and mercaptoethanol and left in contact with air to permit reoxidation of the sulfhydryl groups to form disulfide bonds. When ribonuclease is treated in this way, the enzymatic activity does in fact reappear, as first described by Christian Anfinsen and his colleagues. The process is slow, requiring about 6 hours for half of the initial activity to reappear, and 20 hours for virtually complete refolding. At this time more than 90% of the initial activity is present. The refolded material is indistinguishable from the starting ribonuclease. The intrinsic viscosity returns to a value of 3.3, and even the very sensitive criterion of x-ray diffraction indicates that the reoxidized material possesses the same crystal structure as the starting molecule.

The experiments with ribonuclease strongly indicate that primary structure is sufficient to determine tertiary structure. Other enzymes have been studied in detail, including some with several subunits, and in many of these cases similar patterns of refolding take place. However, the refolding process is always a slow one—measured in hours or many minutes. Yet, in the cell, proteins fold in a small number of minutes or even seconds. The implication of this discrepancy of times is that the mechanism in the cell is different from the one in the test tube. Possibly the formation of polypeptide residues one at a time on the ribosomes favors rapid folding by an undetermined mechanism. We will consider this question shortly.

Synthesis of Ribonuclease

The experiments indicating that primary structure determines tertiary structure, based on the disruption of tertiary structure in urea and the reformation when the urea is removed, are not completely convincing. There remains the possibility that even in the urea a small amount of critical structure exists

Cleavage by:
1. Trypsin if R_1 is lysine or arginine
2. Chymotrypsin if R_1 is phenylalanine, tyrosine, or tryptophan
3. Cyanogen bromide if R_1 is methionine, in a chemical process that changes methionine to homoserine, a serine analog with an additional methylene group

FIG. 8-7 *Protein-cleavage Agents.*

that is formed in some manner other than through the spontaneous folding of the primary structure, and it is this kernel of structure that forms a gridwork for the folding of the rest of the molecule when the urea is removed. This possibility was, however, fully eliminated by the complete synthesis of ribonuclease by standard chemical methods. When the primary structure is fabricated in the test tube, one amino acid at a time, the resulting material then takes on the tertiary structure of the native enzyme as isolated from a cell; there is no possibility that a mechanism other than primary structure determining tertiary structure could apply. Such a synthesis was carried out, relying on the prior knowledge of the primary structure of ribonuclease.

In order to obtain the primary structure, shown in Fig. 8-3, several advanced analytical techniques are required. First, it is necessary to degrade the large molecule into smaller pieces that can be handled more easily. The enzymes, trypsin (which cleaves at lysine and arginine) and chymotrypsin (which cleaves at phenylalanine, tryptophan, and tyrosine) are very useful in breaking down the molecule. The chemical cyanogen bromide has also found widespread use in cleaving polypeptides at the C-terminal side of methionine. These various cleavage mechanisms are summarized in Fig. 8-7. Another essential tool in the analysis of primary structure has been the automatic amino acid analyzer. The composition of both the starting enzyme and the smaller peptides obtained by degradative cleavage can be determined by hydrolysis of these molecules, usually in acid, and analysis of the product on the autoanalyzer. The analyzer separates amino acids through ion-exchange chromatography on the basis of their

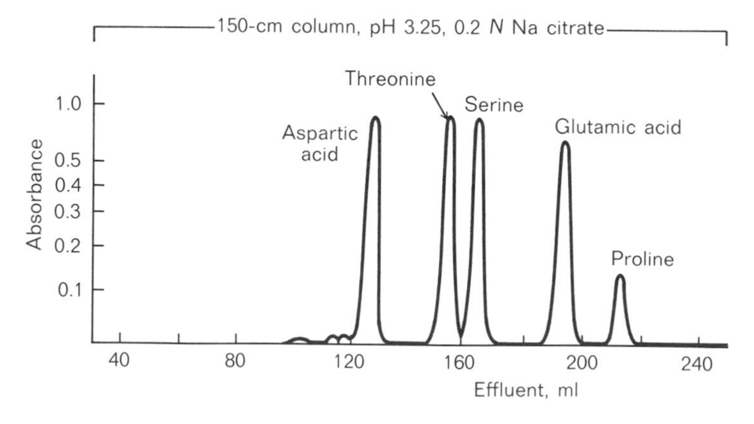

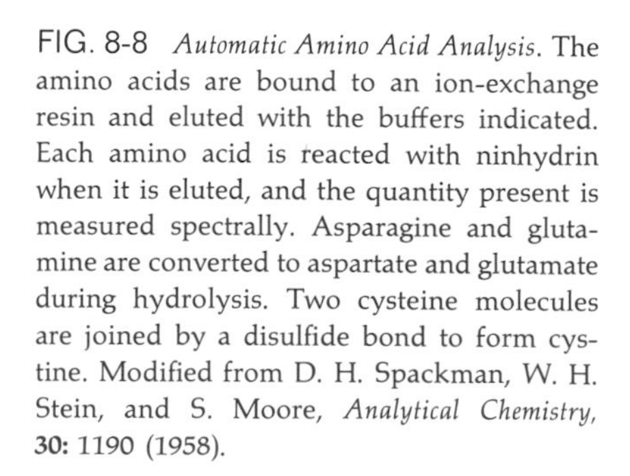

FIG. 8-8 *Automatic Amino Acid Analysis.* The amino acids are bound to an ion-exchange resin and eluted with the buffers indicated. Each amino acid is reacted with ninhydrin when it is eluted, and the quantity present is measured spectrally. Asparagine and glutamine are converted to aspartate and glutamate during hydrolysis. Two cysteine molecules are joined by a disulfide bond to form cystine. Modified from D. H. Spackman, W. H. Stein, and S. Moore, *Analytical Chemistry,* **30:** 1190 (1958).

charge. As the molecules are eluted from the chromatographic resins, they are mixed with small amounts of ninhydrin, which react with the amino groups of the amino acids to form a compound with a characteristic blue color. The absorption of light is measured and recorded on a continuous chart. A typical elution pattern from an analyzer is shown in Fig. 8-8.

While amino acid analysis of the parent molecule and the peptide fragments can provide their amino acid content and some information on the order of the fragments, the exact sequence of amino acids in each fragment is needed. The first, or amino-terminal, amino acid can be determined by one of two chemical methods—the *Sanger* method or the *Edman* method. Both rely on derivatization of the free α-amino group and isolation of the derivatized material after hydrolysis. The derivatized amino acid can then be analyzed as a derivative, or it can be liberated from the derivatizing moiety and characterized directly as an amino acid. These methods are summarized in Fig. 8-9. An important advantage of the Edman method is that it can be recycled, so that after removal of the first amino acid in a sequence and its identification, the second amino acid in the sequence can be reacted and also analyzed in a second step. By using this system and some automated methods, it is possible to determine rapidly the sequence of 10 to 20 amino acids from the amino-terminal end rather easily. (While the FDNB method is less versatile, it was used by Sanger to determine the sequence of insulin, completed in 1956.) Sequence analysis can also proceed from the carboxyl-terminal end, and in this instance the enzymes carboxypeptidase A and B, exopeptidases that release one amino acid after another beginning with the C-terminal end, are very useful. By employing a variety of these methods,

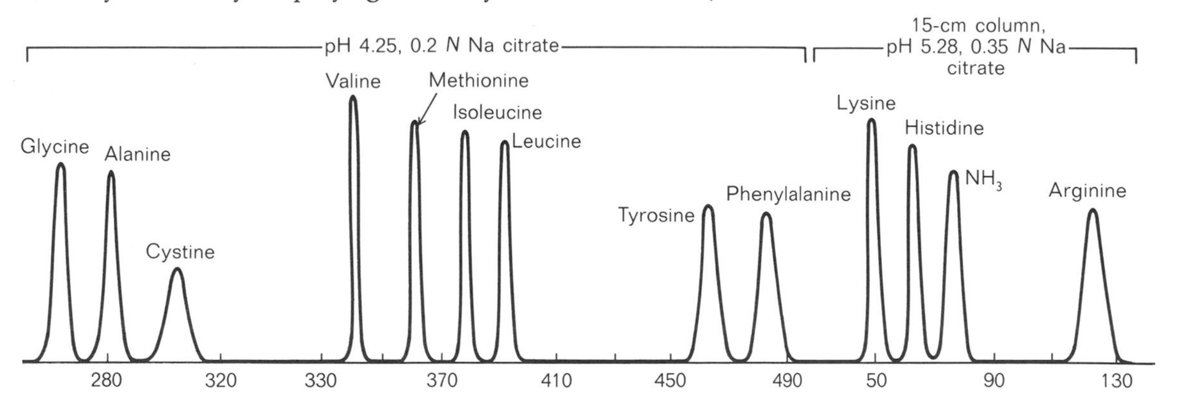

Sanger reagent, fluorodinitrobenzene (FDNB):

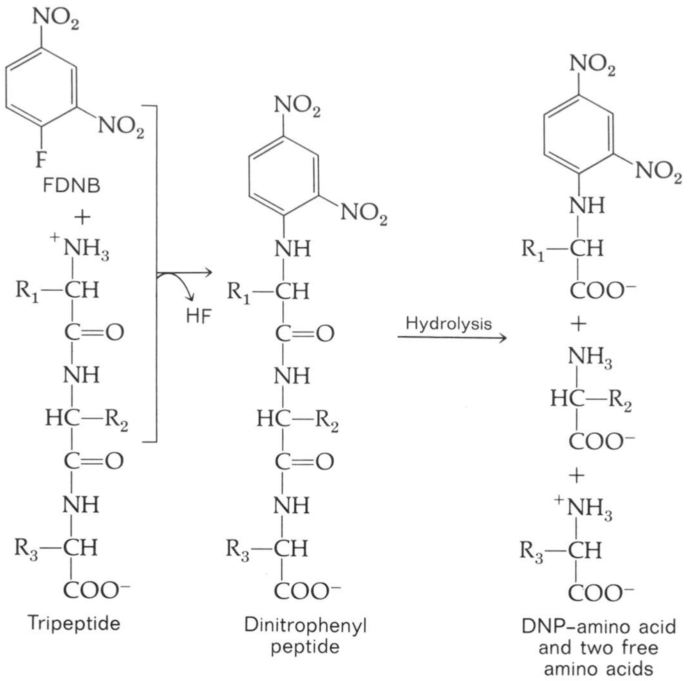

FDNB + Tripeptide → Dinitrophenyl peptide (+ HF) —Hydrolysis→ DNP-amino acid and two free amino acids

Edman reagent, phenylisothiocyanate:

Phenyliso-thiocyanate + Tripeptide → Phenylthiocarbamyl peptide —Mild hydrolysis→ Phenylthiohydantoin derivative of first amino acid and remaining dipeptide

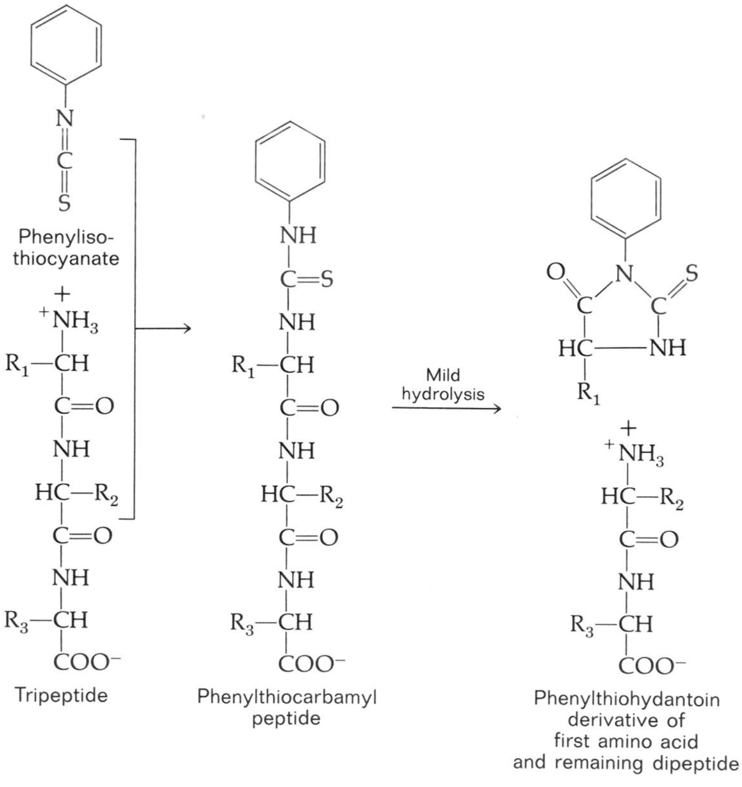

FIG. 8-9 *Sequence Analysis of Peptides.* The Sanger method gives a derivative of the amino-terminal amino acid while the remainder of the peptide is hydrolyzed. The Edman method gives a derivative of the amino-terminal amino acid and leaves the remaining peptide intact for further sequence analysis.

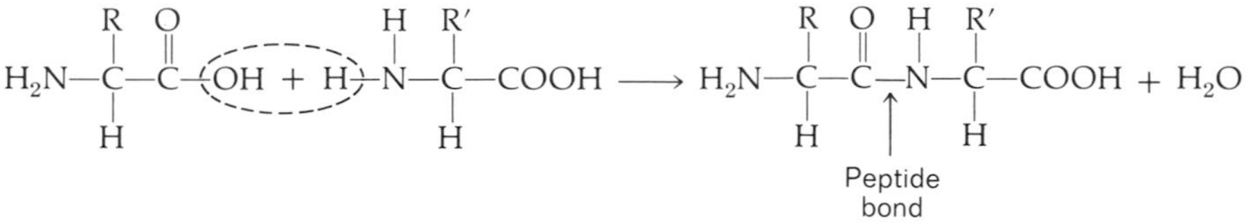

$$H_2N-\underset{\underset{H}{|}}{\overset{\overset{R}{|}}{C}}-\overset{\overset{O}{\|}}{C}-OH + H-\overset{\overset{H}{|}}{N}-\underset{\underset{H}{|}}{\overset{\overset{R'}{|}}{C}}-COOH \longrightarrow H_2N-\underset{\underset{H}{|}}{\overset{\overset{R}{|}}{C}}-\overset{\overset{O}{\|}}{C}-\overset{\overset{H}{|}}{N}-\underset{\underset{H}{|}}{\overset{\overset{R'}{|}}{C}}-COOH + H_2O$$

Peptide bond

FIG. 8-10 *The Peptide Bond.*

the sequence of ribonuclease shown in Fig. 8-3 was determined.

With knowledge of the sequence of ribonuclease in hand, it is possible to consider synthesis of the molecule. The strategy of peptide synthesis is an old one—protect the groups not to be reacted, apply a suitable system for condensing the amino and carboxyl groups to form a peptide bond (Fig. 8-10), and then unblock the groups protected. The major advance in methodology that permitted the synthesis of ribonuclease to be achieved in a relatively short time was the attachment of the starting amino acid to a solid matrix. The synthesis begins with the carboxyl-terminal valyl residue attached to the solid-phase particle. The penultimate amino acid, serine, blocked at the amino group with *t*-butyloxycarbonyl derivative, is added. Thus the only "free" groups are the amino group of the valine and the carboxyl group of serine. In the presence of the catalyst for condensation, dicyclohexylcarbodiimide, reaction of these two free groups occurs to form a peptide bond. The blocking group on serine can then be removed and the cycle repeated. When the peptide is finished, the terminal group bound to the inert resin particle can also be removed by gentle treatment. The various stages in the cycle synthesis are summarized in Fig. 8-11. Using these methods, Gutte and Merrifield were able to synthesize ribonuclease in an automated apparatus that provided a yield of 98.6% at each step. Since there are 124 residues in ribonuclease, the small discrepancy in the yield from a fully quantitative reaction is multiplied at each stage of the synthesis, but the final yield was still a substantial value, about 17%. In addition much of the "impure" material accounting for the remaining 83% could be removed. When the final product was isolated, a fraction with all the enzymatic properties of ribonuclease was obtained. Thus the structure required for activity had been achieved. The complete synthesis of an enzyme such as ribonuclease from its constituent amino acids thus proves beyond any doubt that primary structure determines tertiary structure, at least for ribonuclease.

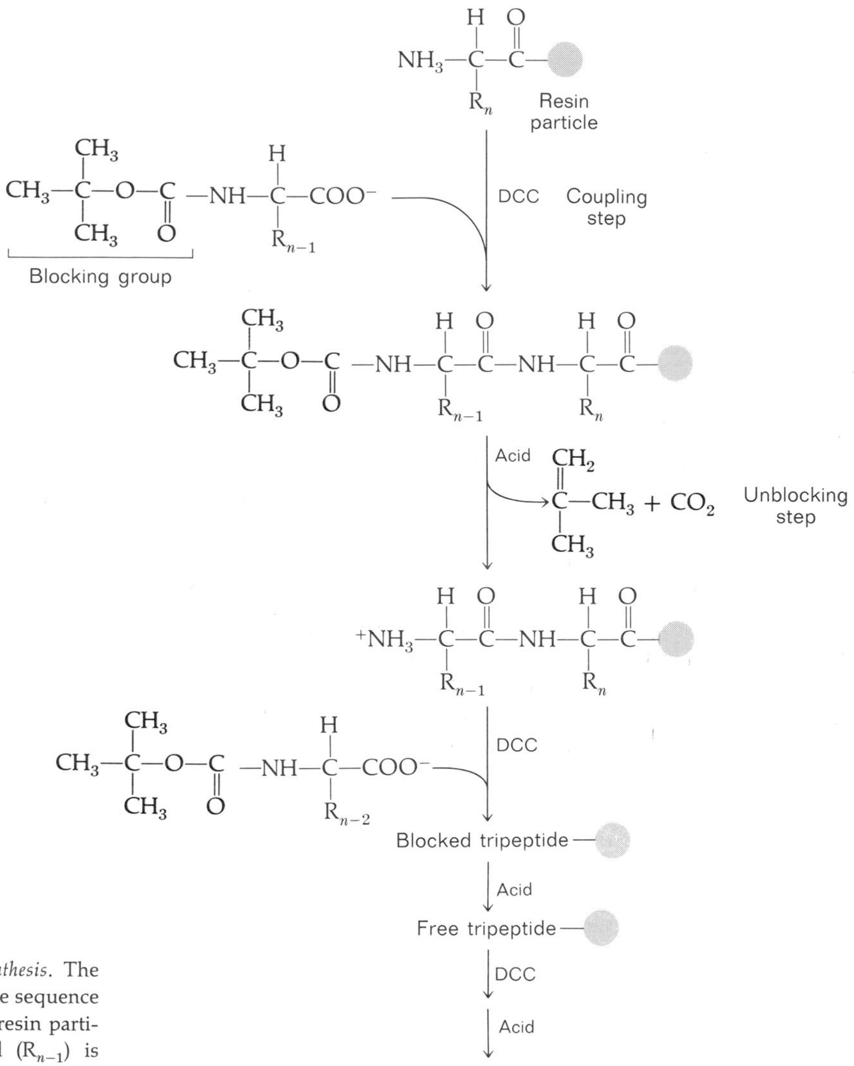

FIG. 8-11 *Solid-phase Peptide Synthesis.* The carboxyl-terminal amino acid of the sequence to be synthesized is attached to a resin particle. The next-to-last amino acid (R_{n-1}) is added in the coupling step with its amino group blocked. In the presence of the catalyst DCC (dicyclohexylcarbodiimide), a peptide bond is formed. The coupling step is followed by an unblocking step to remove the blocking group by acidification and eliminate the volatile products. The next blocked amino acid can then be added and the processes repeated until the desired sequence is obtained. Then the polypeptide is removed from the resin particle. Based on B. Gutte and R. B. Merrifield, *Journal of Biological Chemistry,* **246:** 1922 (1971).

Principles of Folding

With the pattern established that primary structure determines tertiary structure, we now wish to know how this determination actually occurs. What are the principles and mechanisms that actually govern the folding? First, we must consider the nature of amino acids and peptide linkages. A very important point in this respect is that proteins contain only L-amino acids. Since the α-carbon of amino acids is symmetric, both D- and L-enantiomorphs occur. These two forms of amino acids are pictured in Fig. 8-12. For reasons that we do not understand, only the L-enantiomorph is found in proteins of biological origin. Some D-amino acids occur in cell walls and other peptide systems, but all the enzymes and structural proteins with which we are familiar contain only L-amino acids. Presumably the presence of L-amino acids can be traced to the origin of life itself. At that time L-amino acids were incorporated in the first living organisms, and the stereospecificity of biological reactions then required that continuing biological systems employ L-amino acids. Perhaps chance alone determined that L- rather than D-amino acids should be used. Or possibly some cosmic factors were responsible, such as polarized light which preferentially enhanced the formation or destruction of L- or D-amino acids.

COO^- | COO^-
$H_3\overset{+}{N}$—C—H | H—C—$\overset{+}{N}H_3$
CH_3 | CH_3
L-Alanine | D-Alanine

FIG. 8-12 *D and L Forms of Alanine.* All the amino acids except glycine are optically active; others such as threonine and isoleucine possess two centers of asymmetry (see also Appendix III).

Given the presence of L-amino acids, we might imagine that the bonds of the α-carbon to the amino group and to the carboxyl group could enjoy virtually unlimited freedom, as could the bonds of the side chains. However, certain restraints are imposed by the physical bulk of the groups found in protein structures and by the chemistry of peptide linkages, severely limiting the range of accessible conformations. First, peptide bonds are *planar.* Since these bonds have a large amount of pi character they will closely hold to a planar structure. Moreover, they generally occur in the *trans* configuration (see Fig. 8-13). It follows from the planarity of these bonds that the peptide backbone will be determined by two angles called ϕ and ψ, depicted in Fig. 8-14. Angle ϕ connects the amide group with the α-carbon and ψ refers to the bond between the α-carbon and the carbonyl group. By definition both of these angles have a value of 0 when the respective peptide bonds are coplanar and the amide hydrogen and carbonyl oxygen are away from the side chain of the α-carbon. In this geometry the side group is below the plane, indicating that we are dealing with L-amino acids.

The freedom of conformations of protein structures is further limited by simple steric hindrance of the component atoms. Steric hindrance of the backbone itself can be expressed in a Ramachandran plot (Fig. 8-15), in which the values from 0 to 360° of the angle ψ are plotted as functions of the angle ϕ, also from 0 to 360°. Certain combinations of these two angles, indicated by the shaded areas in the figure, will be prohibited due to the bulk of the component atoms themselves. For example, with values of ψ near 0°, and values of ϕ at about 180°, very strong contact occurs between the

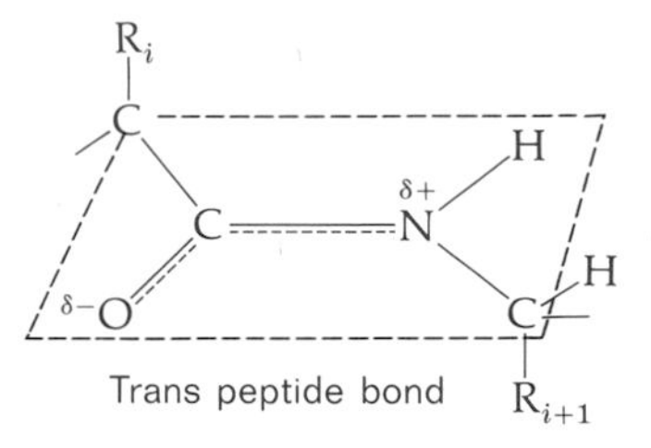

FIG. 8-13 *Planar Peptide Bond.* Atoms of the peptide bond lie in a planar arrangement due to sharing of pi electrons. The configuration is trans with O and H atoms on opposite sides, not cis which would place O and H atoms on the same side.

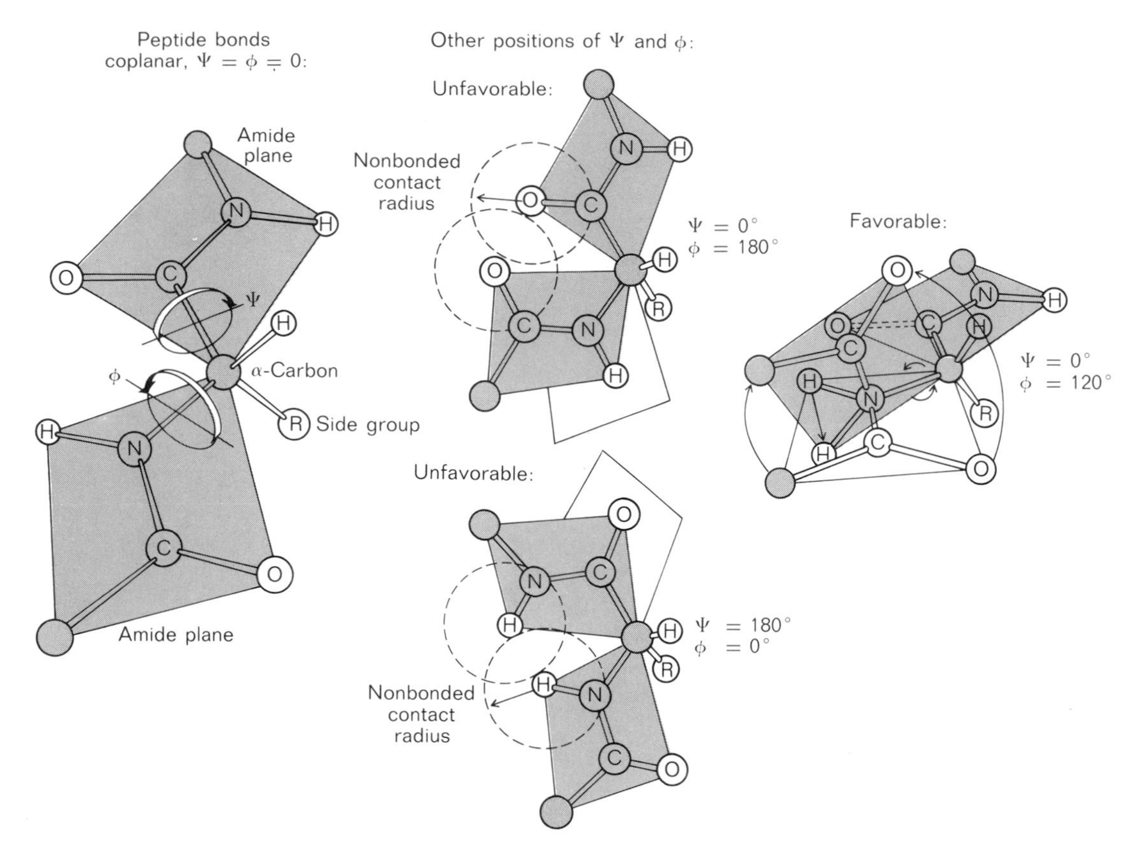

FIG. 8-14 *Backbone Angles.* Two amide planes are joined by the tetrahedral bonds to an α-carbon. Rotation of these bonds can occur (given by angles ψ and ϕ). Certain angles are unfavorable due to contact between atoms (center). One set of angles (far right) is especially favorable because the bulky carbonyl groups are as far as possible from the side chains. Reprinted by permission from R. E. Dickerson and I. Geis, *The Structure and Action of Proteins,* Harper & Row, New York, 1969. Copyright by Dickerson and Geis.

oxygen atoms of adjacent carbonyl groups. Similarly at $\psi = 180°$ and $\phi = 0°$, the hydrogens of adjacent amide groups come into contact. Thus, when the various possible steric hindrances are considered, only a relatively small fraction of the Ramachandran grid is available to the conformation of the polypeptide.

Amino acid residues in one of the conformations determined by a permissible combination of ϕ and ψ from the Ramachandran formulation tend to form two types of regular structures—helices and sheets. The value of $\phi = 120°$ is particularly favorable since this corresponds to the peptide C$=$O bond being as far as possible from the side chain of the next residue. Under these conditions, a number of different helices are formed, depending on the value of ψ. The α-helix, a right-handed helix first deduced by Linus Pauling, is especially favorable and widely encountered in protein structures (see Fig. 8-16). The β-sheet structure occurs with the peptide extended, so that hydrogen bonds can form in a direction vertical to the line of the polypeptide backbone with adjacent polypeptides. This type of structure is depicted in Fig. 8-17 and has also been found in several proteins.

In both the α-helix and β-sheet structures, hydrogen bonds of the type $\rangle$N—H···O=C$\langle$ between adjacent turns of the helix make important contributions to the stability of the structures. Just as in the case of A-T (or A-U) and G-C pairing, hydrogen bonds, although relatively weak chemical forces, nevertheless play a critical role in stabilizing protein structure. Of course not all positions in proteins are occupied by amino acid residues in helices or sheets. Many amino acids occur in conformations we cannot assign to any regular structure. For example, Fig. 8-18 illustrates conformations of all the peptides found in the enzyme lysozyme. We see that while many of them occur at the positions assigned to the helices, particularly the α-helix and the β-sheet structures, many are dispersed elsewhere on the Ramachandran plot. Thus knowledge of helices and sheets alone is not enough to understand structure, and other factors must be considered.

Organized structures of the type found in helices or sheets have an additional property that has been very useful in the analysis of protein conformation. These structures

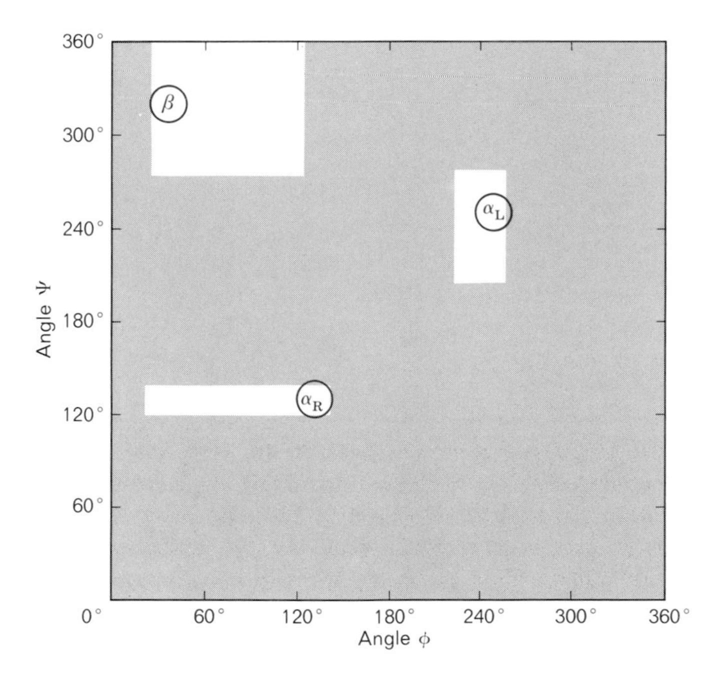

FIG. 8-15 *Ramachandran Grid.* The combinations of angles ψ and ϕ are shown with the regions of steric interference of atoms shaded. The three open areas correspond to the angles of ψ and ϕ where steric hindrance is relieved. Abbreviations are α_R for right-handed α-helix, α_L for left-handed α-helix, β for β-sheet structure. For additional details, see G. N. Ramachandran and V. Sasisekharan, *Advances in Protein Chemistry,* **23:** 283 (1968).

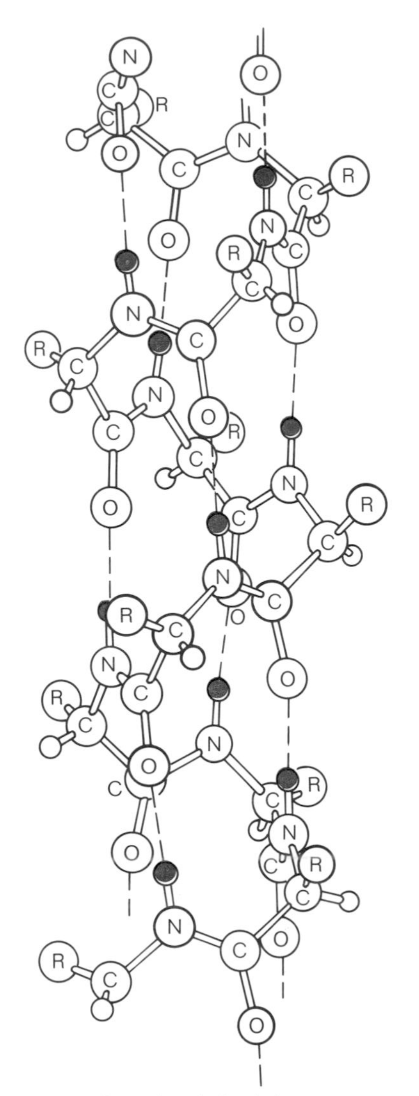

FIG. 8-16 *The α-Helix.* The helix is stabilized by hydrogen bonds (indicated by dashed lines) between amide and carbonyl groups on adjacent turns of the helix, 3.6 residues or thirteen atoms apart. From B. W. Low and J. T. Edsall, in D. E. Green, ed., *Currents in Biochemical Research,* Interscience, New York, 1956.

introduce an optical asymmetry into proteins beyond that present due to the asymmetric α-carbon atoms, and optical rotatory methods can be used to detect their presence. Optical rotatory dispersion (ORD) has been especially useful in analyzing these structures. Much work in the field derives from studies with model polypeptides, first prepared by Efraim Katchalsky and his coworkers, such as synthetic polylysine or synthetic polyglutamic acid. At low pH ($pH < 4$), polyglutamic acid is protonated, neutralizing the negative charge of the glutamate carboxyl groups. At neutral pH these charges cause strong repulsion between neighboring residues. In the uncharged condition the molecule spontaneously folds into a structure that, by optical and hydrodynamic properties, has been identified as an α-helix. Polylysine has a similar behavior at the other end of the pH spectrum. The ε-amino group of lysine has a positive charge in the neutral range, and neighboring residues are strongly repulsed. However, at high pH, about 11, when the ε-amino group is deprotonated, the polylysine will also form a helical structure. Interestingly, an α-helix is formed at relatively low temperatures, but if the material is heated to 40°, a β-sheet structure is formed. The optical rotatory dispersion properties of both the α-helix and β-sheet structures are indicated in Fig. 8-19, as well as the dispersion properties of the random coil as it exists at neutral pH. It is evident from this figure that quite distinct optical properties are correlated with the sheet and helix, and both of these are in marked contrast to the rotation observed with the random coil. Optical rotatory dispersion studies of naturally occurring proteins can thus provide some indications of the presence of α-helix or β-sheet structures if spectra with these features are found.

Based on this analysis of polypeptide structures, we can now examine more closely the structure of ribonuclease. Close analysis of the ribonuclease structure (Fig. 8-20) reveals both α-helix and β-sheet structures. The α-helix is found in the N-terminal region of the ribonuclease chain involving residues 2 to 12 and 26 to 33. Extensive β-sheet structures are found beginning at residues 42 to 49, with one sheet laid down; an extensive V-shaped formation of β-sheet structure involves residues 71 to 92, 94 to 110, and 80 to 86, folding back to pair with residues 42 to 49. An additional α-helix is found at residues 50 to 58. We will consider some of the amino acid residues in detail in Chapter 9, when we consider

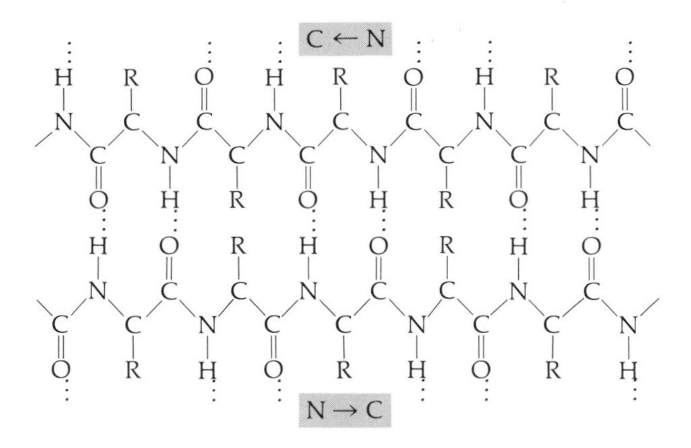

FIG. 8-17 *β-Structure: Antiparallel Pleated Sheet.* Polypeptide chains in an extended conformation run in opposite directions with hydrogen bonds between chains.

the mechanism of ribonuclease action. Ribonuclease is one of the enzymes studied to date which is relatively rich in β-sheet structures. In contrast the protein hemoglobin, which will occupy much of our attention in Chapter 10, on cooperativity, has no β-sheet structure and is made up largely of α-helices.

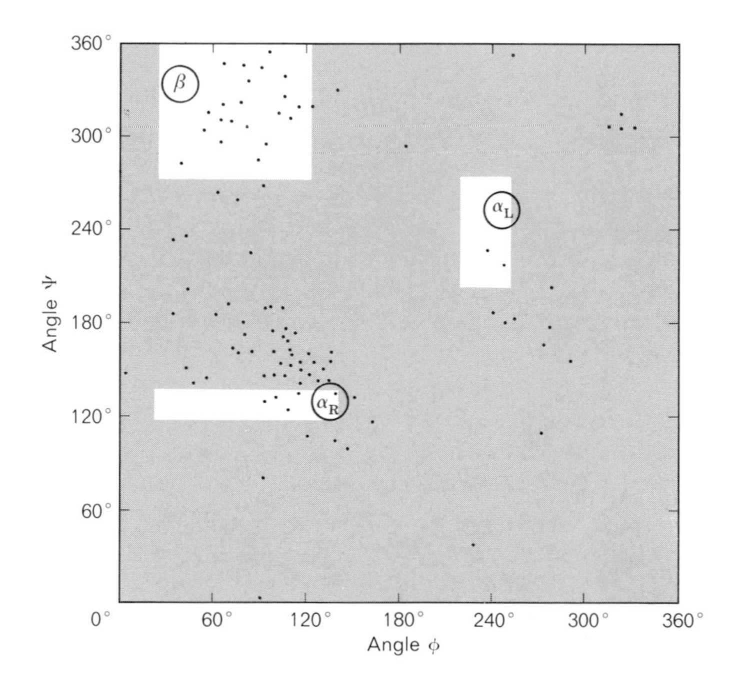

FIG. 8-18 *Values of ϕ and ψ for Each Amino Acid Residue of Lysozyme.* The values cluster about the three combinations of ψ and ϕ with a minimum of steric hindrance. Reprinted by permission from R. E. Dickerson and I. Geis, *The Structure and Action of Proteins,* Harper & Row, New York, 1969. Copyright by Dickerson and Geis.

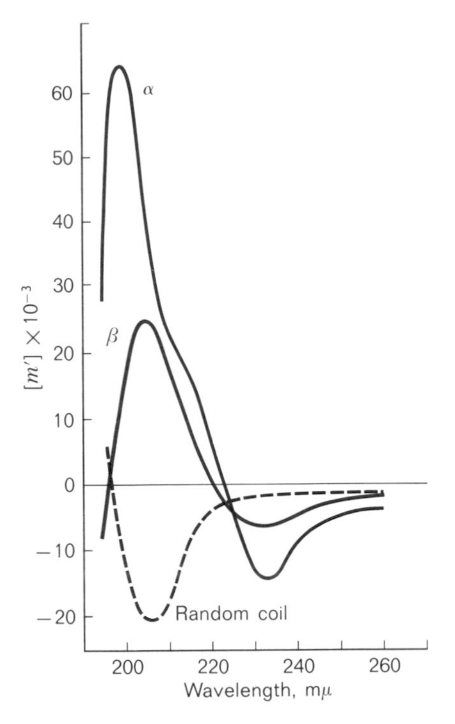

FIG. 8-19 *Optical Rotatory Dispersion of Polylysine.* Ordered structures occur above pH 11: β-sheet structure predominates above 40°C; α-helix is formed at lower temperatures. From P. Chou and H. Scheraga, *Biopolymers;* **10:** 657 (1971).

Dynamics of Folding

A striking feature of protein three-dimensional structure is that hydrophobic groups are often on the inside. These groups, particularly valine, leucine, isoleucine, methionine, tryptophan, and phenylalanine, are less likely to be found in contact with water. For example, if we return to the structure of ribonuclease (Fig. 8-20) and examine the helices along with the amino acid sequence given in Fig. 8-3, we see that in the α-helical regions, hydrophobic residues are found on the side of the helix at the interior side of the molecule, while polar residues are found at the surface in contact with water. This observation can be explained by *hydrophobic interactions.*

Hydrophobic interactions are an entropy-driven process. Any reaction that tends to proceed spontaneously can be thought of as yielding free energy. This free energy can come from two sources, as is well known from the familiar equation

$$\Delta G = \Delta H - T\,\Delta S$$

If we consider the reaction of hydrophobic groups (R),

$$2\,\mathrm{R(H_2O)}_n \rightleftharpoons \mathrm{R \cdot R} + 2\,\mathrm{(H_2O)}_n$$

we have a clear idea of how this reaction proceeds. The reactants are in strong contact with water, which limits the freedom of the water and lowers the entropy of the system. When the R groups come together, water is liberated and the entropy of the system increases. Binding of water by the unpaired R groups gives off some heat that tends to counteract the gain in entropy as the water is released, but the absolute magnitude of the term $T\,\Delta S$, particularly at physiological temperatures, is greater than the magnitude of ΔH, and it is the entropy term that predominates. At lower temperatures, approaching 0°C, the $T\,\Delta S$ diminishes, and under these conditions ΔH may predominate and the reaction may tend to the breaking of hydrophobic bonds. This principle has led to an explanation of certain cold-labile enzymes—enzymes that lose their activity in the cold. A similar description applies to electrostatic interactions, which are also entropy driven in much the same way as hydrophobic bonds. They too may exert a significant role in cold-labile enzymes.

Van der Waals interactions, as described in the discussion of DNA (Chapter 2), also play an important part in the stabilization of protein conformation. In addition hydrogen bonds may contribute to the stability, as is certainly the case

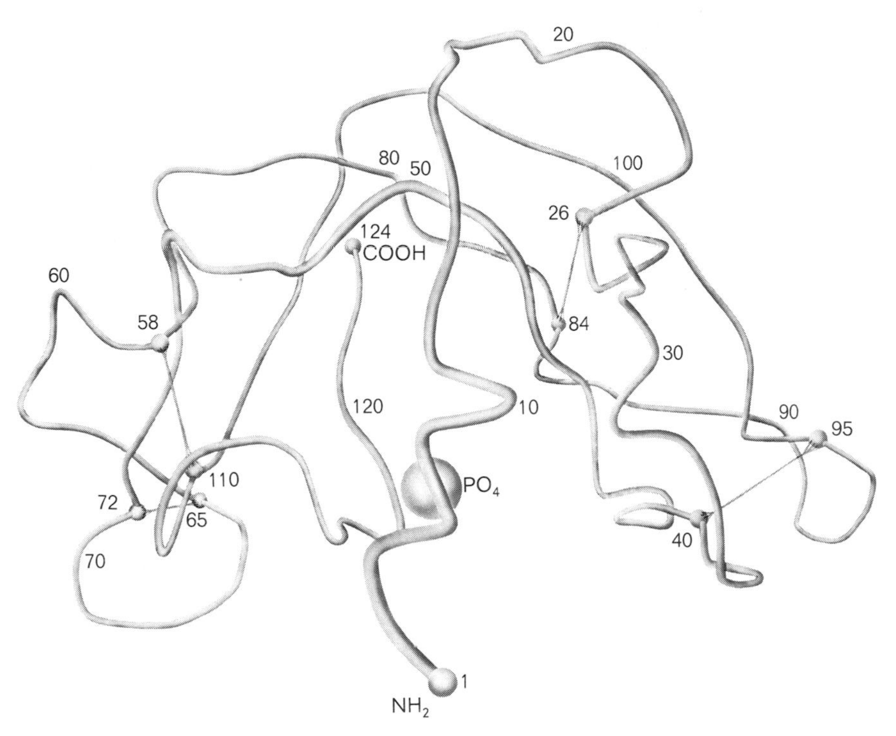

FIG. 8-20 *Structure of Ribonuclease.* From G. Kartha, J. Bello, and D. Harker, *Nature,* **213:** 862 (1967).

in the α-helix and β-sheet structures. In a number of instances hydrogen bonds other than the amide-carboxyl type are also important. Hydroxyl groups of serine, threonine, and tyrosine, as well as the amino groups of asparagine and glutamine, are capable of donating hydrogen atoms for the formation of hydrogen bonds. Similarly, the carbonyl groups of asparagine and glutamine can act as hydrogen acceptors. Many of these residues in various combinations actually occur in the diversity of hydrogen bonds encountered in protein structure. Simple electrostatic ion-pair interactions, such as $—NH_3^+ \cdots {}^-OOC—$ also make important contributions to structure, especially in the case of hemoglobin.

With the similarity of structural forces between proteins and DNA, especially hydrogen-bond-stabilized helices, it is not surprising that proteins show some of the same stability properties noted with DNA. At elevated temperatures, often above 40°C, proteins also "melt" with the hyperchromic spectral shifts and increased viscosity characteristic of a random coil. Extremes of pH can cause unfolding as well. Most proteins are composed of a mixture of many positively and negatively charged amino acids. The exact proportions vary, accounting for the ability to separate proteins on the basis of net charge. However, at high or low pH the positive or negative charges may be canceled and the electrostatic balance of the protein disturbed, resulting in disruption of structure. Thus, in working with proteins, enzymologists must avoid extremes of pH or temperature, although in some cases, if the protein under investigation is especially stable, such extremes used judiciously can provide simple purification steps.

Prediction of Tertiary Structure from Primary

With knowledge that primary structure determines tertiary, and some indication of the dynamics of the process involving hydrophobic bonds and van der Waals interactions, one could imagine predicting the tertiary structure of an enzyme simply from its primary structure. Such a prediction could be later verified by crystallography. The realization of such a prediction would be a great triumph of protein chemistry, and such efforts in this direction are now being actively pursued by Harold Scheraga and his colleagues as well as a number of other investigators. The problem, however, is complex. First, the number of possible conformations of a polypeptide of any size, such as is found in an enzyme, is enormous. To attempt to compare all the conformations with one another to arrive at the most stable conformation requires an enormous amount of time, even employing the large high-speed computers available today. Even if a conformation with less energy than any other conformation is found by the computer, it is not certain that this would be the conformation actually assumed by the protein. When a protein is folding, there is a kinetic aspect to the process. Certain groups may form contacts quickly, which would be difficult to break, so that they will predetermine the final structure in such a way as to yield a conformation that is not necessarily the one of absolute minimum energy.

The synthesis of enzymes from the N-terminal group to the C-terminal group in a cell may be an important consideration, as the folding can occur in steps during the synthesis so that only a limited number of conformations are available at any one time. However, it is not likely that proteins fold from one end exclusively. For example, the reconstitution experiments of ribonuclease can be performed on a derivative of ribonuclease known as ribonuclease S, in which a peptide containing the first 20 residues has been removed by enzymatic cleavage. This protein still forms a proper three-dimensional structure after treatment with urea and can recognize the addition of the 1–20 peptide to reconstruct an active enzyme. Thus the N-terminal groups do not appear to be essential in the folding process. All these factors make the effort to calculate the conformation from the structure very difficult. Some progress has been realized in identifying amino acid residues that strongly tend to form helices and others that are "indifferent" or tend not to form helices, and even break helices when they occur in the amino acid sequence, particularly in a string of two or more (see Table 8-2). But the prediction of an enzyme structure is a long way off. Even perfect knowledge of helix-forming and helix-breaking potentials is not sufficient. For example, the protein cytochrome c, composed of 104 amino acids, contains only a small amount of helix, so irregular conformations are the dominant factor in this protein.

In summary then, we can say that the possibility of actually predicting a tertiary structure from a primary structure is at the present time a very lofty peak that has not yet been scaled. Nevertheless, from our experience with mountains on our own globe, we might surmise that in a matter of time it too will be conquered.

TABLE 8-2 Assignment of Helical Character to Amino Acid Residues

Helix Breaker	Helix Former	Helix Indifferent
Glycine	Leucine	Alanine
Proline	Isoleucine	Threonine
Serine	Phenylalanine	Valine
Aspartic Acid	Glutamic Acid	Glutamine
	Tyrosine	Asparagine
	Tryptophan	Cystine
	Methionine	Histidine
		Lysine
		Arginine

Source: P. N. Lewis, N. Gō, M. Gō, D. Kotelchuk, and H. A. Scheraga, *Proceedings of the National Academy of Sciences*, **65:** 810 (1970).

PROBLEMS

1 Using the sequence of lysozyme given in Fig. 7-2, list the peptides that would be produced by the protein cleavage agents summarized in Fig. 8-7. From the structure of the peptides, try to deduce their order in the protein and thereby reconstruct the complete sequence of lysozyme.

2 On the basis of ionization of side chains, answer the following.
   (a) Polymers of which amino acids would be more likely to form helices at low pH (as in the case of polyglutamic acid).
   (b) Polymers of which amino acids would be more likely to form helices at high pH (as in the case of polylysine).

Chapter 9 Frontispiece *Crystal of ribonuclease mounted inside a thin-walled glass capillary in position for x-ray diffraction investigation. The crystal is ½ mm long. Courtesy D. Harker.*

TERTIARY STRUCTURE AND ENZYME ACTION

9

Biochemical reactions are extraordinarily efficient. The reactions proceed at high rates and virtually no "side products" are formed to diminish yields. Credit for this super-chemistry rests with proteins, acting as biochemical catalysts called enzymes. In this chapter we will describe the quantitative aspects of enzyme action. The key to enzyme catalysis is the three-dimensional geometry of a polypeptide chain, its *tertiary structure*. In virtually every example studied, the active sites of enzymes are restricted to single polypeptide chains. Hence the action is in the domain of tertiary structure. *Quaternary structure*, the level of organization involving several polypeptide chains, deals primarily with cooperative interactions between active sites, or tertiary units. The importance of quaternary structure in cooperativity and regulation of enzyme action will be taken up in the following chapter.

General Characteristics of Enzymes

What do enzymes do? Generally speaking, enzymes lower the free energy of activation. Therefore they simply speed the arrival at equilibrium of a reaction. The lowering of activation free energy is represented schematically in Fig. 9-1. The relative free energies of the initial and final states are the same in the presence and absence of an enzyme, but the hypothetical reaction pathway is changed in the presence of an enzyme. This new pathway has a lower activation barrier. Many reasons can be advanced to account for this lowering, and a number of them were summarized in Chapter 7 where the general question of catalysis was briefly considered.

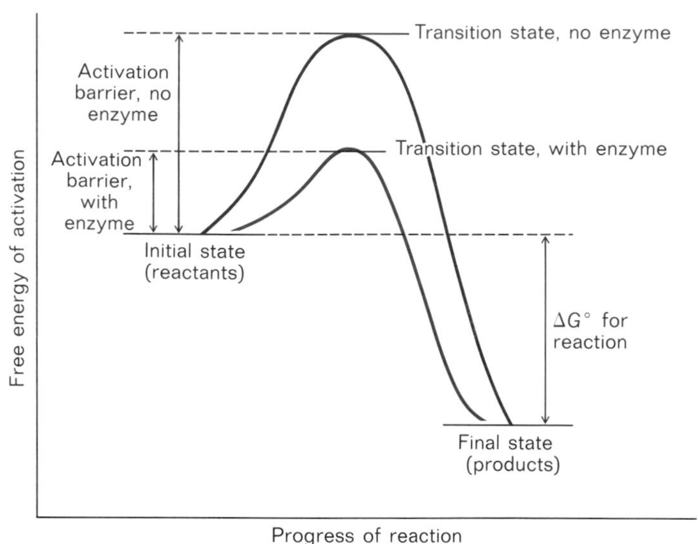

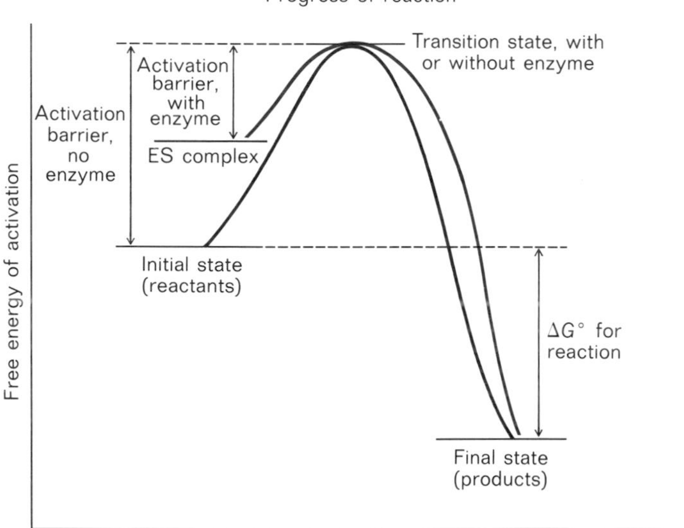

FIG. 9-1 *Free Energies of Activation.* Enzyme can lead to a rate enhancement by lowering the activation energy (logarithm of the rate is inversely proportional to the activation energy). The enzyme can act by lowering the activation energy of the transition state (upper curve) or raising the ground state in the enzyme-substrate complex (lower curve).

In order to go further into characterizing enzymes, we must consider how enzyme activity is measured. Enzyme activity is determined by an *assay,* which measures the production of product with time or the disappearance of the substrate with time. The rate of change, known as the velocity (v), is conveniently measured as the time-dependent appearance of the reaction products, $v = dp/dt$. For any given concentration of enzyme, the dependence of velocity on substrate follows a saturation curve, shown in Fig. 9-2. At substrate concentration at the plateau, where the maximum velocity is reached, the actual velocity will be proportional to the concentration of enzyme and can be used as a direct assay for the enzyme.

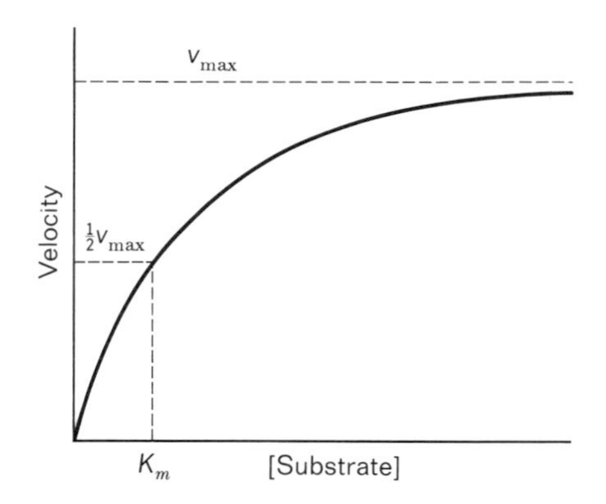

FIG. 9-2 *Velocity-Substrate Profile.* The substrate concentration at $\frac{1}{2}v_{max}$ gives the value of K_m.

Model of Enzyme Dynamics

To explain the saturation type of behavior shown in Fig. 9-2, and to account for a number of other properties of enzymes, a simple model has been developed through the years. It is not a comprehensive model, and many enzymes can only be described by more extensive models, but it nevertheless serves as a starting point for most considerations on enzyme action. The model assumes the three states summarized in Eq. (1):

$$E_0 + S \underset{k_2}{\overset{k_1}{\rightleftharpoons}} ES \xrightarrow{k_3} E_0 + P \tag{1}$$

Enzyme plus substrate can reversibly react to form an enzyme-substrate complex represented by ES. The substrate bound in the complex is then transformed by the enzyme catalytically into the product, which diffuses from the enzyme surface to regenerate free enzyme. Thus, while the substrate is continually depleted and a new product formed, the enzyme recycles and can be reused so that its action is catalytic. In these equations, [E] represents the concentration of total enzyme, [ES] represents the concentration of enzyme-substrate complex, and $[E_0]$ is the concentration of free enzyme, also equal to [E] − [ES]. [S] is the substrate concentration, and since [S] is generally much greater than [ES], the amount of [S] involved in the enzyme-substrate complex can be neglected. Since any reaction $A \underset{k_2}{\overset{k_1}{\rightleftharpoons}} B$ can be described by $d[B]/dt = k_1[A] - k_2[B]$, it follows that

$$\frac{d[ES]}{dt} = k_1[E_0][S] - (k_2[ES] + k_3[ES]) \tag{2}$$

As long as [S] is much greater than [E], the *steady-state* condition occurs; that is, the concentration of ES does not change with time and the amount produced equals the amount being used:

$$\frac{d[ES]}{dt} = 0 \tag{3}$$

Therefore

$$k_1[E_0][S] = k_2[ES] + k_3[ES] \tag{4}$$

or

$$K_m = \frac{k_2 + k_3}{k_1} = \frac{[E_0][S]}{[ES]} = \frac{([E] - [ES])[S]}{[ES]} \tag{5}$$

(where the constant K_m is known as the *Michaelis-Menten constant*)

and

$$[ES] = \frac{[E][S]}{K_m + [S]} \tag{6}$$

Since the measured velocity v of the reaction is proportional to the concentration of the ES complex,

$$v = k_3[ES] \tag{7}$$

When the substrate is *saturating,* so that essentially all the enzyme present is present in the complex, the maximum velocity v_{max} is achieved and

$$v_{max} = k_3[E] \tag{8}$$

Combining Eqs. (6) and (7), we obtain

$$v = k_3 \frac{[E][S]}{K_m + [S]} \tag{9}$$

and normalizing to v_{max},

$$\frac{v}{v_{max}} = \frac{k_3\{[E][S]/(K_m + [S])\}}{k_3[E]} \tag{10}$$

or

$$v = \frac{v_{max}\,[S]}{K_m + [S]} \tag{11}$$

Equation (11) for v in terms of v_{max}, K_m, and [S] is the versatile *Michaelis-Menten equation.*

Consequences of the Michaelis-Menten Equation

1 Substrate concentration at $v = \frac{1}{2}v_{max}$ is equal to K_m. This point can be established by noting that Eq. (11) then reduces to

$$\frac{v_{max}}{2} = \frac{v_{max}[S]}{K_m + [S]}$$

or

$$K_m = [S]$$

Thus, simply by establishing a substrate profile of v versus [S], and measuring the velocity at half of the maximum, we have obtained a value of K_m.

2 In certain cases the value of K_m is often an approximate measure of the substrate-binding constant K_s, where

$$K_s = k_2/k_1$$

from Eq. (1). This relationship applies only when k_3 is the rate-limiting step in the reaction and is appreciably smaller than k_1 or k_2; the K_m from Eq. (5) then reduces to k_2/k_1.

3 The Michaelis-Menten equation can be transformed to provide other equations more suitable for graphical analysis. The most commonly employed is the *Lineweaver-Burk* equation derived from the reciprocal of the Michaelis-Menten equation. Thus

$$\frac{1}{v} = \frac{K_m + [S]}{v_{max}[S]} = \frac{K_m}{v_{max}[S]} + \frac{1}{v_{max}}$$

This form is the common straight-line representation, where graphical analysis of $1/v$ versus $1/[S]$ yields K_m and v_{max} as intercepts and K_m/v_{max} as the slope. A typical graph of this equation is given in Fig. 9-3.

4 Many times in the study of enzymes the interplay of enzymes and inhibitors is encountered. Two types of reversible inhibition can be distinguished—*competitive* and *noncompetitive*. Competitive inhibition is usually caused by a small molecule that resembles the substrate—a substrate analog—but does not fulfill the requirements for a substrate and cannot actually participate in the reaction catalyzed. In such cases the addition of an inhibitor increases the apparent K_m of the reaction system, although the v_{max} is left unchanged. Therefore competitive inhibition is most generally recognized by examining the Lineweaver-Burk plot and noting a family of curves at different concentrations of inhibitor, all passing through the intercept at the point corresponding to the reciprocal of the v_{max} but intersecting the abscissa at different values. Typical competitive inhibition behavior is shown in Fig. 9-4.

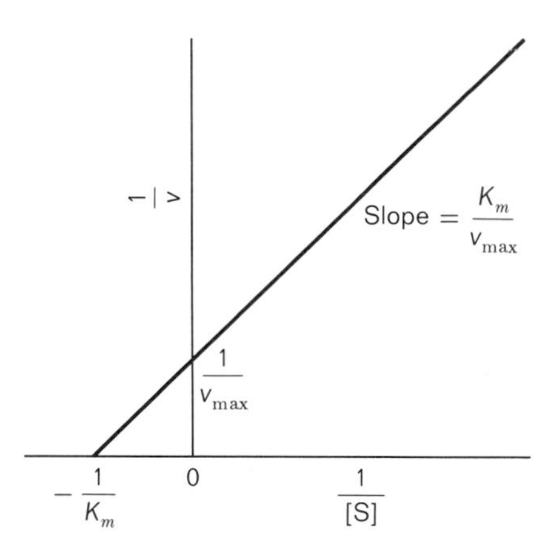

FIG. 9-3 *Lineweaver-Burk Plot.* In this linear form the v_{max} is given by the reciprocal of the intercept on the ordinate and the K_m is given by the reciprocal of the intercept on the negative abscissa.

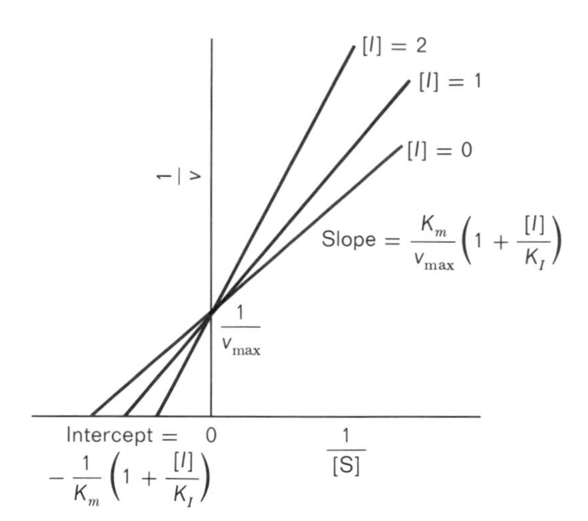

FIG. 9-4 *Competitive Inhibition.* A series of lines with the same intercept on the ordinate is obtained for different concentrations of inhibitor.

Noncompetitive inhibition can involve a general enzyme antagonist, such as a metal ion. In this type of inhibition a fraction of the enzyme molecules is removed from participation in the catalytic events, and the apparent maximum velocity changes. However, since there is no direct interference with the binding of substrate, the K_m of the system remains constant. Noncompetitive inhibition is shown in the Lineweaver-Burk formulation in Fig. 9-5. Irreversible inhibition also is widely encountered in studies with enzymes. Commonly, a reactive amino acid, such as a serine or a cysteine, is present at the active site, and reaction with a general acylating or alkylating agent may derivatize this group and completely and irreversibly inactivate the enzyme. An example is the *covalently* bound inhibitor produced with chymotrypsin by reacting the active serine with an acylating agent such as diisopropylfluorophosphate (DFP) (see Fig. 9-6). Studies along these lines are especially useful in delineating the residues at the active site, and we will consider an example of such studies more thoroughly when we discuss the active site of ribonuclease.

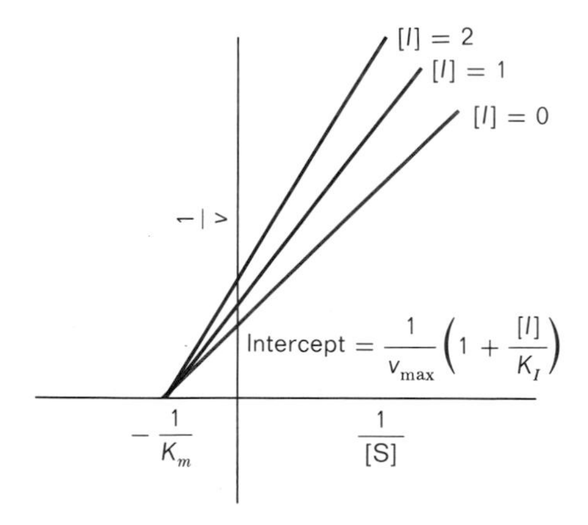

FIG. 9-5 *Noncompetitive Inhibition.* A series of lines with the same intercept on the negative abscissa is obtained for different concentrations of inhibitor.

General Properties of Enzymes

So far we have considered the important parameters K_m and v_{max} in formal terms. It is now useful to note a few common ranges for these numbers. The value for K_m of typical enzymes is about 10^{-4} M. This value is only an average, and

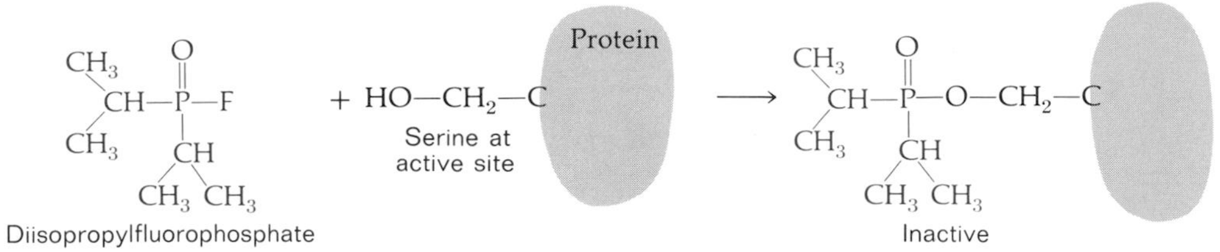

$$(CH_3)_2CH{-}P(=O)(F){-}CH(CH_3)_2 + HO{-}CH_2{-}C\text{(Protein)} \longrightarrow (CH_3)_2CH{-}P(=O)(O{-}CH_2{-}C\text{(Protein)}){-}CH(CH_3)_2$$

Diisopropylfluorophosphate

Serine at active site

Protein

Inactive

FIG. 9-6 *Covalently Bound Inhibition.* Reaction of diisopropylfluorophosphate with serine hydroxyl.

some enzymes are found with a value of K_m well below 10^{-5} and even 10^{-6} M. At the other extreme, enzymes are rarely encountered with values of K_m above 10^{-3} M, and substrates with values of K_m in this range or above are generally considered to be poor substrates.

Values of v_{max} also vary widely among enzymes. The maximum velocity is conveniently considered in terms of the *turnover number.* The turnover number is defined as the molecules of substrate catalyzed per molecule of enzyme per minute. From knowledge of the measured maximum velocity of an enzyme and the *specific activity*—that is, the rate of product formation per milligram of enzyme—and finally, the molecular weight of the enzyme, one can calculate the turnover number. Carbonic anhydrase has a turnover number of over 10^7—this is one of the highest known. More commonly, enzymes of glycolysis and intermediary metabolism show turnover numbers of about 10^3. This still represents a very large number and demonstrates the value of enzymes as catalytic units. Just a small quantity of enzyme can quickly transform a large population of molecules.

In our consideration of the quantitative aspects of enzyme action, we only included one enzyme-substrate state. However, in many enzyme systems, multiple enzyme-substrate complexes are present. Each represents the partial transformation of the substrates along the route to product formation. In many cases multiple substrates are utilized in the reaction, such as reactions involving transfers or synthesis of a product from two substrates. In addition a large number of reactions to be considered in intermediary metabolism involve a *cofactor*—a low-molecular-weight molecule that participates along with the enzyme in carrying out a particular chemical step. In these cases of multiple substrates or reactions with cofactors, complexes are formed that involve several compounds, and in some cases the order of binding of the substrate can be determined. Finally, in a number of instances the enzyme-substrate complex is actually a covalent linkage. A notable example of such an enzyme-substrate complex is the acyl intermediate discussed for chymotrypsin in Chapter 7.

Enzymes catalyze an extremely diverse set of chemical reactions. We have already considered many types of chemical transformations in the synthesis of nucleic acids and in the steps of protein synthesis. Many more enzyme reactions will be encountered in the later chapters dealing with intermediary metabolism. Categories of enzyme action that can be delineated include bond cleavage (which is by hydrolysis), group-transfer reactions, isomerization reactions, addition reactions to double bonds, linkage reactions involving the formation of bonds with ATP cleavage, and oxidation-reduction reactions in which a change of redox potential is the critical point of the reaction. Each of these reaction types will be illustrated in the chapters on metabolism. However, to give a deeper insight into enzyme catalysis we will now consider the mechanism of one enzyme, ribonuclease, in detail.

Mechanism of Ribonuclease

Having described the structure of ribonuclease in Chapter 8, we might now attempt to formulate the way in which this structure catalyzes the hydrolysis of RNA. One of the first clues came from the observation that the reaction proceeds via the cyclic 2′,3′-phosphate intermediate. The structure of such an intermediate is shown in Fig. 9-7 and implies the overall reaction summarized in the same figure. Other important evidence on the role of ribonuclease in the reaction came from the observation that treatment of the enzyme with iodoacetate led to inactivation. Iodoacetate (ICH_2COOH) reacts strongly with SH groups (cysteine) and less well with imidazoles (histidine) and amine groups (lysine). In each case the amino acid side chain displaces the iodine, and carboxymethyl derivatives are formed (Fig. 9-8). Further characterization of the inactive enzyme indicated that the iodoacetate can abolish activity by reacting at either of two positions on the molecule, histidine 12 or histidine 119. These experiments suggested that both histidines were involved at the active site of the enzyme and hence were presumably in close proximity in spite of the fact that they are separated in the primary sequence by over 100 residues. Lysine 41 was also implicated by chemical studies.

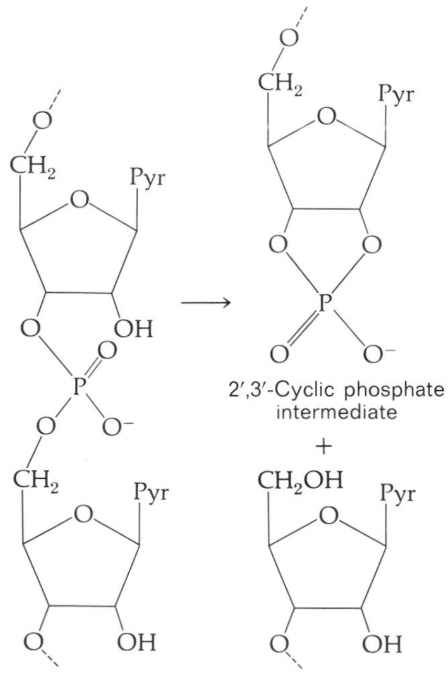

The identification of the two histidine residues and a

2′,3′-Cyclic phosphate intermediate

FIG. 9-7 *First Step in the Ribonuclease Reaction.*

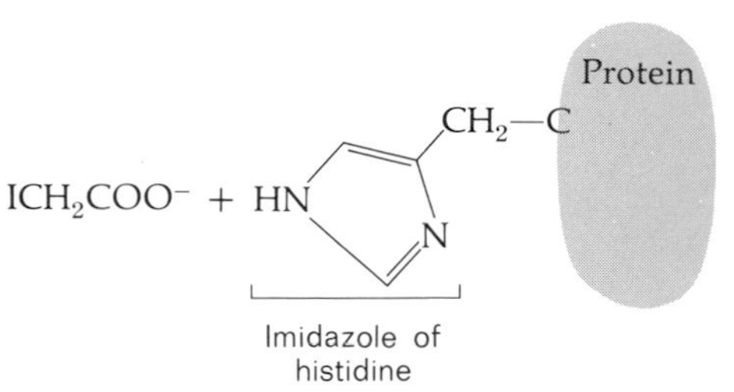

$$ICH_2COO^- + \text{HN(imidazole)}\text{—}CH_2\text{—C(Protein)} \longrightarrow {}^-OOC\text{—}CH_2\text{—N(imidazole)}\text{—}CH_2\text{—C(Protein)}$$

Imidazole of histidine

FIG. 9-8 *Reaction of Iodoacetate with Histidine.* For other reactions of amino acid side chains, see G. E. Means and R. E. Feeney, *Chemical Modification of Proteins,* Holden-Day, Inc., San Francisco, 1971.

consideration of the actual reaction catalyzed suggested that both may be participating—one as an acid and one as a base. This point of view was given further strength by the observation that the pH profile of the reaction for the hydrolysis of 2′,3′-cyclic phosphate shows very steep fall off on either side of pH 7 (see Fig. 9-9). This type of behavior is just what would be expected for acid catalysis and base catalysis. Deprotonation of an acid group at high pH leads to the right half of the curve, while protonation of a basic group at low pH leads to the left half. All of these findings led to the mechanism now supported by the x-ray crystal structure that finds histidines 119 and 12 and lysine 41 in close proximity. The mechanism is summarized in Fig. 9-10. According to this mechanism, histidine 12 acts as a base, abstracting a proton from the 2′-hydroxyl of the sugar moiety, while histidine 119 acts in the positively charged acid form and attracts the phosphate oxygen. Formation of the 2′,3′-phosphate intermediate occurs with release of an RNA fragment and effects the transfer of a proton from one histidine through the phosphate to the

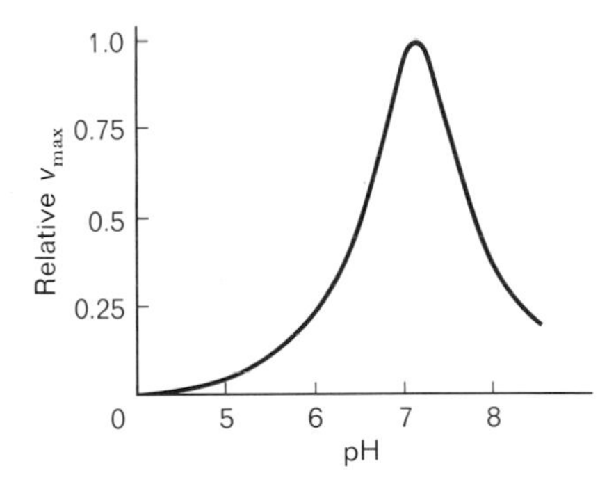

FIG. 9-9 *pH Dependence of Ribonuclease.* The dependence of v_{max} on the pH for the hydrolysis of nucleoside 2′,3′-cyclic phosphate by ribonuclease. From S. Bernhard, *The Structure and Function of Enzymes,* copyright © 1968, W. A. Benjamin, New York, 1968.

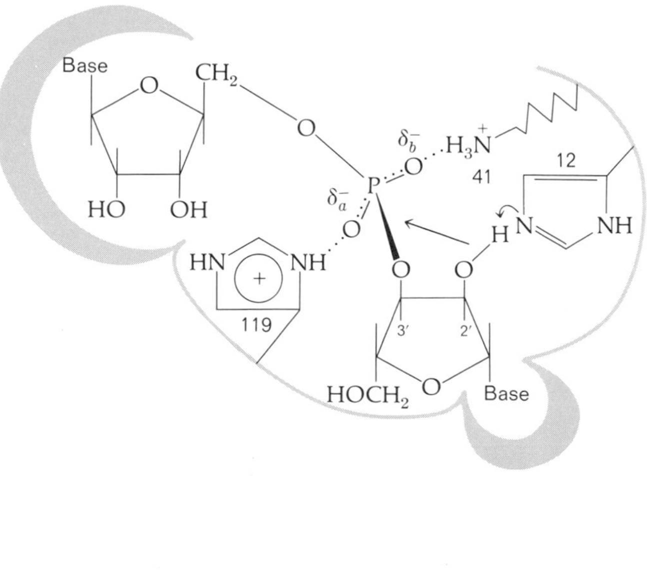

FIG. 9-10 *Substrate at Active Site of Ribonuclease.* From E. A. Dennis and F. H. Westheimer, *Journal of the American Chemical Society,* **88:** 3431 (1966).

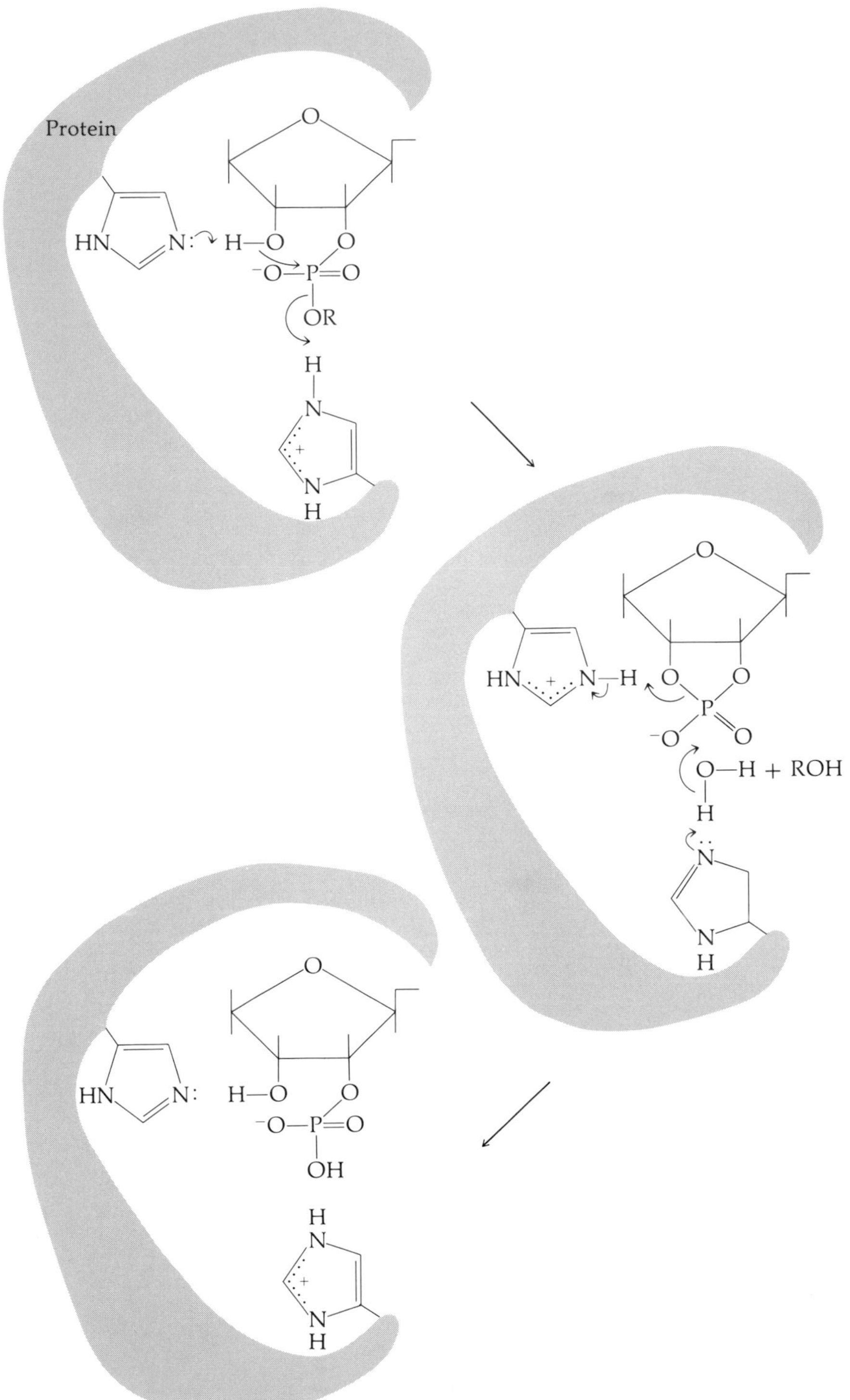

FIG. 9-11 *Mechanism of Ribonuclease.* From S. Bernhard, *The Structure and Function of Enzymes*, copyright © 1968, W. A. Benjamin, New York, 1968.

other. This transfer reverses the roles of the two histidines as acid and base. However, the initial states are reestablished as water attacks to cause final hydrolysis of the cyclic phosphate. These various stages are summarized in Fig. 9-11 and provide an excellent example of the collaboration of x-ray crystallography and organic chemistry to reveal the mechanism of action of a protein.

The specificity of the catalytic mechanism of ribonuclease for the pyrimidine bases (shown in Fig. 4-5) is also understood now in terms of the three-dimensional structure of the enzyme. Hydrogen bonds are formed involving the residues threonine 45 and serine 123, which can accommodate either uridine or cytosine bases. The formation of these bonds is summarized in Fig. 9-12. In contrast to transfer RNA-synthetase, which probably involves mimicry of a protein by the transfer RNA (see Chapter 4), the nucleic acid–ribonuclease recognition described here involves mimicry of a nucleotide by the protein to give a complex suggestive of a base pair in DNA. Thus we see that both types of mimicry have their usefulness in biochemical reactions. It is also worth noting that once again weak hydrogen bonds appear in a critical biochemical role. We have encountered hydrogen bonds at the heart of the double helix, in the transcription and translation processes, and now in the mechanism of an enzyme's specificity. We will see that they also play a part in stabilizing interactions between protein subunits in the quaternary structure of hemoglobin, the subject of the following chapter.

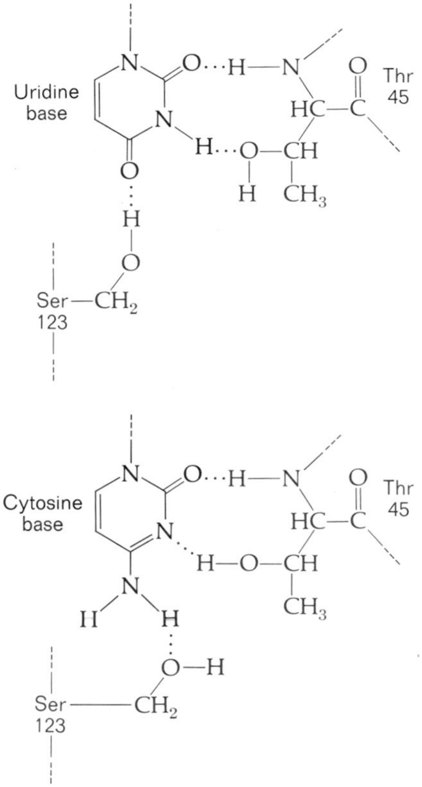

Uridine base — Thr 45 — Ser 123

Cytosine base — Thr 45 — Ser 123

FIG. 9-12 *Ribonuclease-Pyrimidine Complexes.* From F. M. Richards, H. W. Wycoff, and N. M. Allewell, in F. O. Schmitt, ed., *The Neurosciences: Second Study Program,* Rockefeller University Press, New York, 1970.

PROBLEMS

1 An enzyme with a molecular weight of 10^5 has a specific activity of 10^{-2} moles of product formed per minute per milligram of enzyme. What is its turnover number?

2 From the shape of the pH profile for ribonuclease, calculate probable pK values for the histidine acting as an acid and the histidine acting as a base.

Chapter 10 Frontispiece *Crystals of horse hemoglobin. Hexagonal plates (bottom) of deoxyhemoglobin are converted to needles of oxyhemoglobin (middle) by oxygen from an air bubble (top). Courtesy F. Haurowitz.*

QUATERNARY STRUCTURE AND COOPERATIVE INTERACTIONS

10

We have followed the fate of the information in genes through the stages of protein synthesis into a polypeptide sequence. We have noted how the polypeptide folds spontaneously into a specific three-dimensional structure that can catalyze a metabolic reaction. Now we will consider the special properties that arise when several polypeptide chains are present in one protein. With the isolation and characterization of many enzymes in recent years, the presence of more than one polypeptide chain in a protein has been widely encountered. Frequently the polypeptide chains are identical and associate spontaneously through noncovalent forces—the same types of forces that prevail in maintaining tertiary structure. Proteins with a quaternary structure, that is, more than one polypeptide chain, are known as *oligomeric* proteins. In this section we will describe how one establishes the presence of subunits in proteins and explain why most oligomeric proteins contain two, four, or if neither two nor four, usually an even number of subunits. With this background we will go on to consider the major functional advantage of quaternary structure, which is the mediation of cooperative interactions between substrate molecules or other ligands. To illustrate cooperativity we will study in detail the classical example of this phenomenon, the cooperative binding of oxygen by hemoglobin, and also examine a regulatory enzyme, aspartate transcarbamylase, which exhibits many of the important cooperative properties.

Several additional examples of cooperativity will be considered in Chapters 15 and 16 in the realm of regulation of metabolism and the control of protein synthesis.

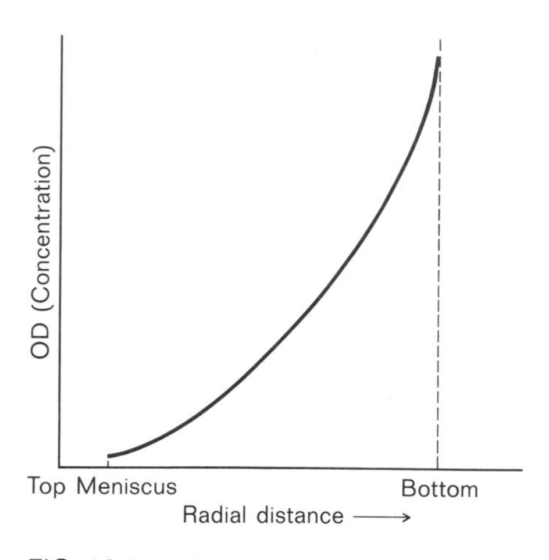

FIG. 10-1 *Sedimentation Equilibrium.*

Occurrence of Subunits

Several lines of evidence are used in biochemistry to determine the presence of subunits in proteins. The conclusion that a particular native enzyme is not the minimum structural unit of the protein generally requires establishing the size or molecular weight of the enzyme first. At the present time, molecular weight is often determined by an ultracentrifuge technique known as sedimentation equilibrium. (Other methods such as osmometry and light scattering could also be employed.) Protein solutions are examined in an analytical ultracentrifuge at moderate rotor speeds for relatively long times, 10 to 20 hours. In these times, if the volume of solution is small (approximately 0.1 ml), the protein solution can achieve *sedimentation equilibrium*. The macromolecules reach a steady state that satisfies the centrifugal force generated by the spinning rotor and the opposing tendency of diffusion. At equilibrium the protein concentration at different points in the cell is invariant with time. The distribution at sedimentation equilibrium can be observed in the analytical ultracentrifuge by, for example, the scanning absorption optical system, which measures the optical density at each point in the spinning centrifuge cell to give a profile of the type seen in Fig. 10-1. The redistribution of the macromolecule can be described by

$$M = \frac{2RT}{(1 - \bar{v}\rho)\omega^2} \frac{d \ln C}{dr^2} \tag{1}$$

This logarithmic equation suggests a simple graphical method for determining the molecular weight as the slope of a straight line of $\ln C$ versus r^2 (see Fig. 10-2). The slope gives $M(1 - \bar{v}\rho)\omega^2/2RT$. With known values of R as the gas constant, T the absolute temperature, ω the speed in radians per second, ρ the density of the solution and $\bar{v}$ the partial specific volume of the material, the molecular weight can be calculated simply from this relationship. Frequently, enzymes of intermediary metabolism are found with molecular weights in the range of about 100,000 to 200,000.

With the molecular weight of the native enzyme in hand, any of the following methods might reveal the presence of subunits.

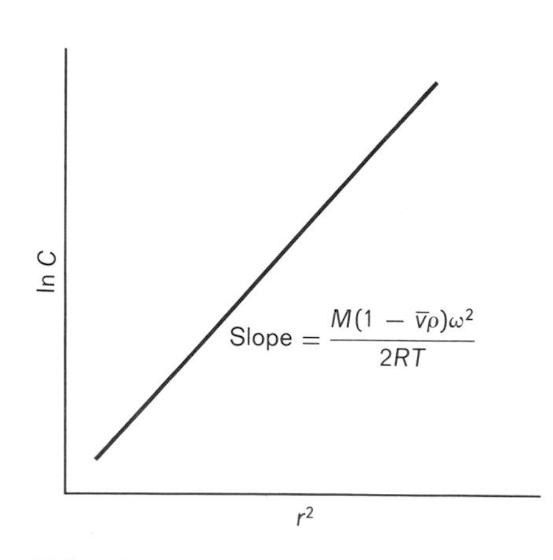

FIG. 10-2 *Analysis of Sedimentation-Equilibrium Data.*

1 *Primary structure redundancy.* If the amino acid sequence of the enzyme is determined, or some fraction of it, each amino acid in the sequence may be represented some integral number of times. For example, if the C-terminal

amino acid is determined, one may find two or more copies of the same amino acid for each molecule of enzyme. In this way the enzyme aldolase was concluded to have multiple subunits when four tyrosine molecules were found to be released for each molecule of enzyme on digestion with carboxypeptidase. When N-terminal analysis is performed (see Fig. 8-9), multiple N-terminal residues may also be revealed.

2 *Decrease in molecular weight in dissociating solvents.* When molecular weights are measured in solvents that disrupt structure, such as guanidine or urea, for an enzyme with quaternary structure, the observed molecular weight decreases, although no covalent bonds have been broken. Thus when aldolase, normally with a molecular weight of 155,000 is examined in guanidine hydrochloride, and certain corrections are made for the presence of the guanidine, a molecular weight of about 38,000 is obtained, providing further evidence for the presence of four subunits.

3 *Spontaneous dissociation with dilution.* In certain cases proteins are in equilibrium with their free subunits. Since the law of mass action requires that dilution favors dissociation, the dependence of molecular weight on concentration can often reveal the presence of subunits. Phenomena of this type have been observed extensively with hemoglobin, for example, where measurements of molecular size by sedimentation velocity revealed a decrease in the sedimentation constant, from about $s = 4.3S$ characteristic of the native tetramer in the millimolar range to values approaching 2.8S (characteristic of half-molecules or dimers), in the micromolar range. (Typical data are shown in Fig. 10-3.) In experiments of this type, dissociation can be observed even in the absence of dissociating solvents such as guanidine hydrochloride.

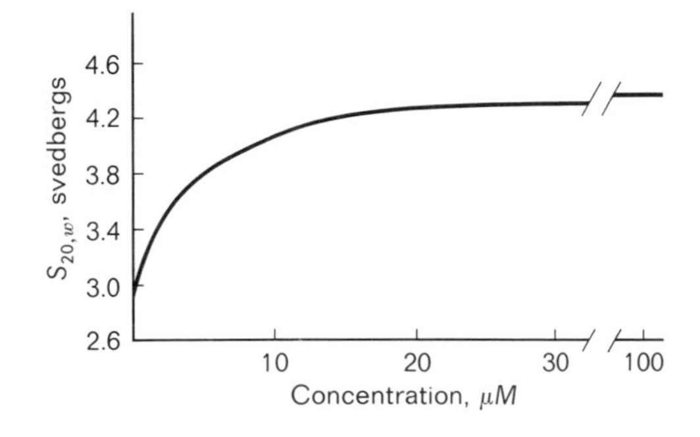

FIG. 10-3 *S Versus Concentration for Carboxy Hemoglobin.* From S. J. Edelstein, M. J. Rehmar, J. S. Olson, and Q. H. Gibson, *Journal of Biological Chemistry,* **245:** 4372 (1970).

4 *Hybridization experiments.* Many enzymes of higher organisms occur as *isozymes.* Isozymes are enzymes that catalyze the same reaction in different tissues, and have slight differences in structure and charge that can be detected by sensitive methods such as electrophoresis, which separates proteins on the basis of mobility in an electric field. A significant observation with isozymes is that hybrids occur. For example, the enzyme aldolase is found in two forms, known as A and C. A commonly occurs in muscle, whereas C is found as the predominant form in brain. However, in many other tissues both A and C are found, as well as three species with intermediate properties. These molecules form a five-membered set, which is the number expected from the binomial distribution of two types of enzyme, each containing four subunits, if the subunits are randomly mixed. The presence of this pattern of behavior has been clearly demonstrated when the middle members of the set are isolated, dissociated in mild structure-breaking conditions, and permitted to reassociate. When reassociated, all members of the five-membered set appear. This behavior is summarized in Fig. 10-4 and represents one of the simplest

Electrophoresis of brain aldolase on cellulose acetate at pH 8.6 shows five isozymes, A, C, and three intermediate species:

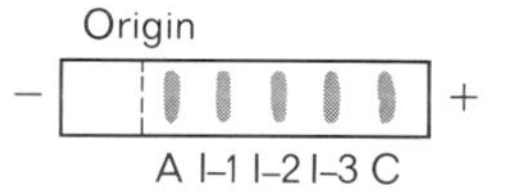

Isolate the five species:

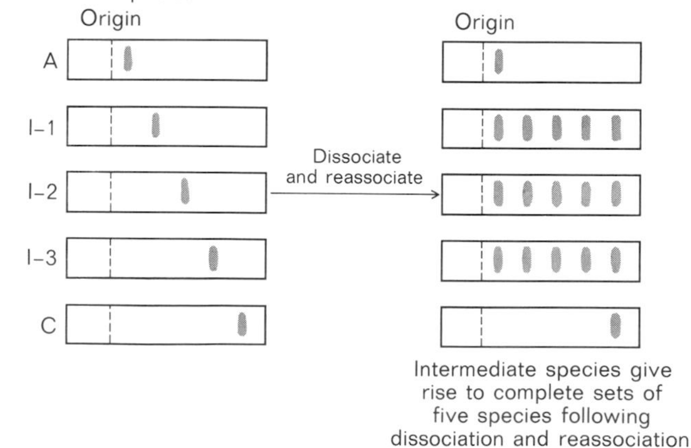

FIG. 10-4 *Aldolase Isozymes.* The five-membered set of electrophoretic bands occurs with a ratio of 1:4:6:4:1 for the concentration of proteins in the bands. Modified from E. Penhoet, M. Kochman, R. Valentine, and W. J. Rutter, *Biochemistry,* **6:** 2940 (1967).

and most conclusive methods for establishing the presence of subunits.

5 *Axes of symmetry.* When proteins are studied by x-ray crystallography, analysis of the x-ray distribution patterns reveals the *unit cell* of the protein. The unit cell is defined as the smallest repeating unit in a crystal. In a crystal of the enzyme ribonuclease, the smallest repeating unit is a structure identified with the entire enzyme itself. However, in a number of cases the size of the unit cell is smaller than the size of the native protein, estimated, for example, by sedimentation equilibrium. Such is the case for the protein hemoglobin, which crystallizes in forms for which the unit cell corresponds to just one-half the native molecule. Therefore two identical units must be present in the native structure, suggesting an axis of symmetry and a structure most easily reconciled with the presence of subunits associated noncovalently.

6 *Observation in the electron microscope.* Individual protein molecules can be seen as tiny spots in the electron microscope. When visualization is enhanced by specific staining agents such as phosphotungstic acid, individual subunits can sometimes be resolved. However, a number of pitfalls, such as artifacts of staining and alterations during drying, are always present with this method and it is therefore generally used in conjunction with other techniques.

7 *Electrophoresis in sodium dodecyl sulfate (SDS).* SDS, an anionic detergent, is an effective structure-disrupting agent for proteins. Even at concentrations below 1% it causes proteins to unfold and oligomeric proteins to dissociate into subunits. Since the SDS molecules form an envelope around the protein, the only charges from this SDS-protein complex exposed to the solvent are those from the sulfate groups of the SDS. These groups will cause the SDS-protein complex to move in an electric field. Electrophoresis experiments with SDS are conveniently performed on polyacrylamide gels. A typical SDS-gel electrophoresis experiment is shown in Fig. 10-5. Mercaptoethanol may also be added to dissociate subunits held together by disulfide bonds. Mobility of molecules in electrophoresis usually depends on both size and charge. SDS-protein complexes all have a negative charge related to the number of SDS molecules bound, which

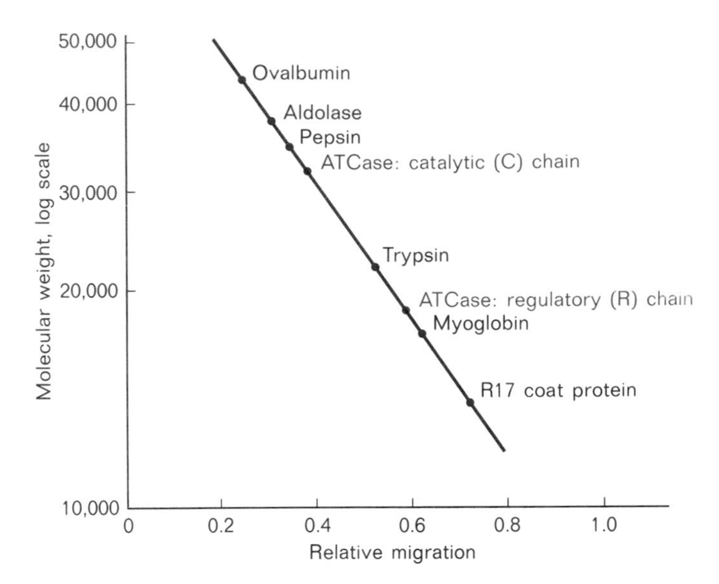

FIG. 10-5 *Sodium Dodecyl Sulfate–Gel Electrophoresis.* Data from K. Weber, *Nature,* **218:** 1116 (1968).

depends on the size of the protein or subunit in question. In this case migration in the electric field is a function of size and a standard curve can be constructed (Fig. 10-5) based on single-chain proteins or protein subunits with known molecular weights. Unknown samples can be studied simultaneously and in the example shown here results on the molecular weights of the two types of subunits (called R and C) of aspartate transcarbamylase were obtained. The significance of R and C subunits is discussed toward the end of the chapter.

Acquisition of Quaternary Structure

Like tertiary structure, quaternary structure arises as a consequence of the primary sequence of the enzyme. This point was established in experiments similar to those with ribonuclease, but with the multichain enzyme aldolase. As seen in Table 10-1, the native structure with a molecular weight of 150,000 can be dissociated in urea into subunits of approx-

TABLE 10-1 Dissociation and Reconstitution of Aldolase

	Molecular Weight	Specific Activity
Water	150,000	13
Urea	38,000	—
Reconstituted in Water	150,000	11.5

Source: E. Stellwagen and H. K. Schachman, *Biochemistry,* **1:** 1056 (1962).

imately one-quarter the size of the parent molecule. As with ribonuclease, the intrinsic viscosity rises from a low value of 4, in accord with the globular structure of the native enzyme, to a considerably higher value of 18 for the enzyme in urea. Under these conditions the specific activity diminishes from a value of 13 for the native enzyme to a level below detection. When the urea is removed, reconstitution occurs to yield the tetramer with its characteristic intrinsic viscosity and a restoration of a large part of the enzymatic activity.

From thermodynamic studies on protein stability and the direct observations on the proteins with subunits studied by crystallography, the nature of the forces holding subunits together can be described. With very few exceptions the forces holding subunits together are hydrophobic or electrostatic in character rather than covalent. Since electrostatic (or ion-pair) bonds, as well as hydrophobic interactions, are favored by entropy factors, it is not surprising that some proteins tend to dissociate into subunits in the cold. Such cold lability can be reconciled with the thermodynamic equation in Chapter 8, which shows that an entropy-driven reaction will tend to be less favored at lower temperatures. Behavior of this type is especially dramatic with an enzyme of oxidative phosphorylation from mitochondria, the adenosine triphosphatase (ATPase) (see Chapter 12). At 20°C the enzyme has a molecular weight of about 280,000 but when the solution is lowered to 5°C, the material dissociates into smaller units with an average molecular weight of only 47,000. The cold-labile dissociation is readily followed by sedimentation equilibrium (Fig. 10-6).

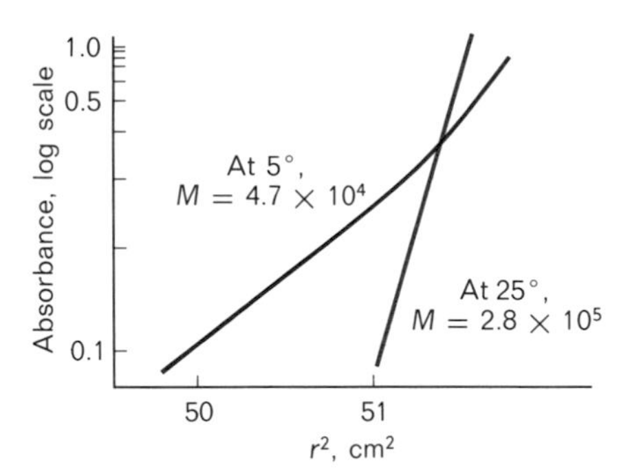

FIG. 10-6 *Cold Lability of Mitochondria Adenosine Triphosphatase* shown by sedimentation equilibrium in an ultracentrifuge. From G. Forrest and S. J. Edelstein, *Journal of Biological Chemistry,* **245:** 6468 (1970).

Symmetry in Quaternary Structure

When the subunit structure of many proteins is compiled, a striking feature is observed (see Table 10-2). A large fraction of the proteins is composed of two, four, or at least an even number of subunits. Why? The answer lies in the symmetry of structures derived from identical units. Two types of bonding are possible in subunit structures composed of identical subunits. These types are summarized in Fig. 10-7 and are known as *isologous* and *heterologous* bonding. In isologous binding, the bonding surfaces contributed by each subunit to a given domain are identical. This type of bonding leads to *dimers,* closed structures of *dyad symmetry,* that is a twofold rotational axis. Two dimers can also associate by isologous

TABLE 10-2 Subunit Composition

No. of Subunits	Known Proteins with This Structure	Selected Examples
1	Many	Chymotrypsin, Myoglobin, Cytochrome c
2	About 50	Liver Alcohol Dehydrogenase, Enolase
3	1	Aspartate Transcarbamylase Catalytic Unit
4	About 60	Aldolase, Lactate Dehydrogenase
5	2	Arginine Decarboxylase, Hemocyanin
6	7	Asparaginase
7	None	None
8	3	Hemerythrin
9	None	None
10	None	None
11	None	None
12	2	Glutamine Synthetase

Source: I. M. Klotz, N. R. Langerman, and D. W. Darnall, *Annual Review of Biochemistry,* **39:** 25 (1970).

binding to form a tetramer, but larger fully closed, isologous structures are geometrically impossible. Thus the prevalence of dimers and tetramers is explained through isologous binding and the limitation of isologous binding to dimers and tetramers. The advantages of isologous binding are that the structures formed are completely closed. Once the bonding surface of one subunit has associated with the identical surface of its partner, the two surfaces are no longer free to interact with other subunits, and nonspecific association with various other protein subunits cannot occur.

With heterologous bonding, the surfaces contributed by subunits to a single interface are different. In a sense, one provides the lock, the other provides the key. Since the structures are identical, the subunit that provided the lock must have a key protruding elsewhere, and the subunit that provided the key must also contain a lock. These unassociated locks and keys are then free to bond with other subunits. Thus heterologous binding can form linear, open structures, in contrast to the closed dimers and tetramers of isologous binding. Such open structures lead to simple linear arrays, circular structures, or helices, depending on the exact angles of bonding. Circular structures resemble isologous structures in that they are closed, but only heterologous-bonding domains are involved. With appropriate dimensions, closed, odd-numbered structures can also be formed, and principles of this type account for the proteins such as arginine decarboxylase and hemocyanin, which are pentamers with a fivefold rotational axis. Tetramers can also be formed with heterologous symmetry if each bonding contact approximates a

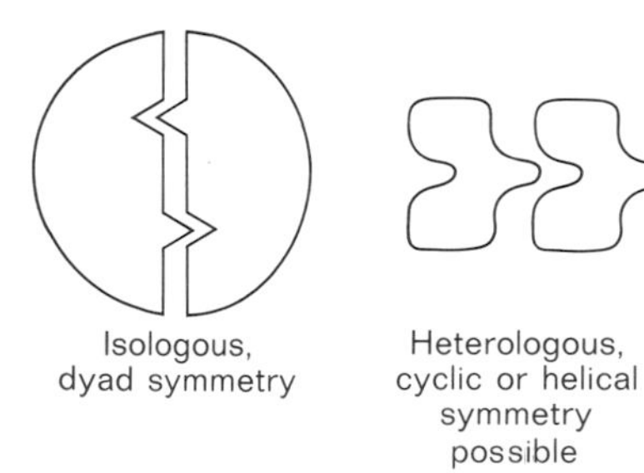

FIG. 10-7 *Subunit-bonding Modes.* From J. Monod, J. Wyman, and J.-P. Changeux, *Journal of Molecular Biology,* **12:** 88 (1965).

right angle. The enzyme tryptophanase appears to be represented best by a structure of this type. In some cases both heterologous and isologous modes of association may be present in a single protein. Thus glutamine synthetase with twelve subunits appears to be constructed through the heterologous association of six isologous dimers. The final structure is thus a six-membered ring, where each member contains two subunits related by a dyad axis of symmetry.

With symmetry principles in mind, we can, with hindsight, understand why tetramers and dimers are prevalent—because of isologous symmetry—but little can be said concerning why certain proteins have four subunits and others have six. In addition many proteins contain no subunits at all. In surveying the large number of proteins that have been studied, however, it is possible to note that the monomeric examples are often found among secreted proteins. In addition the great majority of proteins with a quaternary structure contain no disulfide bonds, whereas many monomers do contain disulfide bonds. These observations suggest that secretions may place some limitation on the size of proteins, and in addition small proteins may require disulfide bonds in order to maintain their stability. The presence of multiple subunits in a protein is a stabilizing force, just as crystallization tends to stabilize structure. However, the most obvious advantage for the presence of subunits is the capability of cooperative interactions between the ligand- or substrate-binding sites, and we will now consider these interactions.

Hemoglobin-Oxygen Equilibrium: Cooperativity Without Catalysis

In Chapter 7 we noted that hemoglobin binds oxygen cooperatively, as indicated by a sigmoidal saturation curve, with important physiological advantages. While many of the enzymes we will consider also show sigmoidal properties in the velocity versus substrate curve, hemoglobin is a convenient case for detailed study of the mechanism of cooperativity, since it is uncomplicated by catalysis. In addition, as one of the first proteins isolated and studied extensively, a great deal of information has been accumulated on hemoglobin. With the added development of an atomic model from x-ray crystallography, we can relate its structure to function in considerable detail.

One of the first explanations for cooperativity in oxygen

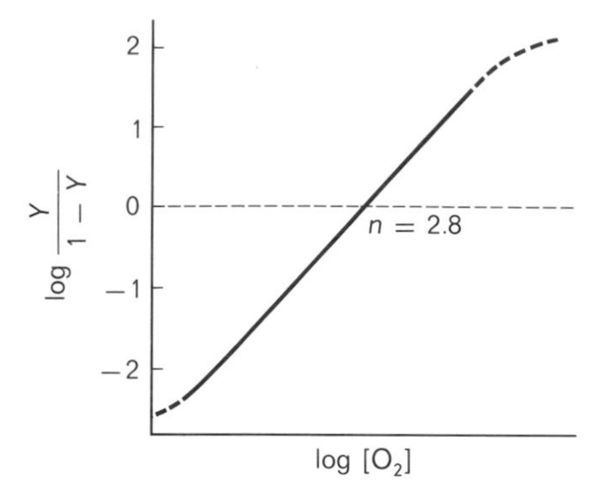

FIG. 10-8 *Plot of the Hill Equation.* The slope approaches a value of unity at the extremes.

binding by hemoglobin, proposed by Hill in 1910, was simply that the reaction with oxygen or other ligands is a high-order reaction. If the binding sites possess some property that requires all of them to act at the same time, the reaction can be written as

$$\mathrm{Hb} + n\mathrm{O_2} \rightleftharpoons \mathrm{Hb(O_2)}_n \tag{2}$$

The equilibrium constant for this reaction is given by Eq. (3),

$$K = \frac{[\mathrm{Hb(O_2)}_n]}{[\mathrm{Hb}][\mathrm{O_2}]^n} \tag{3}$$

$$\frac{[\mathrm{Hb(O_2)}_n]}{[\mathrm{Hb}]} = \frac{Y}{1-Y} = K[\mathrm{O_2}]^n \tag{4}$$

and it can be rearranged to give Eq. (4), where Y represents the fractional saturation and can be determined quite conveniently by spectrophotometric methods. Thus the order of the reaction, or the coefficient n, can be determined by plotting $\log [Y/(1-Y)]$ versus $\log [O_2]$, as seen in Fig. 10-8. The slope of the plot is n. When hemoglobin is analyzed in this way, a value of 2.8 is obtained.

The view that cooperative ligand binding by hemoglobin represents simply a high-order reaction is rendered very unlikely by the fact that the value of n is not integral, and secondly, that it is not identical with the number of sites. Since hemoglobin is known to contain four subunits, a fully ordered reaction requires each subunit to bind oxygen simultaneously and yields a value of $n = 4$, not 2.8 as is observed. Moreover, careful analysis of the extremes of the curve of $\log [Y/(1-Y)]$ versus $\log [O_2]$ indicates that the slope tends to unity at both very high and very low values of Y. For a high-order reaction, the slope should remain constant. Thus these arguments have refuted the high-ordered mechanism, but the formulation and the graphical analysis of the data, as shown in Fig. 10-8, have remained very useful, and n, known as the Hill coefficient, is a convenient measure of cooperativity. However, another explanation of cooperativity must be found, and we now know that this explanation lies in the concept of conformational change.

Evidence for Conformational Change

When x-ray crystallography was first beginning with hemoglobin, workers noted that crystals of deoxyhemoglobin tend to shatter when brought into contact with oxygen. This observation implied that the oxygen was causing a conformational change of the protein that rendered it incapable of maintaining the same crystal structure. As the x-ray studies by Max Perutz progressed to a detailed analysis of the structure of hemoglobin, the basis for the different crystal forms was revealed. The transformation from deoxyhemoglobin to oxyhemoglobin involves a rotation of the subunits with respect to one another, so that the distances between the subunits is altered, and some minor rearrangements of structure occur within the subunits. Hemoglobin contains two α-chains and two β-chains, as summarized in Fig. 7-9, and the major movement occurs between the β-chains. The distance from iron to iron in the heme groups of these chains decreases from 39.9 Å in deoxyhemoglobin to 33.4 Å in oxyhemoglobin. Thus a conformational change is clearly present, but other methods were required to relate the conformational change to cooperativity.

Evidence for functional significance of the conformational change came from kinetic experiments with hemoglobin. Since the oxygenated and deoxygenated forms of hemoglobin display different spectral properties (see Fig. 10-9), it is very simple to follow the change from one form to another. Spectra

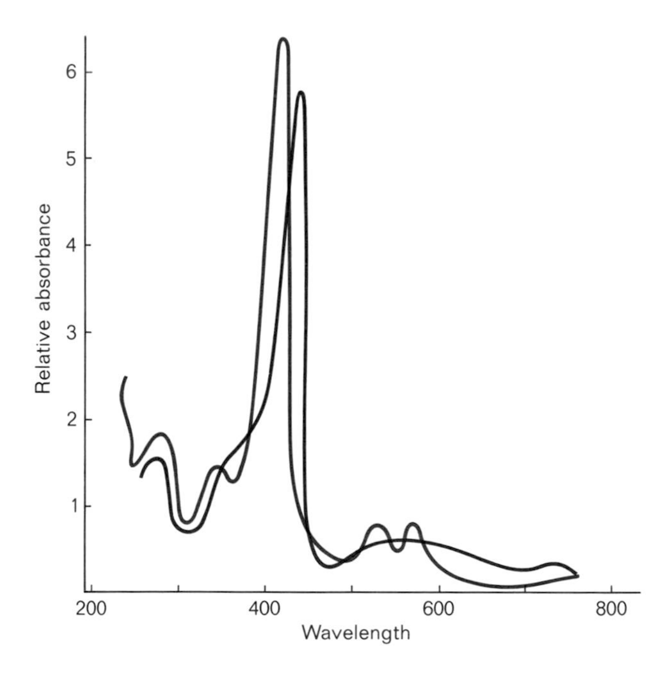

FIG. 10-9 *Spectra of Hemoglobin:* oxyhemoglobin (colored) and deoxyhemoglobin (black).

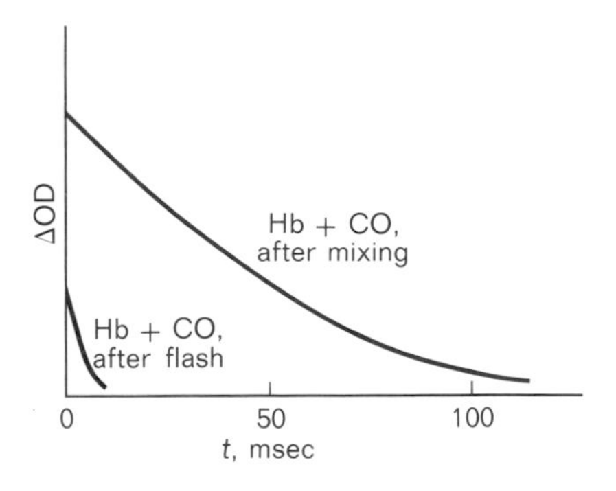

FIG. 10-10 *Reaction of Hemoglobin with Carbon Monoxide.* Change in OD is given for reaction of hemoglobin with CO on mixing and after partial flash photolysis. Based on data from Q. H. Gibson, *Progress in Biophysics and Biophysical Chemistry,* **9:** 1 (1959).

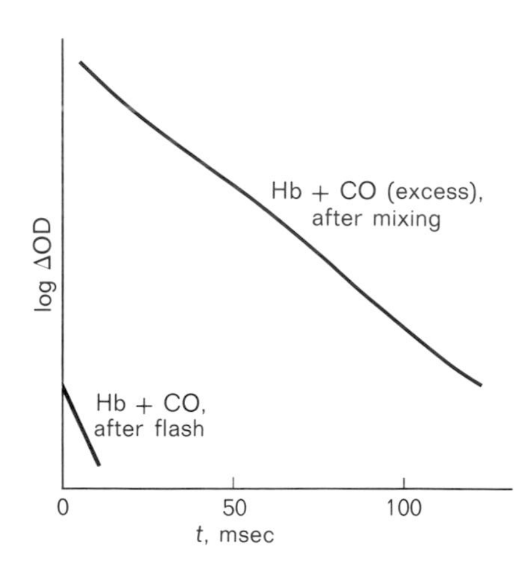

FIG. 10-11 *Logarithmic Plot of Data for Reaction of Hemoglobin with Carbon Monoxide.* The rates from the slopes indicate that the rate of combination of CO with hemoglobin following partial flash photolysis is about 30 times greater than when hemoglobin and CO are mixed.

of this type are found commonly with proteins containing heme prosthetic groups and will be encountered again in the heme-containing proteins of oxidative phosphorylation. Although oxygen is the principal ligand associated with hemoglobin, it also binds carbon monoxide (CO). In fact, CO is bound about 200 times more tightly than oxygen, accounting for its potency as a poison. For biochemical studies, however, the reaction of CO with hemoglobin (Hb) has the advantage that the product formed (HbCO) is very photosensitive. When HbCO is subjected to a quick flash of light, the CO is photodissociated, and free hemoglobin is generated.

When hemoglobin and carbon monoxide are rapidly mixed and the reaction followed spectrally, one observes a relatively slow rate of reaction, as shown in Fig. 10-10. The change in OD approximates an exponential decay curve, typical of bimolecular reactions. When CO is in excess, the system tends to follow first-order kinetics, and a nearly linear change then is obtained of log ΔOD versus time (as seen in Fig. 10-11) with a slight acceleration in rate with time due to the cooperative nature of the reaction.

When the HbCO is formed and a small amount of the CO quickly flashed off, quite different kinetic properties are observed. A partially liganded hemoglobin, $Hb(CO)_3$, is formed and in this case the CO reacts very rapidly with the few free heme sites available—about 30 times more rapidly than in the mixing experiment where all heme sites on deoxyhemoglobin are available. This differential effect, discovered by Quentin Gibson, implies that the state of the molecule trapped by flashing off a fraction of the CO quickly is different functionally from the state of the molecule starting out in the complete absence of ligand. One state reacts rapidly with CO; the other state reacts slowly. With the existence of two functionally distinct states, it is possible to formulate a mechanism for cooperativity in hemoglobin that satisfies many of the properties of the molecule.

The Two-state Theory of Cooperativity

When two conformational states of a protein are present, as in the case of hemoglobin, cooperativity can be generated with the following conditions. First of all, the equilibrium between the two states—call them R and T, where R stands for relaxed and T for tight—must be free and spontaneous. If described by the constant L, where $L = [T]/[R]$, the equilibrium must lie well in favor of the T state in the absence of ligand. In addition, the two states R and T must have different affinity for ligand, with the T state possessing lower affinity than the R state. (The affinity is given by the dissociation constants K_R and K_T, and a relative affinity term, c, can be defined where $c = K_R/K_T$ and $c < 1$.) If these conditions are met, cooperativity will occur.

This formulation for a protein exhibiting cooperative interactions is known as the *allosteric model,* and was proposed by Monod, Wyman, and Changeux. It represents the simplest explanation for cooperative ligand binding by hemoglobin and many other proteins, and recent work indicates that at least in broad outline, the model is a reasonably good description for hemoglobin. In the absence of ligand, the molecular population is almost entirely in the state indicated by T_0 (Fig. 10-12), which corresponds to deoxyhemoglobin. Since L is large, the equilibrium spontaneously favors the T_0 conformation and only negligible amounts of the R_0 state are present. The subscript indicates the number of ligand molecules bound; hence the subscript 0 indicates the absence of ligand. When ligand is added, the equilibrium will shift gradually to the R state since R has a higher affinity for ligand and hence gains more stability with each molecule of ligand bound, pulling the $T \rightleftharpoons R$ transition to the right. When L is relatively large, as for hemoglobin, the fraction of molecules in the R state will only become appreciable after the binding of the second or third molecule of ligand. When four molecules of ligand are bound, virtually all of the hemoglobin molecules will be in the R state, which corresponds to the conformation of oxyhemoglobin. The partition between R and T states for the various degrees of ligand saturation is given by the coefficients L_i, where $i = 1, 2, 3$, or 4, and L_i can be related to the basic constant L by the equation $L_i = Lc^i$. Thus c represents the relative advantage the R state gains in the R-T equilibrium for each molecule of ligand bound.

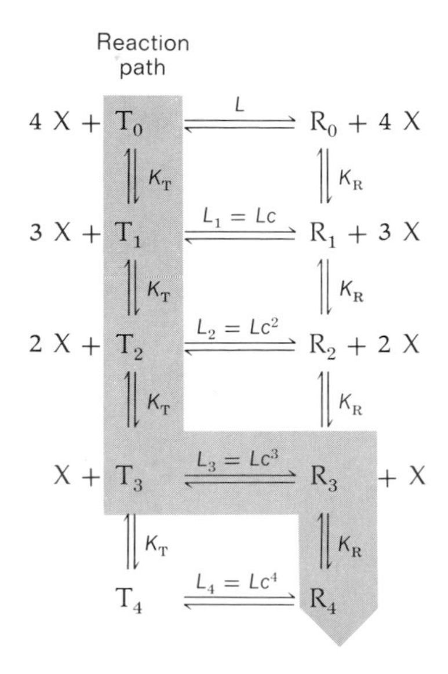

FIG. 10-12 *Details of the Allosteric Mechanism.* The two states T and R are given with subscripts that refer to the number of oxygen molecules (X) or other ligand bound. The predominant species, T_0, T_1, T_2, T_3, R_3, and R_4, are indicated by the reaction path. Model based on J. Monod, J. Wyman, and J.-P. Changeux, *Journal of Molecular Biology,* **12:** 88 (1965).

When the properties of the allosteric model are analyzed in greater detail, as shown in Fig. 10-13, many characteristics needed to describe the properties of hemoglobin are found. In the representation of log $[Y/(1 - Y)]$ versus log $[O_2]$, we see that the predicted behavior includes regions at the extremes of the curve, where $n = 1$, corresponding to the properties of the T state at the lower extremity and of the R state at the upper extremity. When these regions of $n = 1$ are extrapolated to log $[Y/(1 - Y)] = 0$, corresponding to 50%

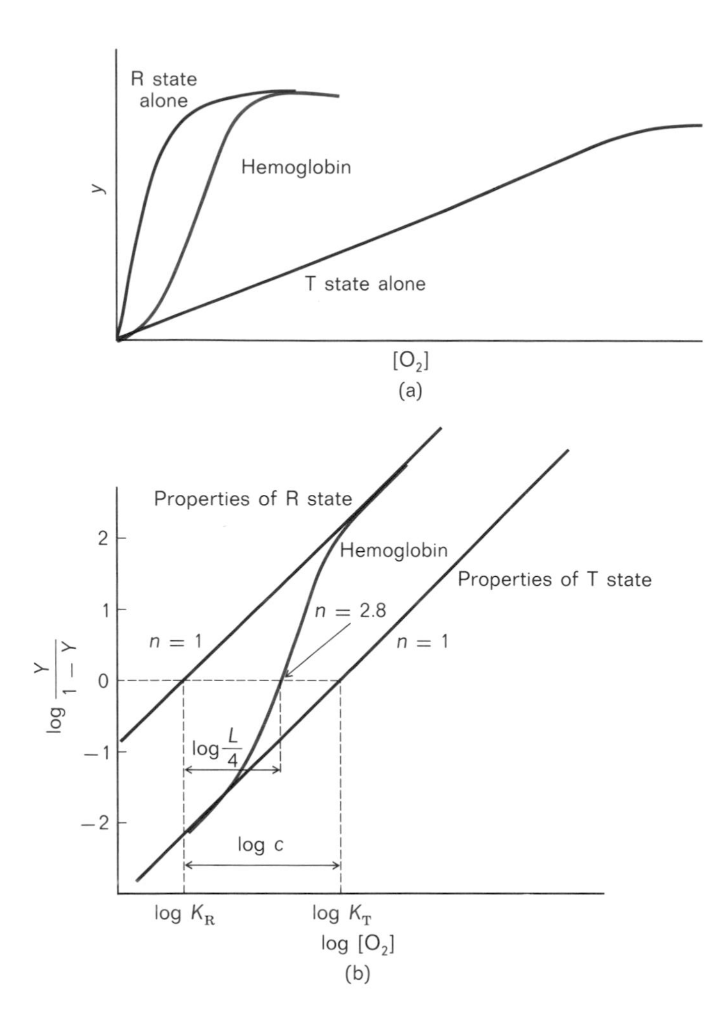

FIG. 10-13 *Implications of the Allosteric Model.* Properties in terms of (a) a saturation plot (Y versus O_2) and (b) a Hill plot are given for hypothetical behavior of the R state and the T state, as well as for hemoglobin.

saturation, the value of log K_R and log K_T are obtained. The difference between them gives log c. In the central region of the curve, the steepness increases to give a value corresponding to $n = 2.8$. Thus cooperativity ($n > 1$) appears as a concerted transition from the low-affinity T state to the high-affinity R state.

Further deductions can be made concerning the value of L. While K_R and K_T are fixed by the affinity of the R and T states, the value of log $[O_2]$ at half-saturation will depend on L. Conversely, if K_R is known independently and the value of the oxygen concentration at half-saturation can be measured, the magnitude of L can be calculated. This relationship can be appreciated intuitively by noting that the more stable the T state, that is, the larger L, the more ligand must be added before half-saturation can be reached and the farther to the right the saturation curve lies. Since four molecules of oxygen are present on the molecule of hemoglobin, the linkage between oxygen binding and the value of L takes the simple form $L = [O_2]^4_{1/2}$ for values of $[O_2]_{1/2}$ in the usual range. (If L is extremely small or extremely large, it tends to become independent of $[O_2]_{1/2}$.) In these relationships the units of O_2 are normalized to give a value of $[O_2]_{1/2} = 1$ for the binding constant to the R state. The term $[O_2]_{1/2}$ refers to the concentration of oxygen at half-saturation ($Y = 0.5$) and provides a convenient measure of overall affinity.

Consequences of the Two-state Model

When cooperativity arises as a consequence of the equilibrium between two conformations R and T of a protein, the value of the conformational equilibrium L must be in an optimal range for cooperativity to occur. If L is too low, the R state is already favored, even in the absence of ligand, and the addition of ligand simply populates the R state with no shift in conformational equilibrium to cause cooperativity. The Hill coefficient will be at its basal level, $n = 1$. At the other extreme, if L is very large, the T state may be so stable that the binding of ligand, even though it favors the R state, is not sufficient to cause the conformational transition. Therefore the addition of ligand simply populates the T state; cooperativity is absent and again the Hill coefficient observed is unity. In considering the overall dependence of cooperativity, we thus arrive at a bell curve, as seen in Fig. 10-14. Cooperativity will be at its maximum value (highest Hill coefficient) at some optimal value of L. This value turns out to be $L = c^{-i/2}$, where i is the number of ligand-binding sites. As c approaches 1, corresponding to a decrease in the difference in affinity for ligand of the R and T states, bell curves with lower maxima are produced.

One of the important features of the allosteric model is that it also explains the *Bohr effect*. As the pH of hemoglobin solutions is raised, the affinity of hemoglobin for oxygen also rises. This phenomenon is known as the Bohr effect. However, the Hill n remains constant at about 2.8. When the values of $[O_2]_{1/2}$ in the pH range 7 to 9 are converted into values of L, we find that the values of L place the system at the top of the bell curve, between the points noted in Fig. 10-14.

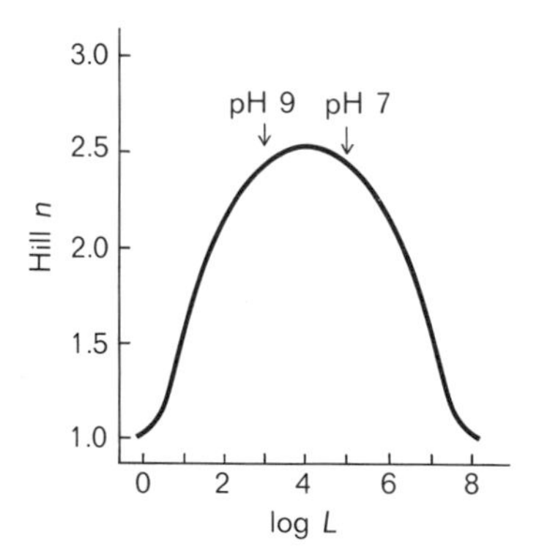

FIG. 10-14 *Bell Curve for Hemoglobin.* Values corresponding to extremes of the Bohr effect, pH 7 and 9, are indicated. Modified from S. J. Edelstein, *Nature,* **230:** 224 (1971).

Thus the marked change in oxygen affinity with pH, coupled with invariance of the Hill *n*, is clearly consistent with the allosteric model.

Further evidence for the allosteric formulation comes from hemoglobins that lie on the slopes of the bell curve. Some genetic variants of hemoglobin as well as derivatives altered by chemical means show a high affinity for O_2, which places them on the left slope of the bell curve and a decrease in the Hill *n* with increasing pH is found in these cases. One form, cat hemoglobin, possesses low affinity for oxygen, placing it on the right side of the bell curve. It displays an increase in Hill *n* with pH as predicted by the model. These observations extend the body of data adequately represented by the allosteric formulation.

Basis of the Bohr Effect

With the determination of the three-dimensional structure of hemoglobin, the formal analysis of cooperativity and affinity can be directly related to the atoms in the molecule. This connection to structure is especially clear when the basis of the Bohr effect is considered. The Bohr effect can be represented by considering protons as allosteric inhibitors that bind preferentially to the T state. When protons bind, the T state gains free energy and the R-T equilibrium is shifted in favor of the T state. Hence lowering pH results in an increase in *L* and diminished affinity for oxygen. Through this mechanism the Bohr effect aids in the delivery of oxygen to the tissues by providing hemoglobin with an extra push to unload its oxygen. In the vicinity of actively metabolizing tissues, CO_2 levels are likely to be high (see Chapter 11), and CO_2 dissolved in water forms carbonic acid. The effect of CO_2 is thus to lower pH and favor the T state or shift the O_2 sigmoid saturation curve to the right. Hence the apparent affinity of hemoglobin for oxygen is lowered and the oxygen tends to be released more readily. Conversely, at the lungs, CO_2 is removed from the blood, and the pH rises with a corresponding increase in the affinity of hemoglobin for oxygen. This factor helps the hemoglobin load up on oxygen to its level of full saturation.

Crystallographic studies on hemoglobin have revealed the atomic basis of the Bohr effect and also suggest reasons why the T state is favored in the absence of ligand. The pivotal difference between deoxyhemoglobin (T state) and

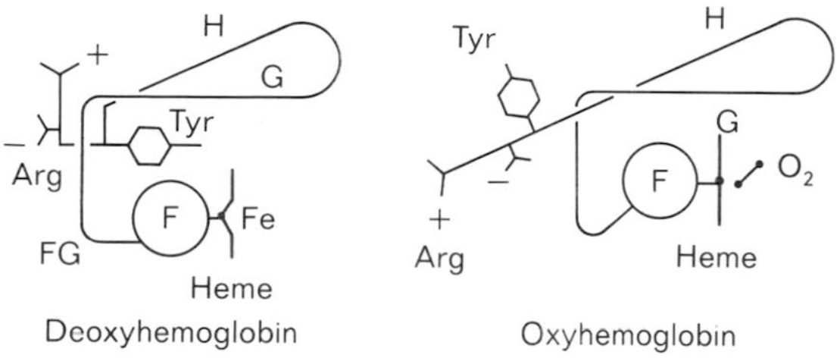

FIG. 10-15 *Changes in Iron and C-terminal Residues of Hemoglobin upon Reaction with Oxygen.* The iron atom moves into the plane of the heme upon reaction with oxygen and triggers rearrangement of the C-terminal arginine and penultimate tyrosine residues. The penultimate tyrosine may actually oscillate in and out of the heme pocket in oxyhemoglobin. The letters F, G, and H refer to helical regions of the molecule. Structures for the α-subunit are shown. From M. Perutz, *Nature,* **228:** 726 (1970).

oxyhemoglobin (R state) concerns the position of the iron in the heme. When oxygen is bound to iron, the electrons are paired and the radius of the iron atom (0.55 Å) is just the right size to fit in the center of the heme. However, when oxygen is removed, the electrons in the iron become unpaired and the iron expands slightly to a radius of 0.70 Å. This increase is enough to force the iron out of the plane of the heme by about 0.8 Å. In a chain reaction of molecular movements, the displacement of iron results in formation of a pocket nearby and the insertion of a tyrosine residue that alters the location of the other carboxyl-terminal residues. These changes are shown for α-chains in Fig. 10-15.

The movement of the tyrosines of the two α-chains (α140) and the rearrangements of the carboxyl-terminal residues bring groups into proper alignment for ion-pair formation. The α-amino groups of the two amino-terminal valine residues of the α-chains form salt bridges with the carboxyl groups of the carboxyl-terminal arginine residues of the "opposite" α-chains (Fig. 10-16). In the β-chains similar movements occur in response to the bulging of the iron atom and the relocation of tyrosines (β145). In this case positively charged imidazole rings of the carboxyl-terminal histidine residues of each β-chain (β146) form salt bridges with aspartates (β94) of the same β-chains (Fig. 10-16). The histidines are also stabilized in this position by salt bridges between their carboxyl groups and ε-amino groups of lysines on the α-chains (α40). Thus the protonated forms of the α-amino groups of the α1 valines and imidazole groups of the β146 histidines are stabilized by salt bridges in the deoxy (T) state. This stabilization is reflected by a higher pK for these groups in deoxyhemoglobin than oxyhemoglobin. When oxygen binds to hemoglobin, the carboxyl-terminal amino acid residues are dislodged (Fig. 10-15), the salt bridges are broken, protons are released, and the molecule assumes the R state. We also see that when the pH is increased slightly, the amino

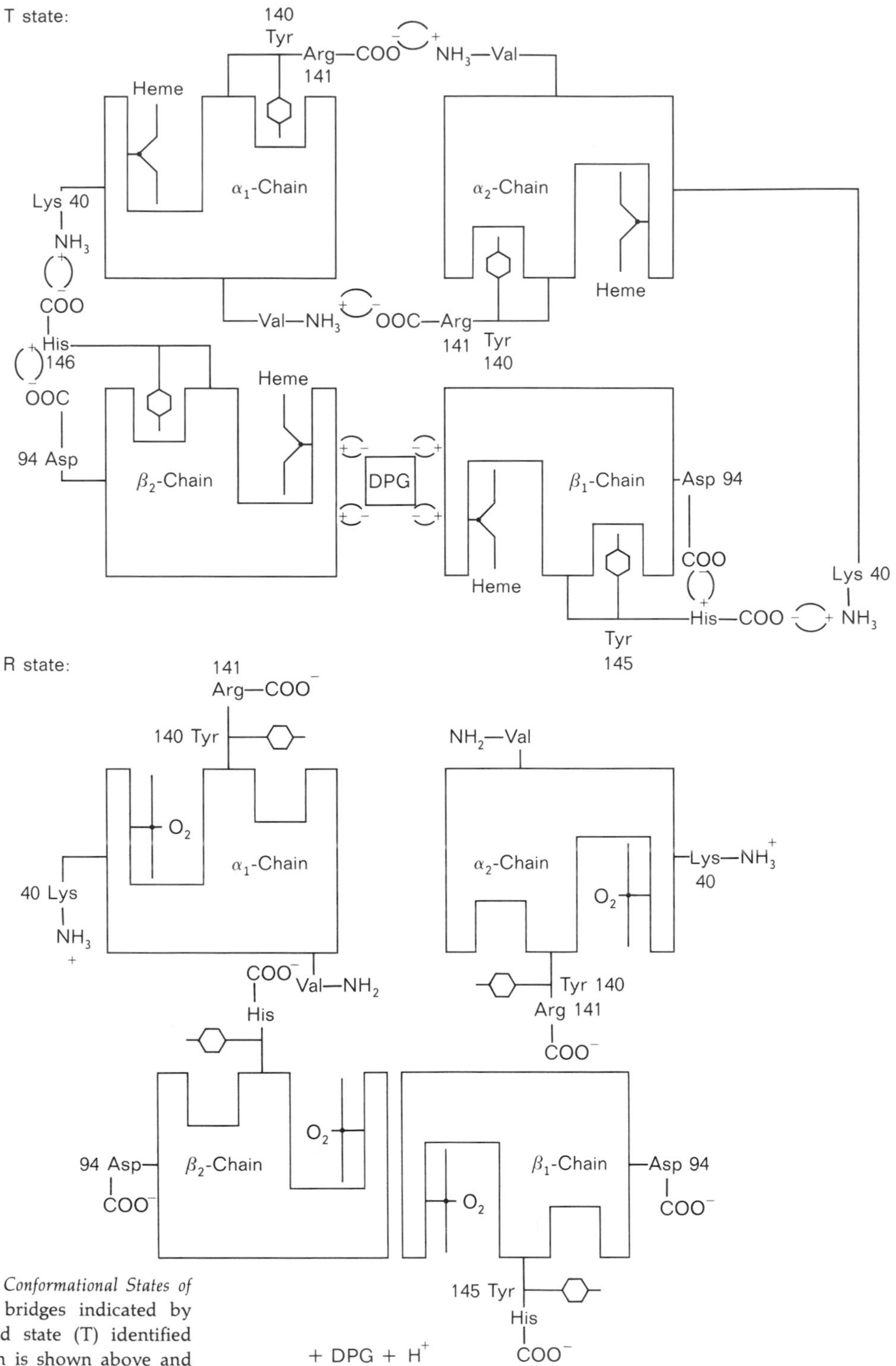

FIG. 10-16 *The Two Conformational States of Hemoglobin,* with salt bridges indicated by color. The constrained state (T) identified with deoxyhemoglobin is shown above and the relaxed state (R) identified with oxyhemoglobin is shown below. Crystal structures of the two states are shown in the Chapter 10 frontispiece. Modified from M. Perutz, *Nature,* **228:** 726 (1970).

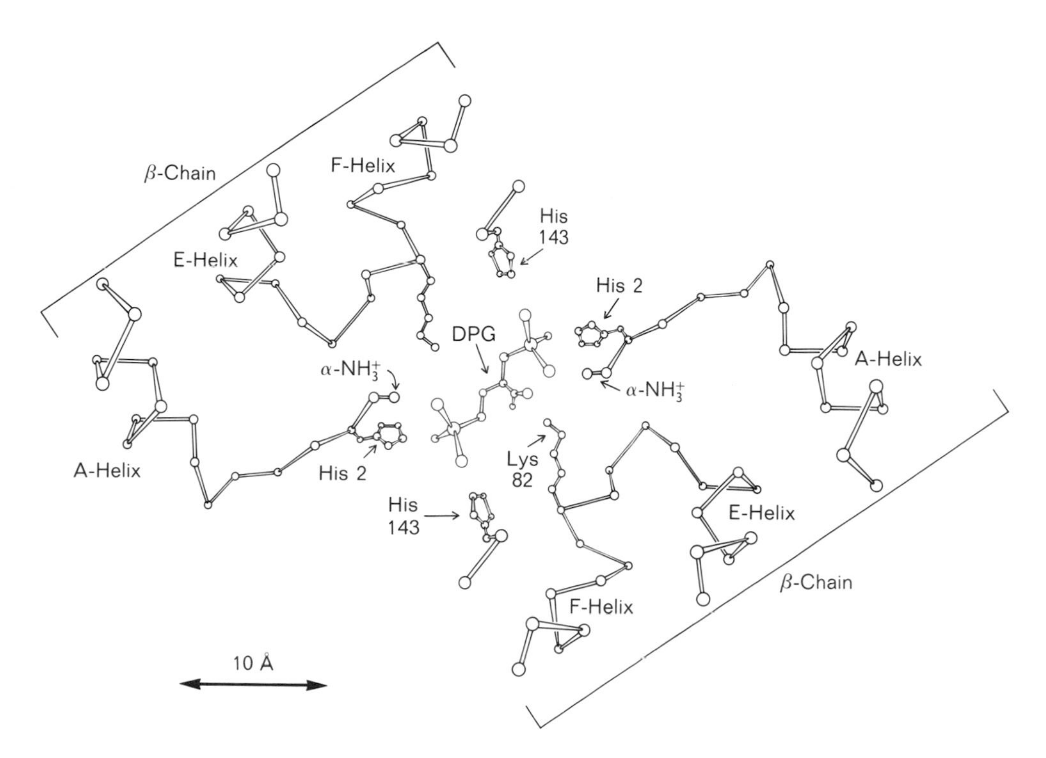

FIG. 10-17 *The Binding of 2,3-DPG to Human Deoxyhemoglobin.* The negatively charged DPG binds to the positive residues, valines 1, histidines 2 and 143 of both β-chains and lysine 82 of one of the β-chains. From A. Arnone, *Nature,* **237:** 146 (1972).

and imidazole groups in the salt bridges of the T state will be partially deprotonated, with a weakening of the salt bridges. In this case, the T state becomes less stable, L is decreased, and the affinity for oxygen rises. These interactions account for at least a large part of the Bohr effect.

Additional stability for the T state and a small contribution to the Bohr effect also result from the binding of a small molecule, diphosphoglycerate (DPG), in a pocket between the β-chains. Amino acid residues histidines 2 and 143 and lysine 82 are involved (Fig. 10-17). In the R state the β-chains are in close proximity (Fig. 10-16) and DPG is displaced. The contribution to the Bohr effect derives from the stabilization of the histidine residues in the protonated form by the negative charges of DPG. When DPG is released, the stabilization is lost and the protons dissociate.

One other aspect of the structural differences between oxy- and deoxyhemoglobin concerns hydrogen bonds and provides an interesting continuation of the focus on hydrogen bonds in nucleic acid structure, protein synthesis, and protein action. The reorientation of subunits in changing from oxy- to deoxyhemoglobin involves changes principally in the α_1-β_2

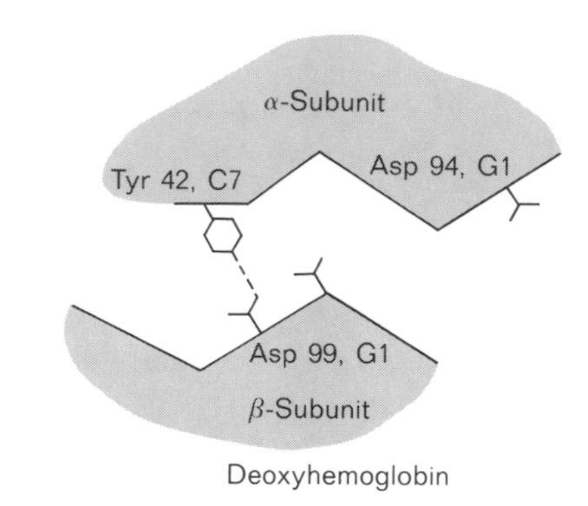

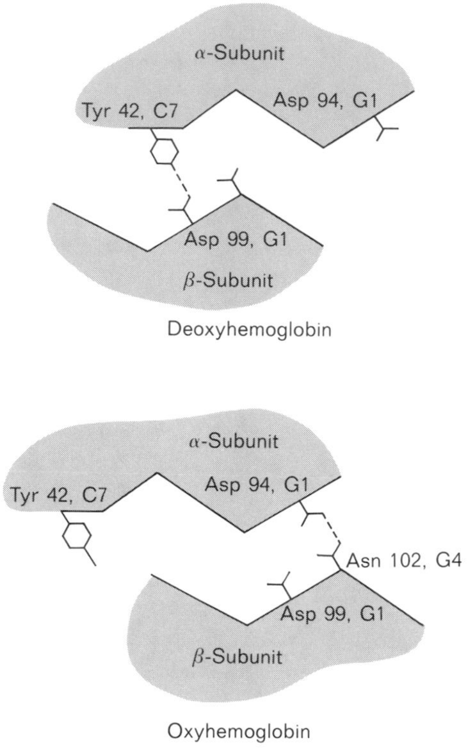

FIG. 10-18 *Subunit Bonding in Hemoglobin.* The transition from the deoxy to the oxy conformation involves a shift in the amino acid residues in contact at the α_1-β_2 subunit interfaces. From M. F. Perutz and L. F. TenEyck, *Cold Spring Harbor Symposium of Quantitative Biology,* **36:** 295 (1971).

subunit interface (Fig. 7-9). As one subunit slides across the other, one hydrogen bond at each α_1-β_2 interface present in the deoxy structure is broken and another hydrogen bond is formed, as shown in Fig. 10-18. The importance of just one pair of hydrogen bonds in the delicate balance between conformations of hemoglobin is revealed by mutant forms of human hemoglobin. In hemoglobins Yakima and Kempsey, the aspartate at β99 is replaced by amino acids that cannot make these hydrogen bonds. The T state is thereby destabilized and L decreases, resulting in abnormally high oxygen affinity. In hemoglobin Kansas, the asparagine at β102 is replaced, destabilizing the R state. Therefore L is raised, resulting in low oxygen affinity. Once again we find hydrogen bonds playing a crucial role in biochemical processes.

Cooperative Interactions in an Allosteric Enzyme

While hemoglobin is the protein studied most extensively in terms of cooperative ligand binding, the phenomenon itself is found widely in regulatory or allosteric enzymes that act at pivotal reactions in intermediary metabolism. At these key points in metabolism it is extremely valuable for the cell to be able to regulate the rate of enzyme catalysis, depending on the composition and quantity of other metabolites in the cell. Regulation can occur with high efficiency when the dependence of reaction velocity on substrate concentration is sigmoidal. In this case relatively small changes in substrate concentration or small shifts in the position of the curve by the action of activators or inhibitors exert a substantial influence on the velocity. The leverage or amplification that results from such S-shaped curves is analogous to the advantages of cooperativity in oxygen transport by hemoglobin.

One of the most thoroughly studied regulatory enzymes is aspartate transcarbamylase, abbreviated ATCase. This enzyme catalyzes the first step in the pathway for the synthesis of cytidine triphosphate (CTP), the reaction of carbamyl phosphate and aspartate to yield carbamyl aspartate. Seven further transformations are required before CTP is synthesized. However, the important aspect of this reaction is that the presence of CTP tends to inhibit the reaction itself. The reaction scheme is summarized in Fig. 10-19. Behavior of this type is called allosteric. The term allosteric refers to the fact that the inhibitor CTP bears little or no resemblance to the substrate and presumably reacts at a different site. The designation allosteric can also apply to a protein such as hemoglobin because oxygen at one site has an effect on the binding of oxygen at other sites, even though there is no direct steric contact.

Looking at the reaction of aspartate transcarbamylase (ATCase) in somewhat more detail (Fig. 10-20), we see that the velocity versus substrate curve of the enzyme is sigmoidal, reminiscent of the oxygen equilibrium curve for hemoglobin. In the nomenclature of allosteric enzymes, the sigmoidal behavior in substrate binding is referred to as a *homotropic* interaction—that is, like molecules affect one another—in this case aspartate molecules. When CTP is added, the curve is shifted to the right. Effects involving two types of ligands—substrate and effector (CTP)—are known as *heterotropic* interactions and can be positive or negative. CTP is considered a negative effector since it shifts the curve to the right, or inhibits the reaction. In contrast ATP is an activator and tends to raise the rate of the reaction for a given level of aspartate. One can appreciate that if ATP is in excess, the cell could gain

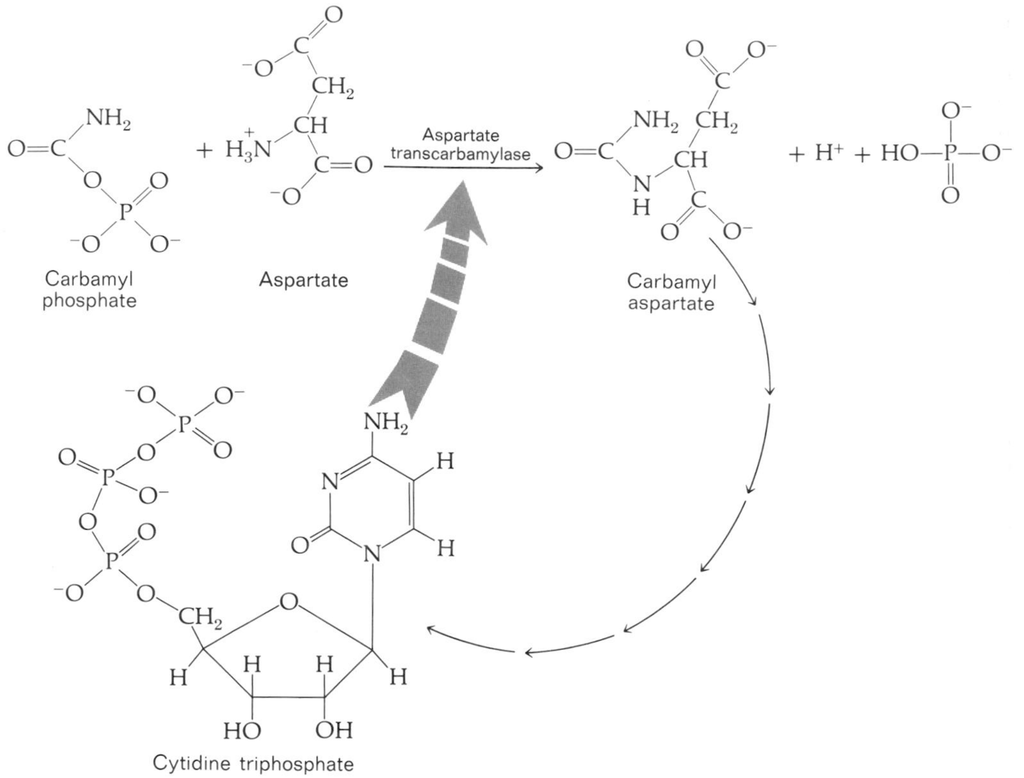

Carbamyl phosphate

Aspartate

Aspartate transcarbamylase

Carbamyl aspartate

Cytidine triphosphate

FIG. 10-19 *Feedback Inhibition of Pyrimidine Pathway.* From J. C. Gerhart and A. B. Pardee, *Cold Spring Harbor Symposium of Quantitative Biology,* **28:** 491 (1963).

an advantage by synthesizing more CTP so that metabolic steps involving both CTP and ATP, such as nucleic acid synthesis, can proceed at full speed. By analogy, we see that the pH, or Bohr effect, for hemoglobin can be considered as a heterotropic interaction in which protons are negative effectors.

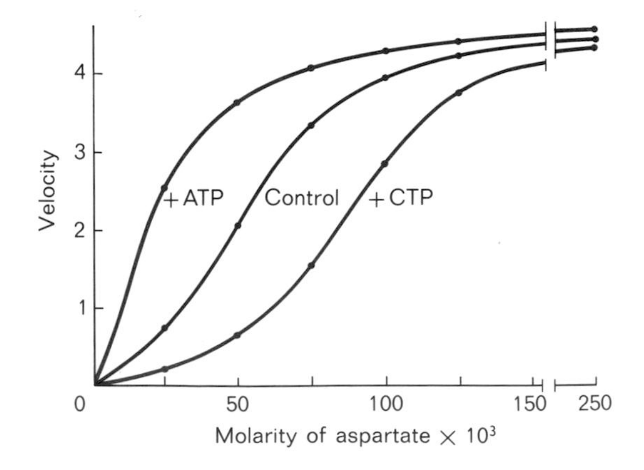

FIG. 10-20 *Role of Effectors on ATCase.* In the absence of effectors, velocity versus substrate (aspartate) profile is sigmoidal (in the presence of carbamyl phosphate). ATP, a positive effector or *activator,* shifts the curve to the left, while CTP, a negative effector or *inhibitor,* shifts the curve to the right. From J. C. Gerhart and A. B. Pardee, *Cold Spring Harbor Symposium of Quantitative Biology,* **28:** 491 (1963).

From our study of the allosteric model, the action of effectors can be readily explained by changes in *L*. Inhibitors bind to the T state, increasing *L*, while activators bind to the R state, decreasing *L*. Thus activators cause movement to the left on a hypothetical bell curve analogous to the one presented for hemoglobin (Fig. 10-14). Therefore it is not surprising that activation of ATCase by ATP results in not only a shift to the left, but a noticeable decrease in the degree of sigmoidality.

Work with aspartate transcarbamylase by John Gerhart and Howard Schachman has provided a fascinating picture on the architecture of the protein. The structure of the molecule is complex. It has a molecular weight of about 300,000, and a sedimentation coefficient of approximately 12S. A striking finding was that in the presence of certain agents, such as mercurials, especially PMB (*p*-mercuribenzoate), the molecule tends to dissociate specifically (Fig. 10-21). Dissociation yields two protein fractions, one with a sedimentation coefficient of 5.8S and the other with 2.8S. The interesting aspect of this splitting of the molecule is that the 5.8S fraction still performs the catalytic function of the enzyme. However, the velocity versus substrate curve is now hyperbolic and not sigmoidal. The facility for regulatory control with CTP appears to be associated only with the 2.8S subunit. Thus, to achieve regulation and sigmoidal velocity versus substrate curves the 5.8S and 2.8S subunits must be mixed (see Fig. 10-22). Experiments show that under these conditions the native-type molecule of molecular weight 300,000 is regenerated. In addition direct measurement of the binding properties of the 2.8S subunit reveal that it does display the affinity for CTP characteristic of the native enzyme. Thus aspartate transcarbamylase is constructed of two different types of subunits—a *catalytic* (c) subunit and a *regulatory* (r) subunit. Somehow the two subunits interact to yield cooperative substrate binding by the enzyme and regulatory features in response to concentrations of nucleoside triphosphates.

Further chemical studies on the enzyme revealed that the native molecule contains six polypeptide chains of the regulatory type (MW 17,000) and six polypeptide chains of the catalytic type (MW 33,000). Each 5.8S catalytic subunit contains three catalytic polypeptide chains and represents one of the few trimers found among proteins. There are two 5.8S-type units in the native structure. Each 2.8S regulatory subunit contains two regulatory polypeptide chains, and three 2.8S-type subunits are present in the native structure. Crys-

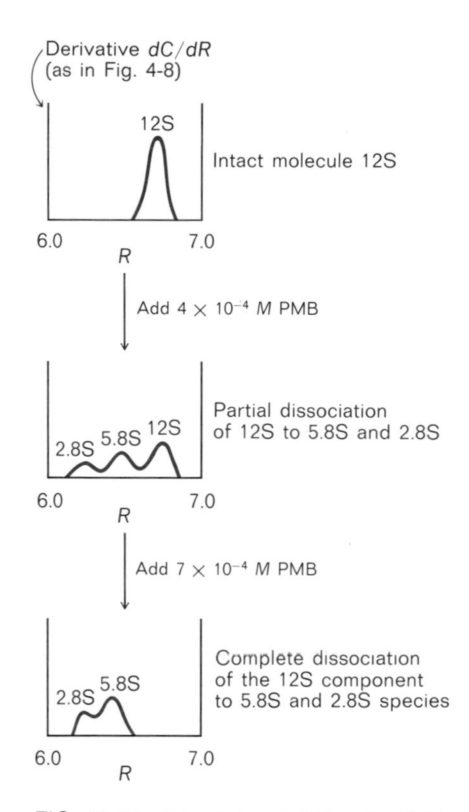

FIG. 10-21 *Dissociation of ATCase by PMB.* The native molecule (12S) dissociates into two components on addition of PMB, a 5.8S species and a 2.8S species. From J. C. Gerhart and H. K. Schachman, *Biochemistry,* **4:** 1054 (1965).

tallographic work on the enzyme by Wiley and Lipscomb reveals that a threefold axis of symmetry is also present in the molecule. Thus we can visualize the native structure of ATCase as containing two trimers, each representing the catalytic subunits, connected by three bridges, each bridge connecting one catalytic unit to another, and each bridge containing two of the regulatory polypeptide chains (see Fig. 10-23).

The mechanism of cooperativity of ATCase is very much like the pattern seen for hemoglobin—two states of different conformation can be recognized. For example, in the absence of ligand, the sedimentation coefficient is 11.6S. However, as ligand is added, the sedimentation coefficient decreases about 3.5%. These results are summarized in Fig. 10-24. A change

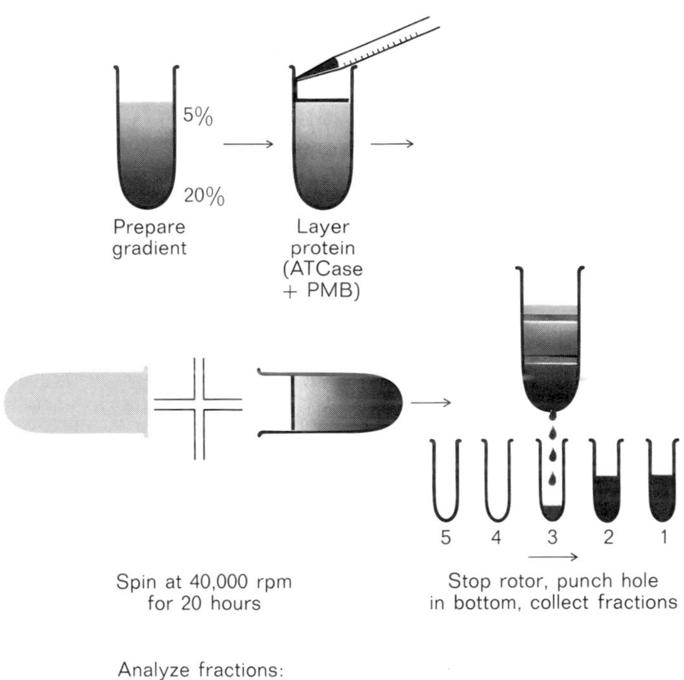

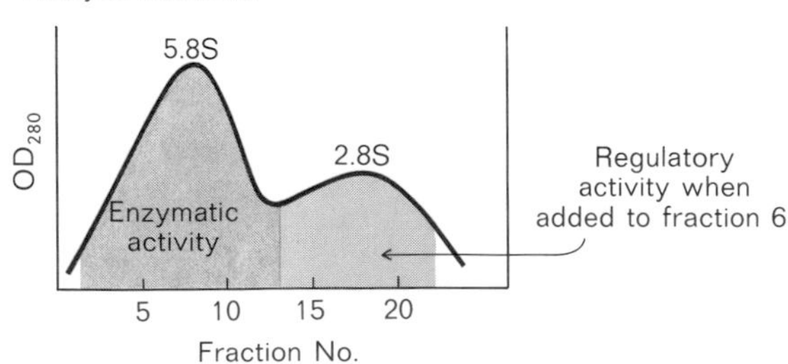

FIG. 10-22 *Separation of ATCase 5.8S and 2.8S Units.* The 5.8S and 2.8S units are sedimented on a preparative sucrose gradient and fractions are collected. Enzymatic activity is associated only with the 5.8S material but the 5.8S unit is "desensitized," that is, insensitive to CTP inhibition unless the 2.8S material is added back. From J. C. Gerhart and H. K. Schachman, *Biochemistry,* **4:** 1054 (1965).

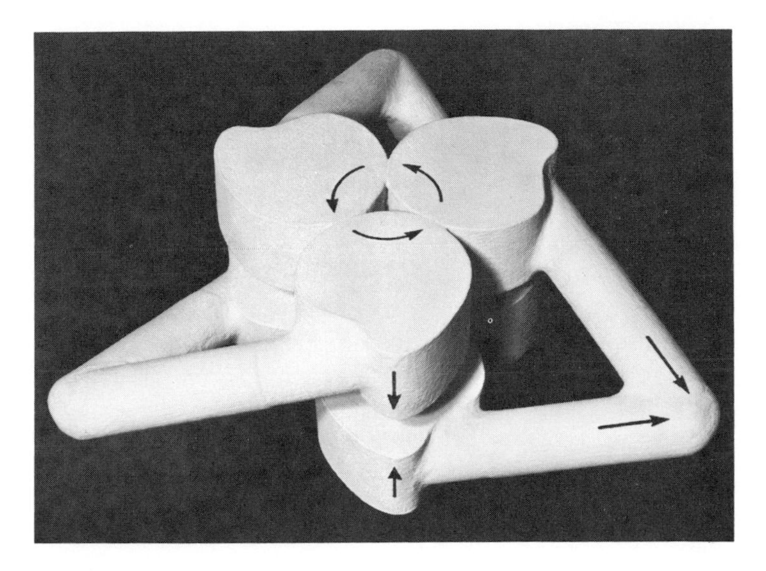

FIG. 10-23 *Model of ATCase.* Electron micrographs and other data lead to structural model of ATCase. Each of the two catalytic subunits (5.8S) contains three polypeptide chains and are seen at the top and the bottom of the model. Catalytic units are connected by the "elbow" joints, each involving two regulatory polypeptide chains. A total of three such joints are present; each comprises a regulatory subunit 2.8S. Photograph courtesy of H. K. Schachman. Further details in J. A. Cohlberg, V. P. Pigiet, and H. K. Schachman, *Biochemistry,* **11:** 3396 (1972).

in sedimentation coefficient indicates that a great deal of molecular rearrangement must occur on binding of ligand since no change in molecular weight occurs. Evidently the state favored in the presence of ligand, the R state, is appreciably less compact than the T state and manifests greater frictional resistance. Various properties of ATCase can be represented satisfactorily with a two-state model. In this case the binding of CTP favors the T state, whereas the binding of ATP would favor the R state. Addition of the principal substrate, aspartate, shifts the equilibrium from T to R. However, in contrast to hemoglobin, the value of L is probably very much smaller—about 10—and the degree of cooperativity achieved is generally less, with a Hill coefficient below 2. The differences in cooperativity between aspartate transcarbamylase and hemoglobin illustrate the variations in detailed expression of the theme of cooperativity in ligand binding. A wide range of allosteric behavior (and quaternary

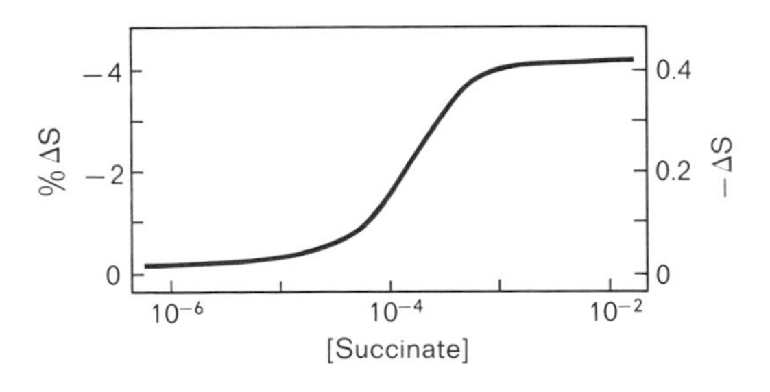

FIG. 10-24 *Conformational Change in ATCase.* The S value decreases on addition of succinate, a substrate analog, in the presence of carbamyl phosphate. From J. C. Gerhart and H. K. Schachman, *Biochemistry,* **7:** 538 (1968).

structures) are found in the numerous examples of enzymes that display control mechanisms based on conformational equilibria and a number of other variations will be encountered in later chapters.

Supramolecular Structures

The same principles and forces that apply in the assembly of proteins at the level of quaternary structure can be extended to even larger molecular aggregates. Examples of such systems encountered elsewhere in this book are the spontaneous formation of ribosome subunits from appropriate RNA and protein components (Chapter 4), large enzyme complexes such as in pyruvate dehydrogenase (Chapter 14), or fatty acid-synthesis enzymes (Chapter 11), and the bacteriophages referred to in several chapters. Even the large T4-type bacteriophages appear to form spontaneously from individual components, although in this case discrete steps can be recognized. Presumably some of the same principles of macromolecular assembly apply as well to other cellular structures, membranes, and organelles. However, we will leave macromolecular structures with the discussion of allosteric proteins to turn to the transformations of small molecules and see how the powerful principles of catalysis and cooperativity are put to use to provide the energy and building blocks of living things.

PROBLEMS

1 Treatment of 10 mg of an enzyme with carboxypeptidase releases 400 μmoles of arginine (1 μmole $= 10^{-6}$ mole). Sedimentation studies on the enzyme reveal a molecular weight of 10^5. What suppositions can be made about the subunit structure of the protein?

2 The catalytic subunits of aspartate transcarbamylase can be prepared in chemically modified form with an electrophoretic mobility distinct from the untreated catalytic subunit. If the two types of catalytic subunits are mixed, dissociated, and reconstituted, how many bands would be expected in an electrophoresis experiment? What would be the ratio of the concentrations of the bands?

3 Ligand bonding to a noncooperative protein gives a Hill constant of $n = 1$; cooperative binding is indicated by $n > 1$. How might a binding curve which gives a value of $n < 1$ on a Hill plot be explained?

Part II

In the first half of our survey of biochemistry we concentrated on nucleic acids and proteins and emphasized the flow of information from DNA, through RNA, to its expression in the catalytic or regulatory activity of a protein. We must now consider the equally fundamental subjects, *metabolism* and *bioenergetics.* These areas of biochemistry have benefited considerably from recent progress, building on important findings from as far back as the nineteenth century. While we have emphasized the reliance of our genetic heritage on the tiny hydrogen bond, we must also acknowledge that ultimately all worldly life depends on a small *electron trickle,* triggered by the light of the sun and captured in the molecule ATP. Just how this electron flow is coupled to ATP synthesis is at present unknown and the solution to this mystery may reveal a "code" for translating electrical energy into chemical energy—which is as much of a testimonial to Mother Nature as the genetic code.

We will approach the design of metabolism by first formulating (in Chapter 11) the fundamental relationships of bioenergetics: photosynthesis, the transformations of the carbohydrates and lipids, and the production of "high-energy bonds" of ATP. The key sequences and the balance of energy in metabolic bookkeeping will be emphasized. From this point we will go on to consider the mechanism of coupling ATP synthesis and electron flow in chloroplasts and mitochondria, the organelles devoted to this task (Chapter 12). The important transformations of nitrogen and nitrogen-containing compounds will be explored in Chapter 13. The vitamins, coenzymes, and some of the intricate chemical mechanisms living systems have evolved will comprise the subject of Chapter 14. With the survey of metabolism completed, we will describe just how the cell regulates all of these pathways and sequences to keep the cell's precursors and energy reserves in proper balance (Chapter 15). One type of control circuit feeds back to DNA itself to influence protein synthesis and this process will be analyzed in detail in Chapter 16. Therefore we will return to our starting point, DNA, completing the cycle: (1) DNA is transcribed and translated into proteins, (2) proteins

METABOLISM

catalyze metabolism, and (3) the metabolites influence transcription of DNA. In Chapter 17 we shall relate our biochemical knowledge to one of the more complex subjects of biochemistry—differentiation—and then go beyond biochemistry in Chapter 18 to weigh its implications in other aspects of human experience.

Chapter 11 Frontispiece *Electron micrograph of thin section of a maize chloroplast. Stacked lamellar structures (thylakoids) are evident. Magnification 26,000×. Courtesy L. K. Shumway.*

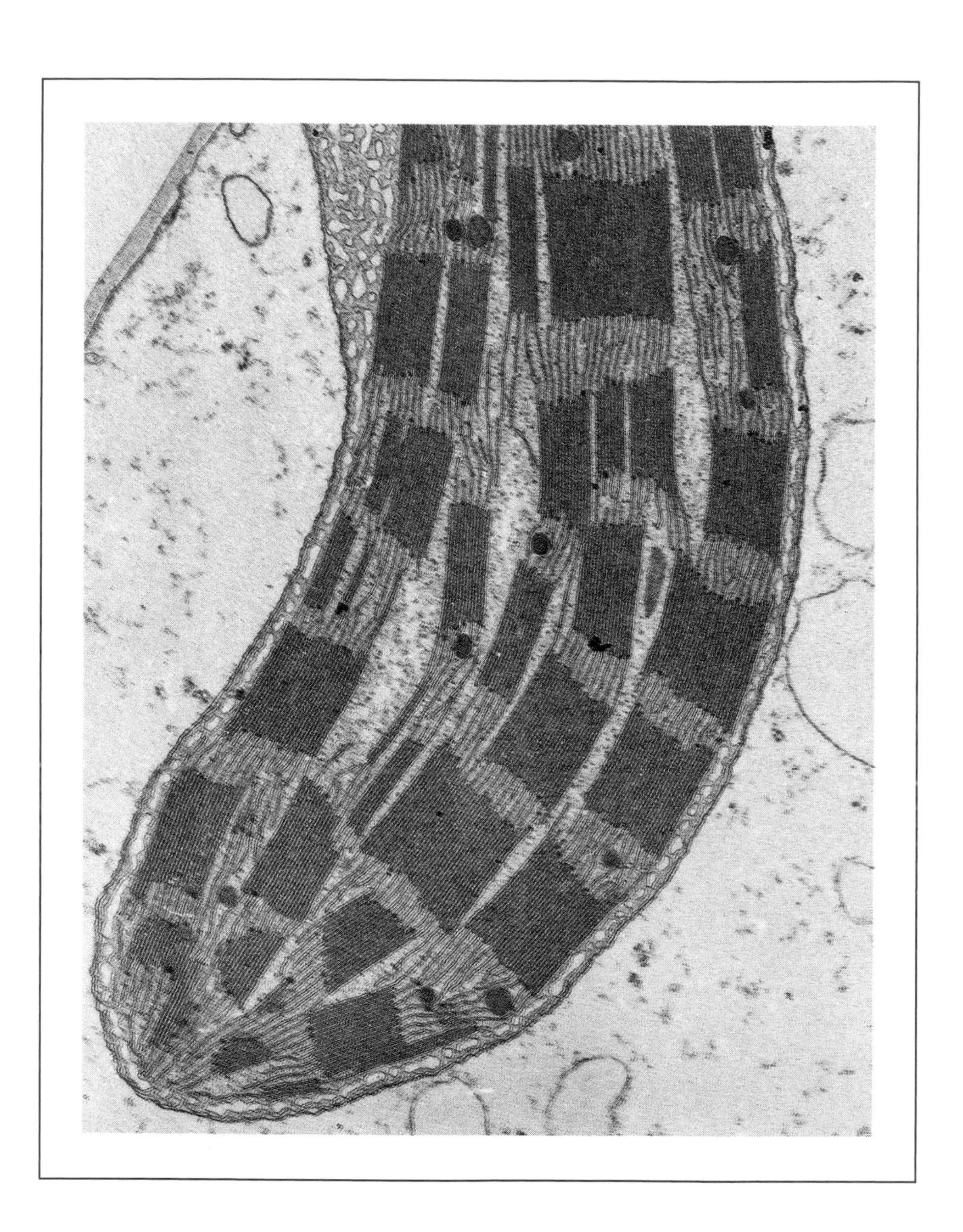

FUNDAMENTALS OF BIOENERGETICS

11

Metabolism in the cell follows laws similar to many of the principles of economics found in the world of commerce. This similarity is illustrated by the story of a prominent corporation president who was asked to expound on the philosophy that governed his activities. A deep, insightful reply was expected. Therefore his answer was almost startling, a bellowing, "Get the money!" In fact this man's daily life consisted of supervising the initiation, development, and distribution of books and other materials for education and seemed far removed from purely monetary considerations. Nevertheless, underlying his work was a simple principle of capitalism. The clarity of this principle may account for his success. The cell is in a closely related position. At first sight its major activity is varying, replicating, and testing its chromosomes. Emphasis is on the cycle of growth: synthesis of DNA for cell division to provide two genes from one and translating and transcribing the nucleic acids to produce proteins for enzymatic activities. Yet we will see that at another level the cell is dominated by its forms of economic principles and seeks to "Get the ATP!"

Metabolic Capitalism

The first principle of economics in a metabolic system is the existence of a *precious commodity*. Wealth in a biochemical economy, as perhaps elsewhere, is associated with the rare. Ice in the summertime is valuable, not in the wintertime—at least not in Ithaca, New York. In a biochemical environment it is the power that is the precious commodity—power to climb a mountain or to synthesize the genes for the next generation. Since the earth contains an oxygen-rich atmosphere and hence provides an oxidizing environment, biochemical systems we study are enlivened by *reducing power*. The abundance of oxygen tends to pull all elements into combination with oxygen. Carbon is oxidized to CO_2. The presence of a reduced, electron-rich, or hydrogenated compound is thus a precious commodity. An example of reduced carbon is methane, CH_4, the opposite of CO_2 in our example. Combining with oxygen, it donates its electrons to electron-poor oxygen in a spontaneous reaction to yield CO_2 and H_2O. This reaction is exergonic and can in principle drive energy-requiring processes. The presence of reducing power in an oxidizing environment can yield work, but only if harnessed by suitable machinery. Otherwise the energy is dissipated as heat. The cell harnesses the reducing power by methods we will describe, and biochemical life proceeds.

Reducing power is effective only by virtue of the oxidizing environment. If life exists on Jupiter, which has a reducing environment rich in hydrogen, methane, and ammonia, bioenergetics on that planet may involve oxidizing power, since there oxygen would be a precious commodity. In fact, a reducing atmosphere may have prevailed during primitive stages of life on earth (Chapter 18), possibly accounting for an anaerobic stage in the metabolism of most cells.

The source of reducing power is energy from sunlight. The energy is captured by green plants and certain microorganisms, through photosynthesis, and is used by them to do work and to reduce organic molecules. The reduced organic molecules, such as glucose, find their way to the animal kingdom and nonphotosynthetic microorganisms, where they are oxidized to release energy. This pattern is illustrated in Fig. 11-1 and provides the essential flow of bioenergetics.

Important Role of ATP

While reducing power thus satisfies the first requirement of a metabolic system, a second important principle must be considered—a *metabolic currency*. We have left unspecified for the moment the process whereby energy is extracted from glucose by animal cells (Fig. 11-1). We will return to it soon but can say that it is very complex. The cell needs a "ready-money" form of energy that does the actual participation in chemical reactions, for the same reasons that a monetary economy has replaced the strict barter system of old. The brickmaker doesn't want to have to carry bricks into a cafe to exchange for a cup of coffee. The same principle applies

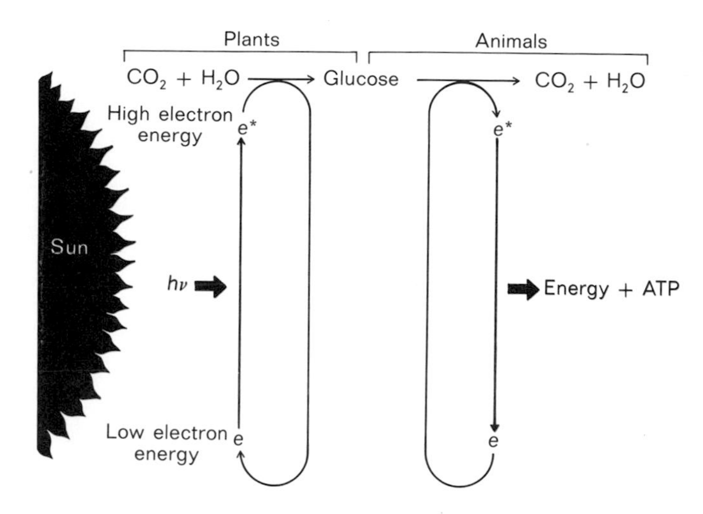

FIG. 11-1 *The Energy Cycle.*

in the cell. Reducing power is difficult to extract from glucose and cannot be conveniently called upon directly each time a peptide bond is synthesized or a muscle contracts. Rather, a *currency* is needed and the gold standard of the cell is ATP, adenosine triphosphate.

ATP, one of the four essential ingredients in RNA synthesis, has evolved as the major direct energy supplier of the cell. Its role stems from certain properties of phosphate anhydrides and is shared by the other nucleoside triphosphates. However, in the majority of cases only ATP serves as the energy provider.

The special property of nucleoside triphosphates and ATP in particular that enables these molecules to act as energy currency is the "high-energy" character of the two terminal anhydride linkages. As first emphasized by Fritz Lipmann, the reaction

$$\text{ATP} \rightarrow \text{ADP} + \text{P}_i \qquad \Delta G^\circ \sim -7000 \text{ cal}$$

occurs with a large release of free energy (ΔG°, see Appendix IV) or, in other terms, is very favorable. Thus, if this reaction is *coupled* to another biochemical process, the free energy change arising from the hydrolysis of ATP can enhance the otherwise unfavorable reaction. We will encounter many examples of such coupling in this chapter. The high-energy character arises from the release of strong repulsive interactions among the four negative charges on the phosphate oxygens (see Fig. 11-2) and the presence of more resonance states in the hydrolysis products than in ATP. The high-

ATP charge distribution:

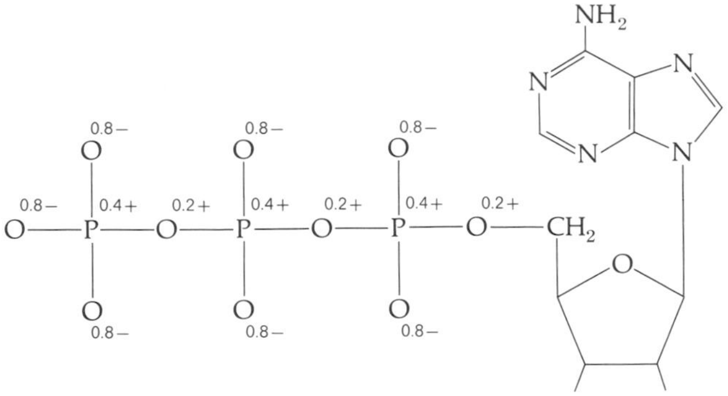

Resonance states of ATP hydrolysis product:

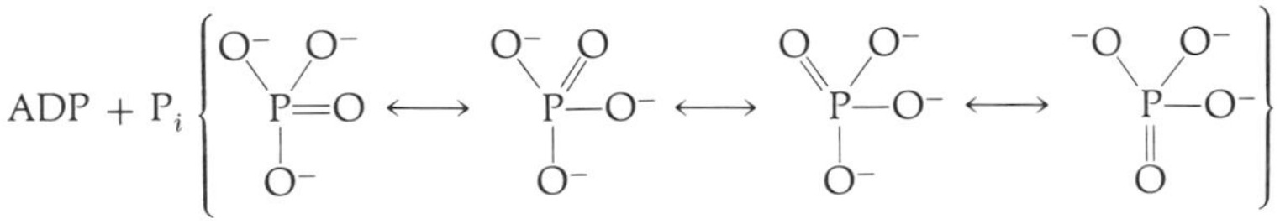

$$\mathrm{ADP} + \mathrm{P}_i$$

Designation of phosphate anhydride linkages as high energy:

A—R—P~P~P

FIG. 11-2 *Properties of ATP.* Another "high-energy" compound, especially important in muscle, is creative phosphate:

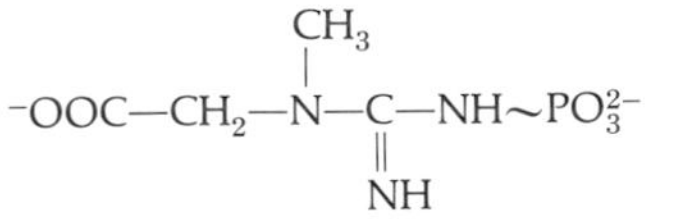

$$^{-}\mathrm{OOC{-}CH_2{-}N(CH_3){-}C({=}NH){-}NH{\sim}PO_3^{2-}}$$

energy character is often represented by a "squiggle" (~) as in writing ATP as AMP~P~P.

With the principles of a precious commodity (reducing power) and metabolic currency (ATP) stated, we can proceed to the actual details involved in metabolism. We will also discover that other compounds, acyl phosphate and phosphoenol pyruvate, can serve as high-energy currency. In addition we will touch on the fundamental process of bioenergetics, the coupling of electron flow, and ATP synthesis, although a thorough discussion will be deferred until the subsequent chapter.

Capturing Energy in Photosynthesis

From our first scheme (Fig. 11-1) we can see that the initial event in the biochemical cycle of life—a true moment of conception—is the capturing of light from the sun. This feat is achieved by the molecule chlorophyll (see Fig. 11-3) in conjunction with other molecular factors. Chlorophyll contains a porphyrin-ring system similar to that found in hemoglobin, although in this case magnesium (Mg) is present as the metal, not iron. By some as yet obscure mechanism, chlorophyll can capture light from the sun and harness the energy to split a molecule of H_2O into oxygen and hydrogen while snatching the electrons (e^-) from the hydrogen. The oxygen is liberated as free O_2 and the stripped hydrogen is left as free protons (H^+).

The captured electrons (two per molecule of water) are elevated by light from a low reducing level to a highly energetic or high reducing state. In this condition the electrons are delivered to one of two hard-to-reduce carrier compounds. The carriers have not been well characterized, but we know they are hard to reduce since they are similar in properties to other compounds that have been studied in detail. Each of the two carriers is reduced by a different *photosystem*. Photosystem I containing chlorophyll a (see Fig. 11-3) delivers an electron via its carrier (Z) to ferredoxin. Photosystem II, containing chlorophyll b in many cases, is activated by shorter wavelengths. It delivers electrons via its carrier (Q) to plastoquinone:

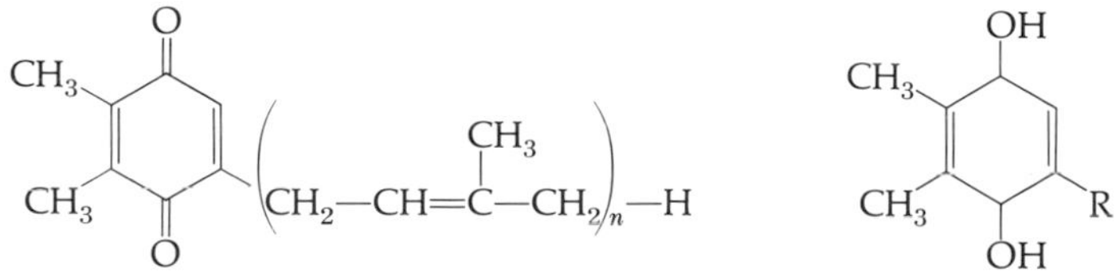
CH3
CH3
O
O
CH3
CH2—CH=C—CH2
n
—H
OH
OH
CH3
CH3
R

Reduced form

It is actually photosystem II which splits H_2O to gain the electrons. After being elevated and partially returned to ground level, the electrons pass on to photosystem I where they are "elevated" a second time.

Many of the steps we have just outlined and some that will be considered shortly are summarized in Fig. 11-4. Several important findings contributing to this scheme were provided by Daniel Arnon. All the compounds are presented at a height corresponding to their oxidation-reduction or *redox* potential, E_0' volts (at pH 7.0). The higher up (i.e., more negative) a compound, the harder it is to reduce. Once reduced, each compound can reduce any other compound below it in the E_0'-pecking order. If X reduces Y, X of course becomes oxidized in the process.

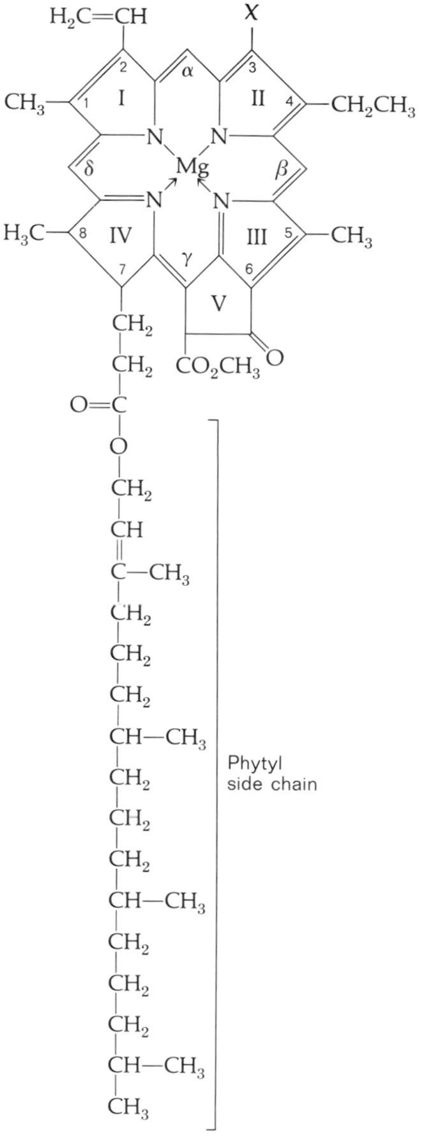
H2C=CH
X
CH3
CH2CH3
H3C
CH3
I
II
III
IV
V
α
β
γ
δ
N
Mg
CO2CH3
O
O=C
Phytyl
side chain

FIG. 11-3 *Structure of Chlorophyll.* The side chain indicated by X is a $—CH_3$ group in chlorophyll a (photosystem I) and a —CHO group in chlorophyll b (photosystem II).

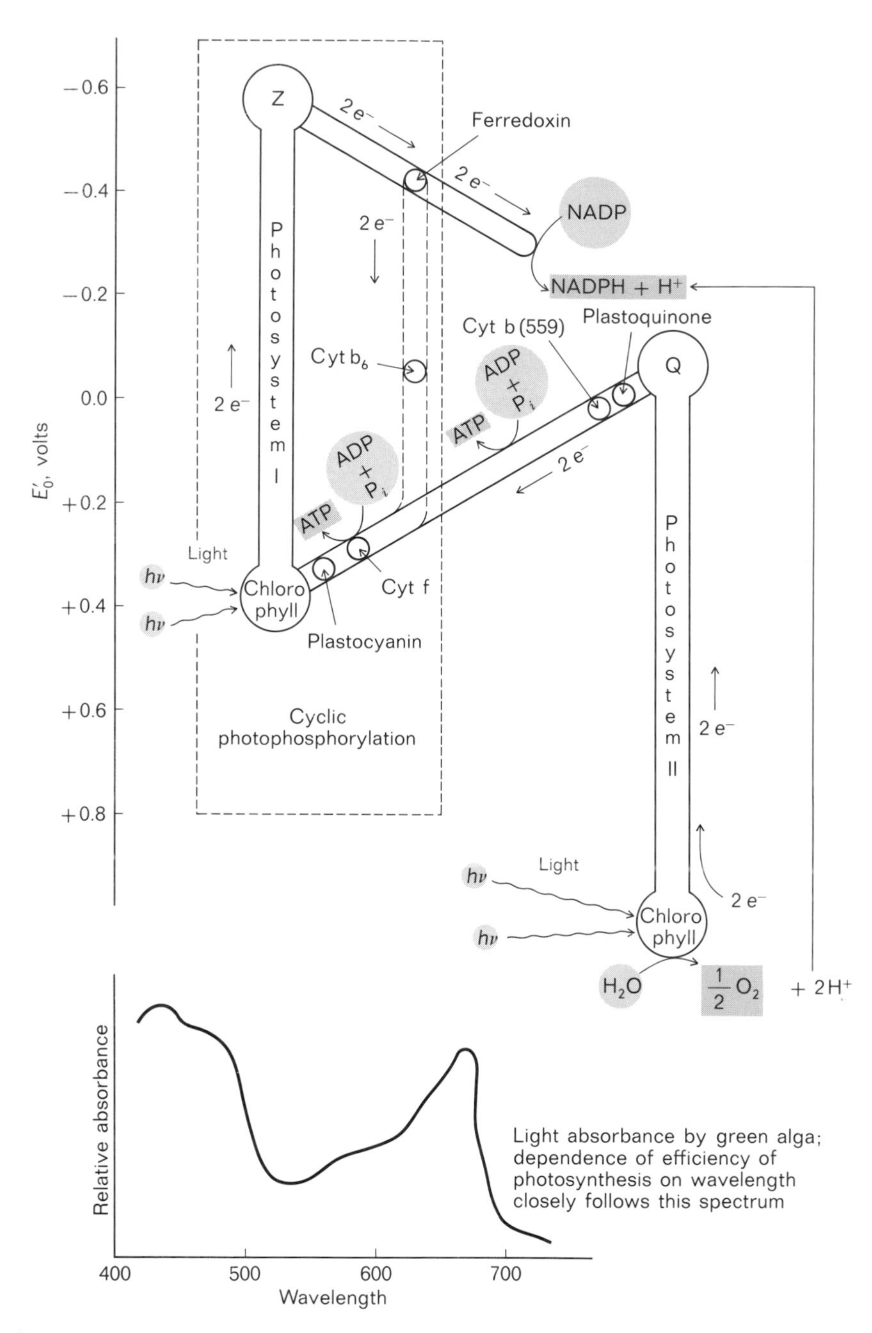

FIG. 11-4 *Photophosphorylation.* Colored circles indicate input, gray rectangles indicate output. Modified from A. L. Lehninger, *Biochemistry,* page 469, Worth Publishers, New York, 1970.

The term E'_0 can be related to the free energy of change of the redox reaction by

$$\Delta G^\circ = -nF\,\Delta E'_0$$

where n is the number of electrons involved, F is the Faraday constant, 23,000 cal/(volt-equiv), and E'_0 is the difference in redox potential between the oxidizing and reducing agents. ΔG° (the free-energy change) is defined in Appendix IV. Thus if reduced ferredoxin (Fd) could donate its electrons to oxygen to yield H_2O and oxidized ferredoxin, the reaction

$$Fd_{red} + \tfrac{1}{2}O_2 \rightleftharpoons Fd_{ox} + H_2O$$

has a $\Delta E'_0$ of (0.8) − (−0.4) or 1.2 volts, and

$$\begin{aligned}\Delta G^\circ &= (-2\text{ equiv})(23{,}000\text{ cal/volt-equiv})(1.2\text{ volts})\\ &= -55\text{ kcal}\end{aligned}$$

Thus the reaction should be extremely favorable.

The captured reducing power is now put to use in two important ways. First, the captured electrons are passed down an *electron-transport* chain (see Fig. 11-4) involving plastoquinone (already noted), cytochrome b (559), cytochrome f, and plastocyanine. For the heme-containing cytochromes, oxidation-reduction involves a change in iron between Fe^{3+} and Fe^{2+}. Finally, the electrons arrive at photosystem I where they can again be "elevated" by light. Somehow, by the mysterious process of electrochemical coupling (to be considered in the next chapter), passage of the electrons down the transport chain is linked to ATP formation from ADP and P_i. This process is called *photophosphorylation*. Two molecules of ATP are apparently formed, at separate sites, for each *pair* of electrons traveling down the chain. The ATP can then go on to contribute to the processes of glucose formation from CO_2 (see below) as well as to many other important energy-requiring reactions.

The second way the high-energy electrons are utilized involves photosystem I. In this case the object is reduction of NADP (structure, Fig. 11-5) via the system-I carrier and ferredoxin, a small (MW 11,600) iron-containing protein. The reaction is catalyzed by the enzyme ferredoxin-NADP oxidoreductase. NADP is a coenzyme, one of the broad class of compounds that facilitates metabolism by filling a precise chemical role. (For more details, see Chapter 14.) As in ATP synthesis, a *pair* of electrons is required for one molecule of NADPH + H^+ to be formed. In the case of NADP, the role

Adenine

This third phosphate is missing in the other common form of the coenzyme, NAD

Nicotinamide

NADP

Substrate + → Substrate + H^+

Oxidized form (NADP or NAD)

Reduced form ($NADPH + H^+$ or $NADH + H^+$)

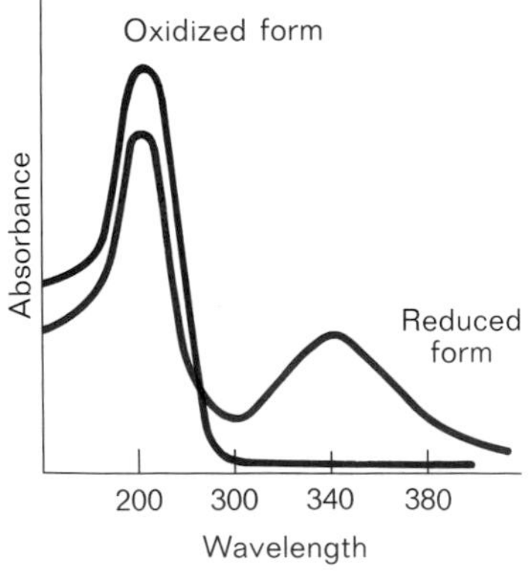

The reduced form of the coenzyme has an absorption peak at 340 nm that is absent in the oxidized form; the difference provides a convenient spectral assay for following reactions involving NAD or NADP

FIG. 11-5 *Properties of Nicotinamide Adenine Dinucleotide Phosphate (NADP).*

is to serve as a mediator of reducing power. Just as ATP aids metabolic processes as a readily available currency in energy-requiring transformations, so NADP, in the reduced form NADPH + H^+ (see Fig. 11-5), provides ready reducing power for reactions that specifically require addition of hydrogen. Both ATP and NADPH + H^+ enter into an enormous variety of reactions and will be encountered often in our discussions of metabolism. In the case of photosynthesis, NADPH + H^+ provides the hydrogens in a form needed for the reduction of CO_2 to form glucose, while ATP provides the driving force.

Before going on to describe the chemical events in photosynthetic glucose formation, one other aspect of the electron-transport chain should be noted. In certain cases the chain of photosystem I can be short-circuited to direct electron transport away from NADPH + H^+ and back into the ATP-generating electron chain from photosystem II (see Fig. 11-4). The connection is made midway down the chain (via cytochrome b_6) so only one molecule of ATP is synthesized on each pass. However, in this process of *cyclic photophosphorylation,* high levels of ATP can be built up without a continuous source of electron donors and acceptors.

Photosynthetic Reduction of CO_2 to Form Glucose

With the availability of ATP and reduced NADP, the formation of glucose from CO_2 can be accomplished. This process, which can occur in the dark, takes place via the *Calvin cycle,* named to honor its discoverer, Melvin Calvin. We will follow the reactions as far as glucose, although plants generally carry the chemistry one step further and store their carbohydrates as polymers of cellulose or starch (see Chapter 15).

Studies on the organic chemistry involved in photosynthesis were greatly aided by the use of a radioactive carbon, ^{14}C. This isotope of carbon is stable (half-life of about 5000 years), and organic compounds containing ^{14}C can be readily monitored by Geiger counters, autoradiography, or liquid-scintillation counters. Studies with ^{14}C distribution have in fact provided the major tool for tracing many of the metabolic pathways. In Calvin's studies simple photosynthesizing algae were exposed to a short pulse of $^{14}CO_2$. The first major compound labeled was 3-phosphoglyceric acid. Chemical analysis localized the ^{14}C in the carboxyl atom and eventually many other compounds were identified in the reactions for incorporating CO_2.

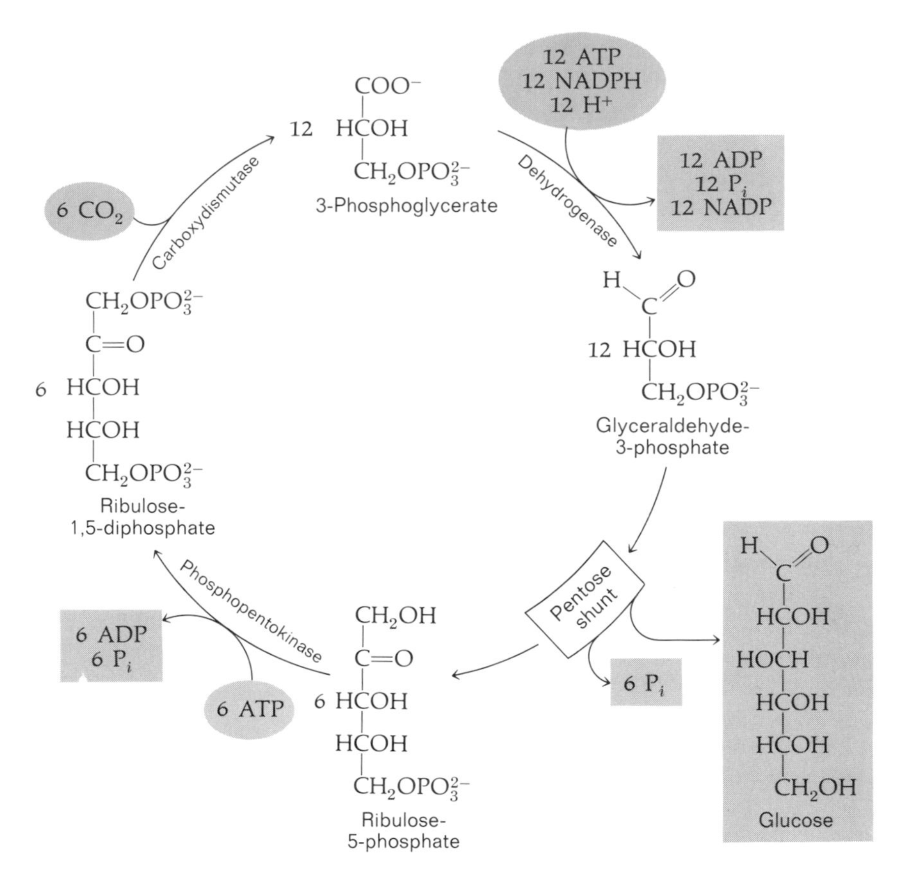

FIG. 11-6 *The Calvin Cycle.* For additional details, see M. Calvin and J. A. Bassham, *The Photosynthesis of Carbon Compounds,* W. A. Benjamin, New York, 1962.

The Calvin cycle is depicted in Fig. 11-6. The critical step is the reaction of free CO_2 with ribulose-1,5-diphosphate, a relative of ribose but with a keto function in the 2 position. The reaction is catalyzed by diphosphoribulose carboxylase, a large enzyme (MW about 0.5 million), and yields two molecules of 3-phosphoglycerate for each molecule of CO_2 and ribulose-1,5-diphosphate. The phosphoglycerate is then reduced to glyceraldehyde-3-phosphate (G3P) by NADPH + H^+ in a reaction requiring ATP for energy. The G3P then enters a complicated scheme that converts the three-carbon glyceraldehyde units into glucose and ribulose. For every twelve molecules of G3P (which arise from six molecules of CO_2 and ribulose-1,5-diphosphate), the shunt produces one molecule of glucose and six molecules of ribulose-5-phosphate (see Figs. 11-7 and 11-8). Rearranging three-carbon units into five- and six-carbon units is a challenging "puzzle," reminiscent of the childhood brain twister of devising a scheme for measuring 1 quart of water with only a 3- and a 5-quart container. Nature has solved the puzzle by a series of reactions involving seven- and four-carbon units, as summarized in Fig. 11-8. Important features include the enzymatic reaction of *aldolase,* which catalyzes aldol condensations of the type

$$\mathrm{H{-}C(R')(OH){-}H} + \mathrm{H{-}C(=O){-}R} \longrightarrow \mathrm{HO{-}C(R')(H){-}C(H)(OH){-}R}$$

and of *transketolase,* which catalyzes transfer reactions of the type

$$\mathrm{CH_2OH{-}C(=O){-}C(H)(OH){-}R} + \mathrm{O{=}C(H){-}R'} \rightleftharpoons \mathrm{CH_2OH{-}C(=O){-}C(H)(OH){-}R'} + \mathrm{O{=}C(H){-}R}$$

A scheme similar to the pentose shunt but "in reverse" will be encountered later in the energy-producing degradation of five-carbon sugars. With the puzzle solved, the cycle can be completed and one molecule of glucose is formed for six molecules of CO_2 captured to give an overall reaction of

$$6\,CO_2 + 18\,ATP + 12\,NADPH + 12\,H^+ \longrightarrow \text{glucose} + 18\,ADP + 18P_i + 12\,NADP^+$$

All the other components listed in Figs. 11-6 to 11-8 are regenerated with each turn of the cycle. They are formally *catalysts* and hence do not appear in the overall reaction.

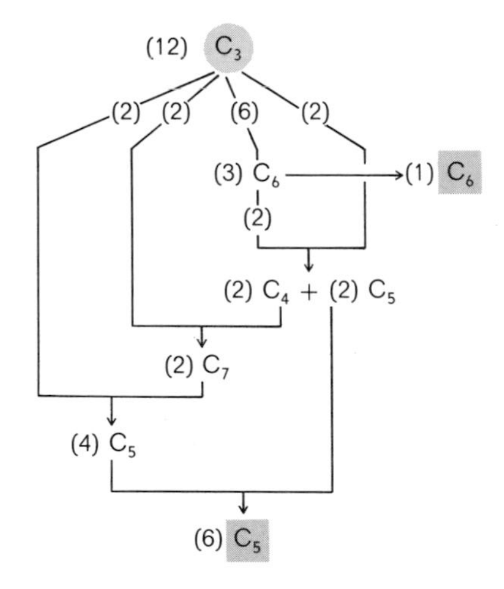

FIG. 11-7 *Pentose Shunt: General Scheme.* The values in parentheses refer to the relative number of molecules that follow each reaction path. Subscripts give the number of carbon atoms in each of the molecules listed.

Summary of Photosynthesis

The synthesis of glucose can now be related to the individual light quanta absorbed by the photosynthesis apparatus. For *each molecule of CO_2 reduced,* two molecules of NADPH + H^+ must be supplied by photosystem I, requiring four electrons to be elevated. Since one quantum of light is required per electron, four quanta are consumed in the process. An additional four quanta are needed to provide the electrons for

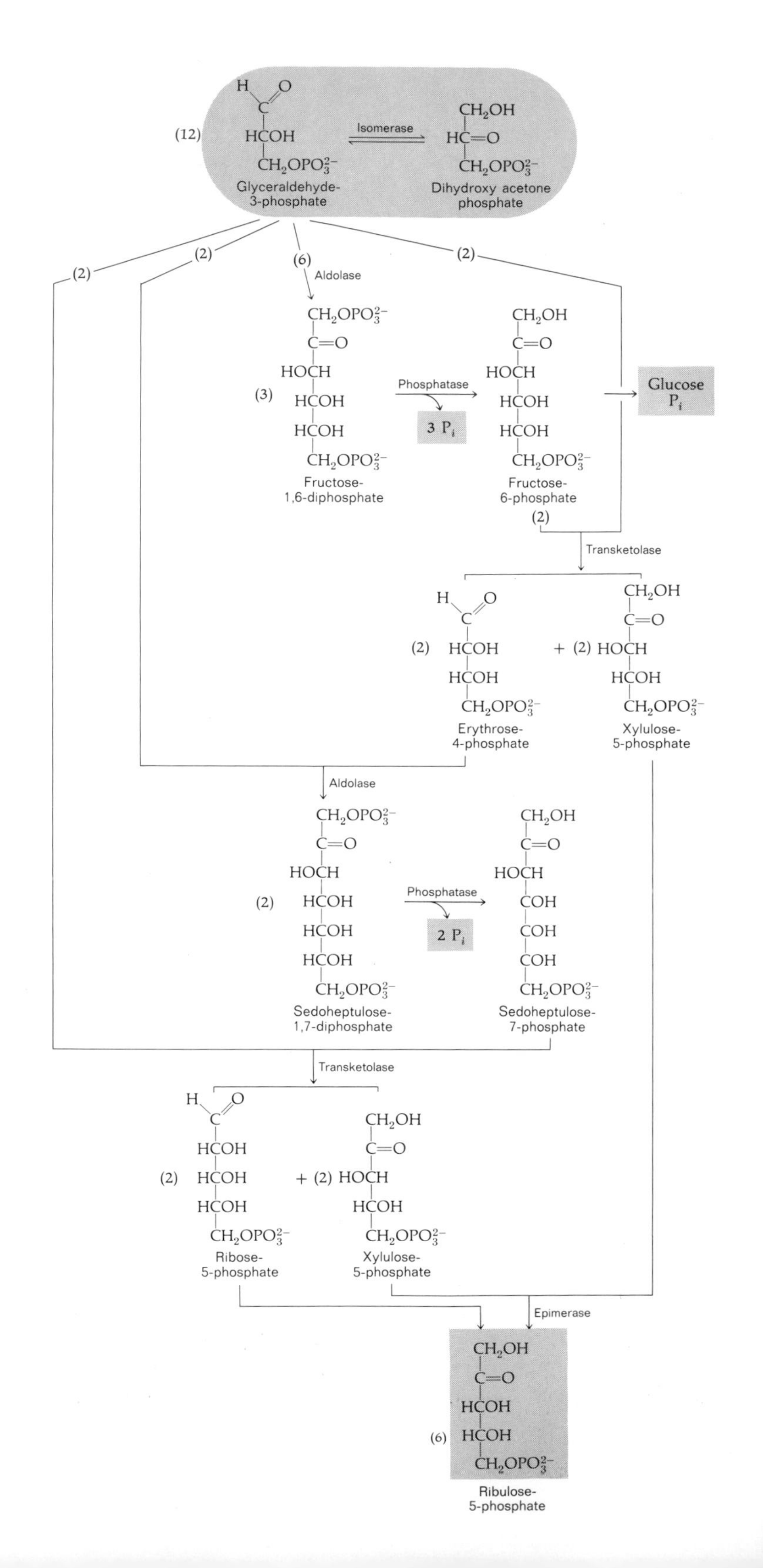

FIG. 11-8 *Pentose Shunt: Structures.*

photosystem I from photosystem II. These electrons probably yield four molecules of ATP along the way. Thus the total reaction described above—involving 6 molecules of CO_2 to form 1 glucose—involves 48 quanta of light. Since light in the green region of the spectrum has roughly 50 kcal/einstein (1 einstein = 1 mole or 6×10^{23} quanta), about 2400 kcal are consumed in producing glucose. With a standard free-energy change ΔG° (see Appendix IV) of about 700 kcal for the combustion of glucose to H_2O and CO_2, we find an overall efficiency of 700/2400 or about 30%. Depending on the exact wavelength of light used, the number may be somewhat higher, possibly approaching 40%. Efficiencies in this range, 30 to 40%, are commonly found in biochemical reactions.

Knowledge of the quantum requirements of photosynthesis also permits the complete reaction of photosynthesis to be specified:

$$6\,CO_2 + 12\,H_2O + 48\,h\nu \longrightarrow C_6H_{12}O_6 + 6\,H_2O + 6\,O_2$$

Experiments by Van Neil demonstrated that all the O_2 liberated in photosynthesis arises from water. Hence 12 molecules of H_2O are required. In more general terms the scheme can be written

$$A + H_2D + n\,h\nu \longrightarrow H_2A + D$$

and many exotic acceptors (A) other than CO_2 and donors (D) other than oxygen of H_2O can be found in nature, particularly among the photosynthetic microorganisms. Hydrogen sulfide, hydrogen gas, and lactic acid are among the donors encountered in nature; nitrogen may be employed as an acceptor to yield ammonia. The more common photosynthetic reactions in plants, involving CO_2 and H_2O, take place in chloroplasts, complex organelles highly specialized for this task (these will be described in detail in Chapter 14). With the production of glucose described, a "precious" reduced commodity, we can now go on to consider how nonphotosynthetic or heterotropic cells cash in this commodity to obtain ATP.

ATP Formation from Glucose Metabolism

Cells obtain ATP from glucose degradation in two ways. The first and minor way is by a relatively simple process termed *substrate-level phosphorylation*. This route is found in the transformations known as *glycolysis* and accounts for only about

5% of the ATP available from glucose. The second and predominant mechanism for deriving ATP from glucose, *oxidative phosphorylation,* is, broadly viewed, a reversal of the process of photosynthesis by which glucose was produced. Metabolic cycles, particularly the tricarboxylic acid cycle (TCA cycle), are entered and transfer the reducing power of glucose breakdown products or metabolites to NAD^+ (a compound similar to $NADP^+$ but lacking a phosphate, see Fig. 11-5), forming $NADH + H^+$. The $NADH + H^+$ then reduces the first compound of an electron chain, similar in some respects to the chain of photosynthesis, and the electrons cascade down a sequence of redox mediators yielding ATP in the process. Here, in a sense, a connection is made between the acceptor of photosystem I of photosynthesis and the electron chain in nonphotosynthetic organisms. The connection involves $NADPH + H^+$ to reduce glucose, transfer of the glucose (in the case of man, for example, by eating), return of the reducing power to a nicotinamide derivative (NAD^+) to close the original "circuit" presented in Fig. 11-1. This major route of ATP formation via the electron chain and oxidative phosphorylation shares many properties with photophosphorylation as will become apparent in the next chapter. We will now describe the substrate-level oxidative phosphorylation of glycolysis and then proceed to NAD^+ reduction via the TCA cycle.

Glycolysis: Token Profit and Preparation for Better Things

Glycolysis is a primitive process. In fact it may be a "fossil" from an early anaerobic phase of earthly development (see Chapter 18). The reactions are complex and, compared to the yield of other pathways, little ATP is produced. Glucose is converted to two molecules of lactate with the concomitant production of two molecules of ATP from ADP + P_i. However, glycolysis does have the important advantage of proceeding without oxygen. In addition two other benefits are realized. (1) The glucose is transformed into compounds suitable for processing in the "high-return" TCA cycle, as well as into compounds needed as intermediates in nucleotide and amino acid biosynthesis (see Chapter 13). (2) In certain organisms, particularly yeast, an intermediate near the end of glycolysis is diverted to form alcohol.

Glycolysis (Fig. 11-9), unraveled by Embden and Meyer-

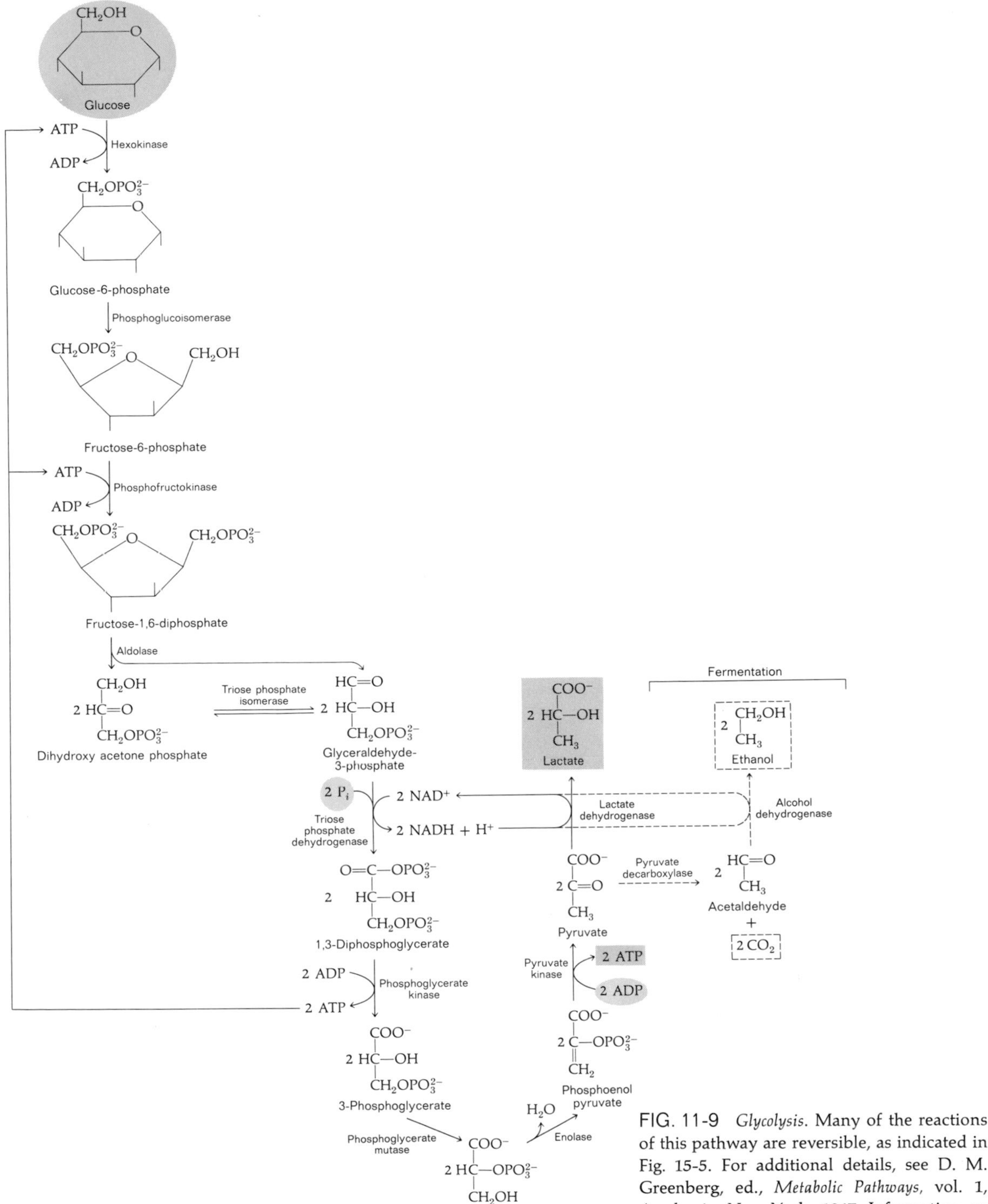

FIG. 11-9 *Glycolysis.* Many of the reactions of this pathway are reversible, as indicated in Fig. 15-5. For additional details, see D. M. Greenberg, ed., *Metabolic Pathways,* vol. 1, Academic, New York, 1967. Information on the structure of several of the enzymes of glycolysis can be found in the series of papers by H. C. Watson and his colleagues beginning with *Nature New Biology,* **240:** 130 (1972). See also Chapter 7 frontispiece.

hof in the 1930s, begins with the input of glucose in the form of glucose-6-phosphate. The glucose is obtained from several sources: the breakdown of starch (see Chapter 15), directly from plant sources, and from interconversion of other sugars such as galactose that may be synthesized by certain photosynthetic organisms or obtained from milk lactose. The phosphorylated glucose is isomerized to fructose-6-phosphate and phosphorylated again to yield fructose-1,6-diphosphate. At this point the familiar aldolase reaction occurs (see Calvin cycle) and three-carbon units are formed. The oxidation of glyceraldehyde-3-phosphate then yields an acyl phosphate, at the 1 position of 1,3-diphosphoglycerate, which in the subsequent kinase reaction is coupled to the formation of ATP. Since two molecules of ATP are formed for each molecule of glucose, glycolysis is now "even," having replenished the two molecules of ATP consumed in the initial phosphorylations.

The three-carbon units are then rearranged to form phosphoenol pyruvate with a high-energy phosphate. In the next reaction ATP is formed (two molecules per glucose), providing the *net* profit of glycolysis. Pyruvate then goes on to form lactate with the oxidation of NADH + H^+, regenerating NAD^+ needed at the triose phosphate dehydrogenase reaction. In yeast and other fermenting organisms the pyruvate is decarboxylated to acetaldehyde and CO_2. The equivalent NADH + H^+ formed in the triose phosphate dehydrogenase reaction is then used in the formation of ethanol.

Each reaction described is catalyzed by a separate enzyme, named in Fig. 11-9. The enzymes are large proteins (MW 1 to 2×10^5) and most contain subunits. Several of the enzymes, particularly those involved with ATP reactions, require the metal magnesium and some interesting allosteric-control properties are found among the glycolytic enzymes (which will be described in Chapter 15).

We noted earlier that glycolysis was not distinguished for its thoroughness in capturing energy from glucose and now we can express this conclusion quantitatively. A comparison of the oxidation energies of glucose and lactate (see Appendix IV) yields a free-energy change $-\Delta G°'$ for glucose to lactate of about 50 kcal. This is a relatively efficient process, since two ATP are formed, which corresponds to a trapping of about 14 kcal/mole or an efficiency of close to 30%—near the value observed in photosynthesis. The efficiency is likely to be even somewhat higher since the calculation was based on standard free-energy changes for ATP (assuming reactants at equimolar concentrations, see Appendix IV). If corrected, the $\Delta G^{\circ\prime}$ for ATP could be somewhat above 7 kcal/mole. However, since glucose contains nearly 700 kcal/mole, only about 7% of the potential energy is released. We can now proceed to the extraction of the remaining energy by the TCA cycle.

The TCA Cycle and NAD$^+$ Reduction

The bulk of the reducing power stored in glucose is released in the TCA cycle, also known as the citric acid cycle or Krebs cycle in honor of its discoverer, Sir Hans Krebs. The purpose of the cycle is to separate carbons (as CO_2) from reducing equivalents, which are concentrated as reduced coenzymes. The cycle begins with a decarboxylation of one of the products of glycolysis, pyruvate, to yield a derivative of acetate and CO_2. The reaction, catalyzed by one of the largest enzyme complexes known—the pyruvate dehydrogenase complex with a molecular weight of about 4 million—has several steps and involves four coenzymes, the familiar NAD$^+$ as well as FAD (flavin adenine dinucleotide, a coenzyme similar in function to NAD$^+$ although slightly easier to reduce), TPP (thiamine pyrophosphate), and CoA (coenzyme A). (The details of the structures and reactions of FAD, TPP, and CoA are described in Chapter 14.) In the case of the pyruvate dehydrogenase reaction we note here that TPP and FAD function in a truly catalytic role (they are not consumed) whereas CoA activates the acetate and is a product of the reaction. An important property of CoA is a terminal sulfhydryl group (for this reason, CoA is sometimes designated CoA-SH). When it is linked in a thioester bond $\text{CoA-S}-\overset{\overset{\text{O}}{\|}}{\text{C}}-\text{R}$, a high-energy compound is formed as in the pyruvate dehydrogenase reaction:

$$\text{Pyruvate} + \text{NAD}^+ + \text{CoA-SH} \longrightarrow \text{acetyl}{\sim}\text{S-CoA} + \text{NADH} + \text{H}^+ + \text{CO}_2$$

Acetyl∼S-CoA is depicted with a "squiggle" (∼) to represent its high-energy character that will provide driving force to the TCA cycle at the point of entry of the two-carbon acetyl unit.

The basic scheme of the TCA cycle (Fig. 11-10) is the

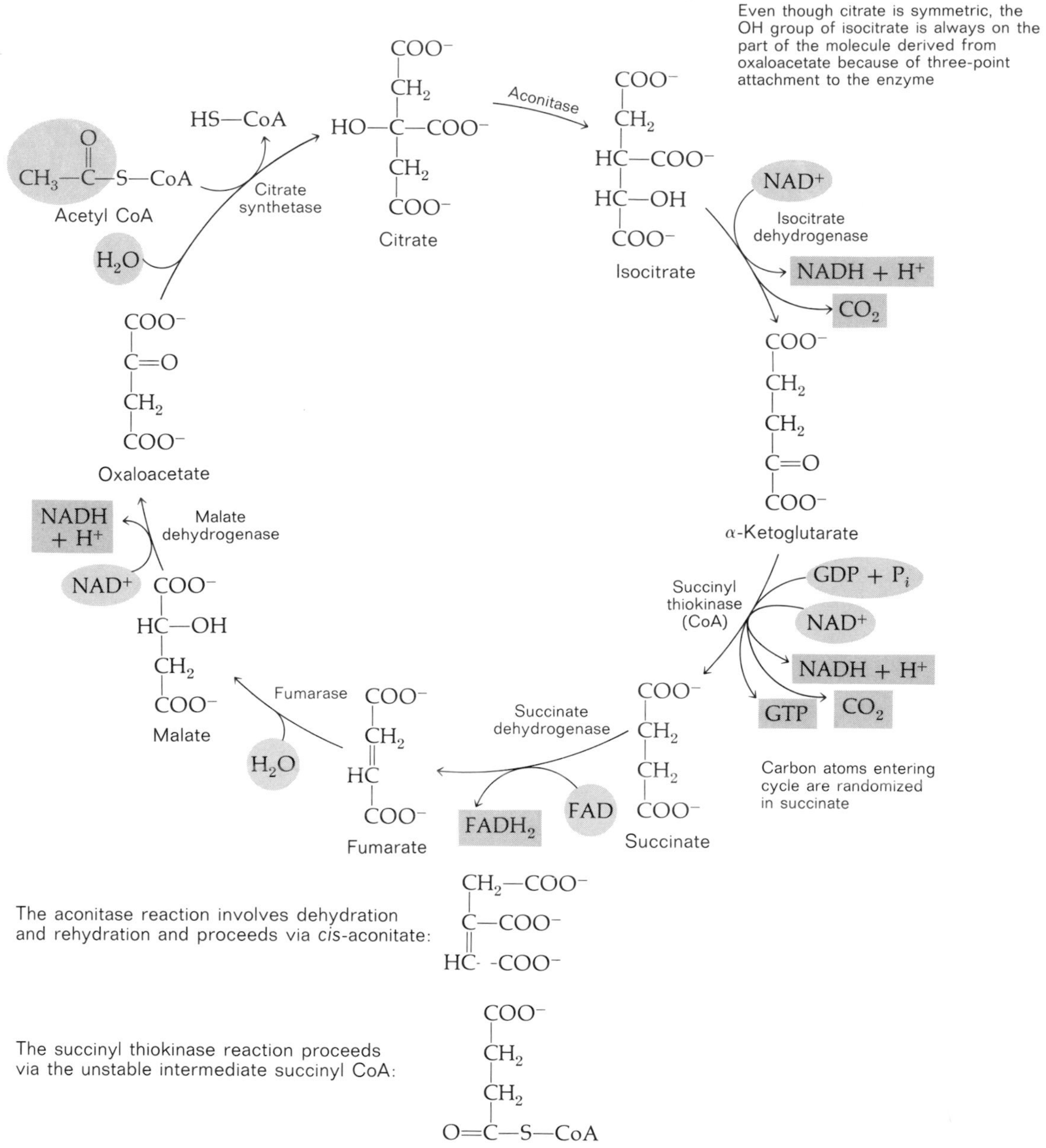
Even though citrate is symmetric, the OH group of isocitrate is always on the part of the molecule derived from oxaloacetate because of three-point attachment to the enzyme
Acetyl CoA
HS—CoA
Citrate synthetase
H_2O
Citrate
Aconitase
Isocitrate
NAD^+
Isocitrate dehydrogenase
$NADH + H^+$
CO_2
Oxaloacetate
α-Ketoglutarate
Succinyl thiokinase (CoA)
$GDP + P_i$
NAD^+
$NADH + H^+$
GTP
CO_2
Malate dehydrogenase
$NADH + H^+$
NAD^+
Malate
Fumarase
H_2O
Fumarate
Succinate dehydrogenase
$FADH_2$
FAD
Succinate
Carbon atoms entering cycle are randomized in succinate
The aconitase reaction involves dehydration and rehydration and proceeds via *cis*-aconitate:
The succinyl thiokinase reaction proceeds via the unstable intermediate succinyl CoA:

FIG. 11-10 *The TCA Cycle.* For additional details, see J. M. Lowenstein, ed., *Methods in Enzymology,* vol. 13, Academic, New York, 1969.

condensation of the two-carbon acetyl units with four-carbon oxaloacetate molecules to yield six-carbon citric acid. Then by a series of transformations the six-carbon units are dehydrogenated and decarboxylated to regenerate the acceptor compound oxaloacetate along with the transfer of reducing power to NAD^+ and FAD. Thus all the organic intermediates are catalytic (nonconsumed). Another important feature is the *incorporation of H_2O* to extract the last of the reducing power in the intermediates by providing additional hydrogens to reduce NAD^+ and FAD and oxygens for CO_2 formation. One substrate-level phosphorylation is also accomplished, at the reaction to form succinic acid, although in this case GTP, not ATP, is formed. The two nucleoside triphosphates are, however, energetically equivalent. Thus the overall reaction, expressed in terms of free acetate (since CoA-SH is regenerated) may be written as

$$\text{Acetate} + 3\ NAD^+ + FAD + GDP + P_i + 2\ H_2O \longrightarrow 2\ CO_2 + 3\,(NADH + H^+) + FADH_2 + GTP$$

We will see, in turning to oxidative phosphorylation, that the reduced NAD^+ and FAD can now be exchanged for the energy currency of ATP.

Oxidative Phosphorylation

The reactions of the TCA cycle and the subsequent coupling of $NADH + H^+$ oxidation to yield ATP, that is, oxidative phosphorylation, take place in the *mitochondria* of animal cells. Mitochondria are complex organelles, the counterpart of chloroplasts in plant cells, and will be discussed more thoroughly in Chapter 14. Here we are concerned with the *accounting* of glucose metabolism and can now complete the saga of reducing power from its starting point at an excited electron in photosynthesis to its final deposition in the reduction of molecular oxygen. It is this last stage, the passage of electrons from $NADH + H^+$ and $FADH_2$ formed in the TCA cycle down the electron-transport or respiratory chain to oxygen that we will consider now. This process of oxidative phosphorylation is similar in concept to photophosphorylation. The major difference, however, is that the highly reducing compound that starts the electron cascade is generated not by light but by the entry of $NADH + H^+$. Then, by sequences of redox reactions (see Fig. 11-11), the reducing power of $NADH + H^+$ is gradually exploited and, at three stages in the chain, electron flow is coupled to form ATP. As in photosynthesis the respiratory chain is composed of several heme-containing cytochromes, some of which have been studied in great detail, especially cytochrome c (see Chapter 7). In addition at least one small molecule, coenzyme Q (Fig. 11-12), is involved as well as two flavin-containing proteins,

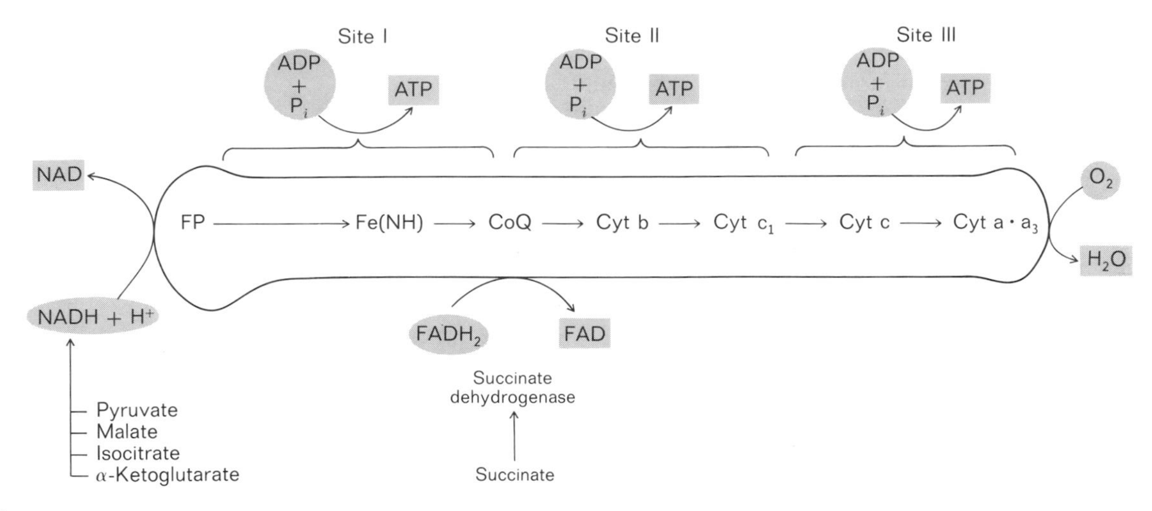

FIG. 11-11 *The Respiratory Chain.* The three sites of phosphorylation are shown.

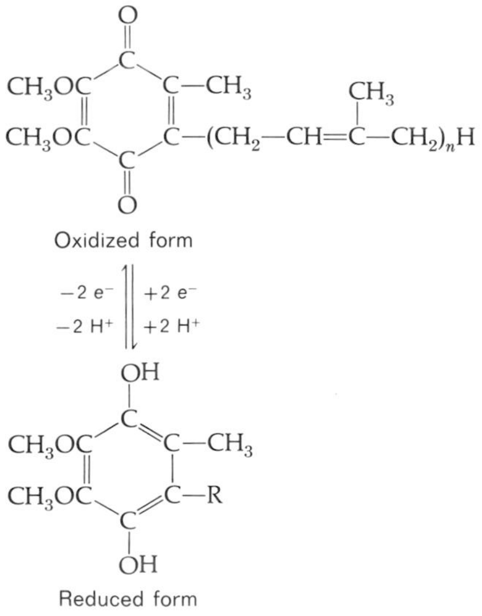

at least two nonheme iron proteins and several other factors. What should be emphasized here is the final output of ATP.

For the passage of a pair of electrons from a molecule of NADH + H^+, three molecules of ATP are formed, one at each of the three sites indicated in Fig. 11-11. Thus the *P:O ratio* (or phosphorylations per O_2 molecule) is 3. The $FADH_2$ bound to succinite dehydrogenase has somewhat less reducing power than NADH + H^+ and can only reduce CoQ. Therefore site I is bypassed and only two molecules of ATP are formed for each molecule of succinate oxidized (P:O ratio = 2). At the end of the chain the electrons react with oxygen to form water. Many of the key steps in the process were discovered by Keilin, Warburg, Chance, and their colleagues using a battery of elegant spectral, chemical, and kinetic methods.

CH_3OC, CH_3OC, C=O, C=O, $C—CH_3$, $C—(CH_2—CH{=}C(CH_3)—CH_2)_nH$

Oxidized form

$-2\ e^-$, $-2\ H^+$ ⇅ $+2\ e^-$, $+2\ H^+$

CH_3OC, CH_3OC, C—OH, C—OH, $C—CH_3$, C—R

Reduced form

FIG. 11-12 *Structure of Coenzyme Q.* Predominant forms are $n = 6$ (CoQ_6) or $n = 10$ (CoQ_{10}).

Summary of Glucose Metabolism

We can now summarize the overall transformation of glucose into energy currency or ATP. From glycolysis, 2 molecules of ATP are formed; and when pyruvate is diverted to the TCA cycle, 2 molecules of NADH + H^+ are left over. Each reduced NAD can then participate in oxidative phosphorylation to yield 3 ATP, but since glycolysis takes place outside the mitochondria, 1 ATP is consumed in transporting the NADH + H^+ inside, leaving a net gain of 2 ATP or 4 per glucose. Pyruvate oxidation yields 4 molecules of NADH + H^+ for each pyruvate or 8 per glucose. These 8 molecules of NADH + H^+ in turn yield 24 molecules of ATP. One $FADH_2$ per pyruvate processed, or 2 per glucose, yields another 4 ATP. In addition, 2 molecules of GTP per glucose are formed by substrate-level phosphorylation at the succinyl thiokinase step. The terminal phosphate of GTP can be transferred to ADP to yield 2 more molecules of ATP. Thus we can summarize the production of a total of 36 molecules of ATP from glucose:

ATP Formed	Source
2	Glycolysis
4	NADH + H^+ (Glycolysis)
24	NADH + H^+
4	$FADH_2$
2	GTP
36	

The overall reaction of glucose oxidation is therefore

$$\text{Glucose} + 6\ O_2 + 36\ \text{ADP} + 36\ P_i \longrightarrow 6\ CO_2 + 36\ \text{ATP}$$

At values for $-\Delta G^{\circ\prime}$ of 7 kcal per ATP hydrolysis and 700 kcal per glucose combustion, an overall efficiency of slightly less than 40% is observed. Water, also produced in the overall reaction, is ignored in the summary.

An interesting point is made if we compare the overall reaction for glucose oxidation with the equation for photosynthetic glucose formation. For each molecule of glucose formed, 18 ATP are consumed along with 12 molecules of NADPH + H^+. NADPH is isoenergetic with NADH and at the going exchange rate (3:1) the NADPH + H^+ corresponds to 36 molecules of ATP for a total of 54 ATP per glucose. Comparing this value with the ATP yield in glucose oxidation gives

Input in Glucose	54 ATP
Return from Glucose	36 ATP
Net Loss	18 ATP

In metabolic trades, energy put into a molecule during formation is never fully regained during degradation. However, as glucose consumers we shouldn't feel short-changed—18 ATP per glucose is a small price to pay for not having to bear leaves.

Fatty Acids: Maximizing Reducing Power

The description of oxidative phosphorylation completes the energy cycle we formulated at the beginning of the chapter (Fig. 11-1) and leaves us with a detailed view of the inner workings of this process. Since economy and efficiency were emphasized in the reporting of this aspect of biochemistry, it is fair to ask if glucose itself is the most appropriate mediator of the transfer of reducing power from plants to animals. We can acknowledge that six-carbon sugars are especially stable configurations among the carbohydrates and in this respect glucose is a good choice; but what about other species of organic molecules? Glucose is clearly not in a fully reduced state. Its hydroxyl and aldehyde functions could be further reduced to methyl or methylene groups. Nature of course has recognized this situation and for maximum reducing power employs *fatty acids,* molecules with the structure

$$^{-}OOC—(CH_2)_n—CH_3$$

Acids rather than fully reduced hydrocarbons are evidently preferred to achieve at least a nominal solubility in water.

On a weight basis, fatty acids store more than twice the energy of glucose:

Food Stuff	Heat of Combustion (kcal/gm)
Palmitic acid	9.3
Glucose	4.2
Proteins	5.6

Proteins also store slightly more energy than glucose, although they are rarely metabolized to completion (see Chapter 13). Thus palmitic acid (see Table 11-1 for structures of fatty acids), on these grounds, would be a better repository of reducing power. While fatty acids do provide an important medium of energy transfer and storage (as will be elaborated below), they possess a number of handicaps. Only molecules of 16 or 18 carbons have the proper solubility. Smaller units are too soluble and disrupt other structures (they are in fact detergents). Larger molecules are too insoluble. Within these limits fatty acids nevertheless play an important role in metabolic transactions, especially in storage as fat. In addition, when coupled to glycerol and combined to form triglycerides and other related molecules known as lipids, fatty acids engage in the indispensable and fascinating process of mem-

TABLE 11-1 Fatty Acids

Common Name	Carbon Atoms	Structure
Saturated Fatty Acids		
Lauric Acid	12	$CH_3(CH_2)_{10}COOH$
Myristic Acid	14	$CH_3(CH_2)_{12}COOH$
Palmitic Acid	16	$CH_3(CH_2)_{14}COOH$
Stearic Acid	18	$CH_3(CH_2)_{16}COOH$
Arachidic Acid	20	$CH_3(CH_2)_{18}COOH$
Unsaturated Fatty Acids		
Palmitoleic Acid	16	$CH_3(CH_2)_5CH{=}CH(CH_2)_7COOH$
Oleic Acid	18	$CH_3(CH_2)_7CH{=}CH(CH_2)_7COOH$
Linoleic Acid	18	$CH_3(CH_2)_4CH{=}CHCH_2CH{=}CH(CH_2)_7COOH$
Linolenic Acid	18	$CH_3CH_2CH{=}CHCH_2CH{=}CHCH_2CH{=}CH(CH_2)_7COOH$
Archidonic Acid	20	$CH_3(CH_2)_4CH{=}CHCH_2CH{=}CHCH_2CH{=}CHCH_2CH{=}CH(CH_2)_3COOH$

brane formation (see Chapter 12). Since fatty acids are such essential components in biochemistry, the synthesis and oxidation of fatty acids will be described, giving particular attention again to the energy accounting in the process.

Fatty Acid Metabolism

Degradation of fatty acids takes place in bites of two-carbon units, thereby paralleling the synthesis of fatty acids in steps of two-carbon additions (see below). This situation leads to a predominance of even-numbered fatty acids at all stages of metabolism. The oxidative breakdown of fatty acids occurs in a spiraling sequence of reactions, summarized in Fig. 11-13. A cycle of (1) activation, (2) dehydrogenation (to form a double bond), (3) hydration, (4) dehydrogenation (to form a keto group), and (5) thiolytic splitting (to cleave acetyl CoA and activate the carbon two positions down the chain) is repeated until the chain is fully degraded. Fatty acid oxidation occurs in the mitochondria and the molecules of acetyl CoA are then processed by the TCA cycle.

The synthesis of fatty acids is slightly more complicated. Although growth of fatty acid chains proceeds in two-carbon steps, additions take place via a three-carbon intermediate, malonyl CoA (see Fig. 11-14). The malonyl CoA is first transferred to an SH-containing acyl carrier protein (ACP). The malonyl—S-ACP then adds onto an acetyl, butyryl, or longer ACP with the release of CO_2. Thus each cycle involves reaction of a three-carbon malonyl derivative with an even-numbered-carbon acyl fatty acid to yield a growth of two

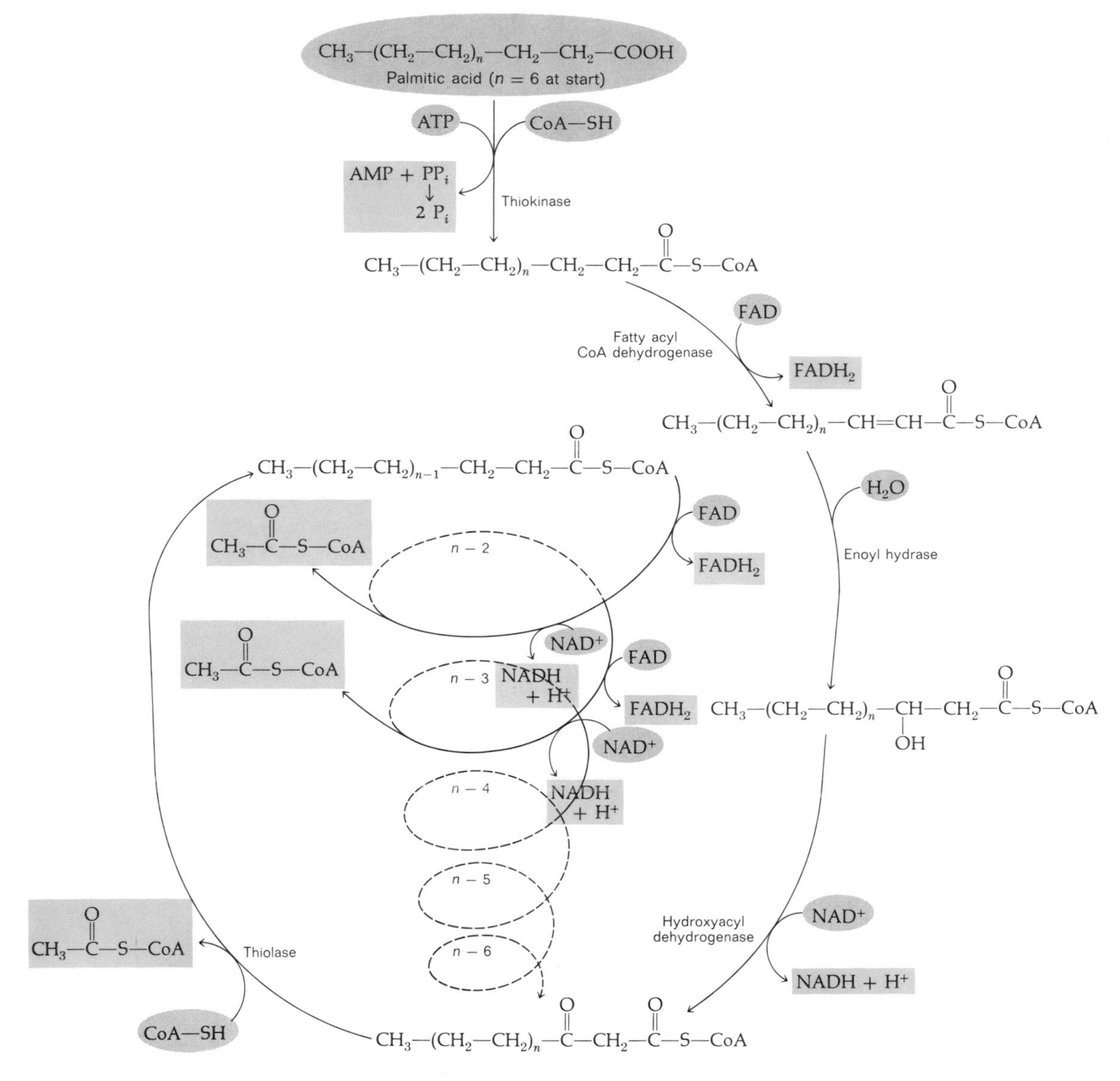

FIG. 11-13 *The Fatty Acid Oxidation Spiral.* Details of the reactions are shown for $n = 6$. Subsequent spirals (for $n - 1$, $n - 2$, etc.) are indicated with details omitted. For additional information, see M. Florkin and E. H. Stotz, eds., *Comprehensive Biochemistry,* vol. 18, American Elsevier Publishing Company, New York, 1967.

carbons and the release of CO_2. The enzymatic reactions of fatty acid synthesis are similar to those in degradation (see Table 11-2) and involve a large complex of enzymes, centered about the carrier protein. Somehow the enzyme complex senses when the cycle has been repeated the correct number of times to form the proper fatty acid, such as palmitic, no

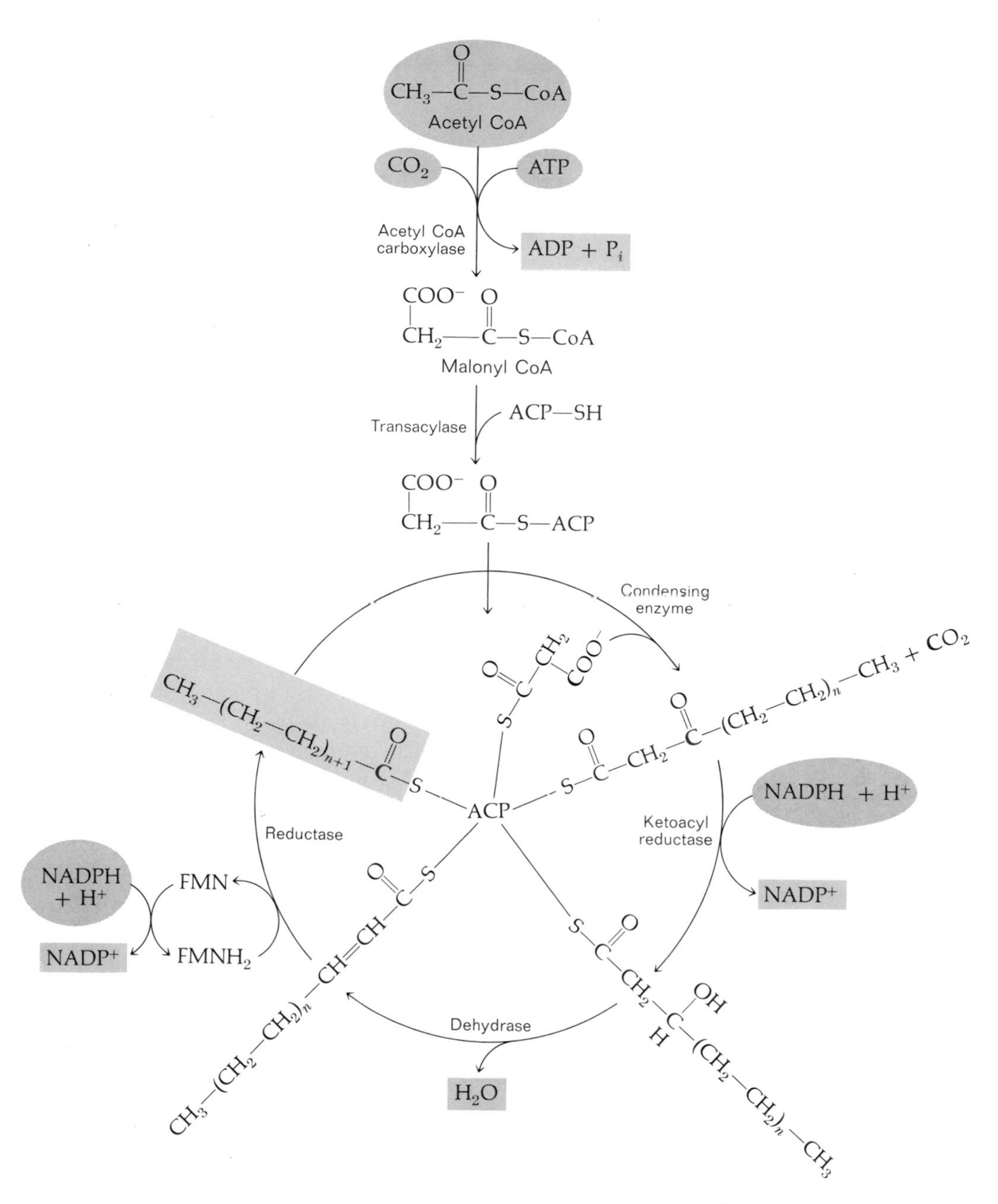

FIG. 11-14 *The Reactions of Fatty Acid Biosynthesis.* ACP stands for acyl carrier protein.

TABLE 11-2 Comparison of Fatty Acid Synthesis and Oxidation

		Synthesis	Oxidation
1	Site	Cytoplasm	Mitochondria
2	Activating Unit	CP—SH	CoA—SH
3	Electron Acceptor for HC=O / HCOH Reaction	$NADP^+$	NAD^+
4	Electron Acceptor for CH=CH / CH_2—CH_2 Reaction	$NADP^+$	FAD
5	Configuration of HCOH Intermediate	D	L
6	Minimum Carbon Unit in Reaction	3; Malonyl CoA	2; Acetyl CoA
7	Additional Coenzymes	Biotin	—

more additions are then made, and the palmityl—S-ACP bond is split to release the free fatty acid.

In plants the reduced NADP required for fatty acid synthesis is available from the reactions of photosynthesis. However, in other organisms a special pathway, the *phosphogluconate pathway* (Fig. 11-15), which resembles the Calvin cycle, provides the major source of reduced NADP for fatty acid synthesis and many other biosynthetic pathways (see Chapter 13). The pathway, discovered by Horecker and Racker, is also known as the hexose monophosphate shunt or the pentose phosphate pathway.

In addition to providing reducing power in the form of $NADPH + H^+$, the cycle provides five-carbon sugars for eventual modification and incorporation in nucleic acids. The discussion of $NADPH + H^+$ also raises a noteworthy point. The triphosphate form of pyridine nucleotide ($NADP^+$) is generally employed in biosynthetic reactions, while the diphosphate form (NAD^+) is almost invariably restricted to degradative, energy-yielding reactions. This distinction will be illustrated a number of times in the subsequent chapters.

Fatty acid metabolism can now be summarized in terms of the yield of ATP per carbon. In terms of ATP produced by a fatty acid, the degradation of palmitic acid is described

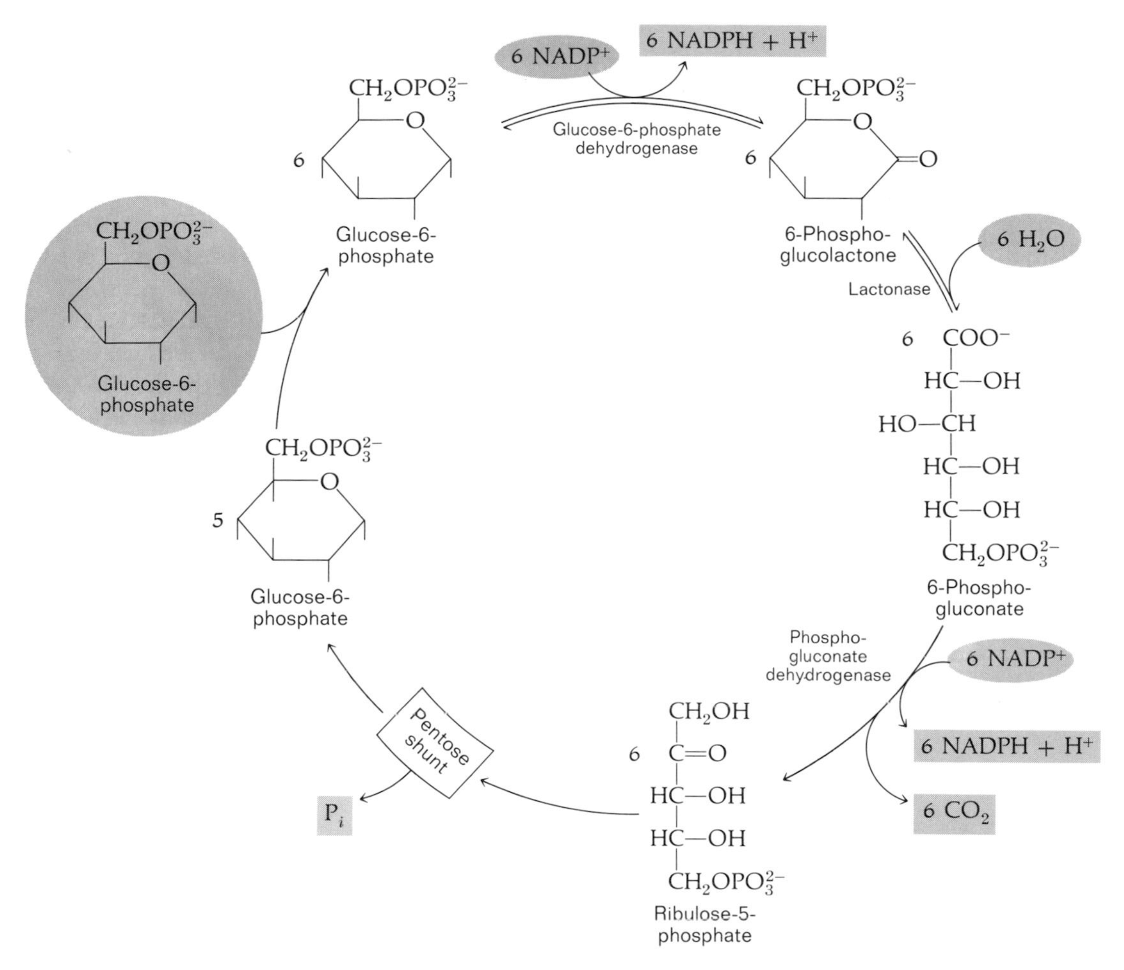

Sum: Glucose-6-phosphate + 12 $NADP^+$ + 6 $H_2O \longrightarrow$ 6 CO_2 + 12 (NADPH + H^+) + P_i

Pentose shunt:

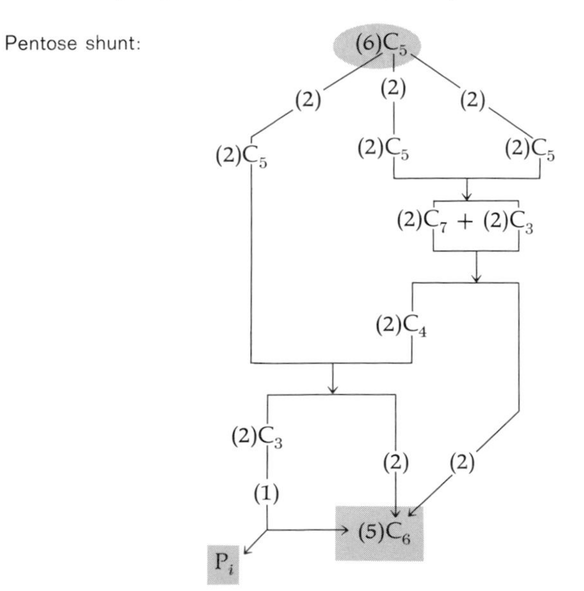

FIG. 11-15 *The Phosphogluconate Pathway.*

in terms of the yield of acetyl CoA:

$$\text{Palmityl CoA} + 7\ \text{CoA} + 7\ \text{FAD} + 7\ \text{NAD}^+ + 7\ \text{H}_2\text{O} \longrightarrow 8\ \text{acetyl CoA} + 7\ \text{FADH}_2 + 7\ \text{NADH} + 7\ \text{H}^+$$

ATP Formed	ATP
From $FADH_2$, 7 × 2 ATP per $FADH_2$	14
From NADH + H^+, 7 × 3 ATP per NADH + H^+	21
	35
	−2*
	33
From 8 Acetyl CoA, 3 NADH + H^+ per Acetyl CoA	72
1 $FADH_2$ per Acetyl CoA	16
1 GTP per Acetyl CoA	8
	96
Total	129

*Minus 2 for palmityl activation by CoA in which ATP is split to AMP and PP_i, with the PP_i further hydrolyzed to 2 P_i.

Thus about 8 ATP per carbon atom are generated by fatty acid oxidation, whereas only 6 ATP per carbon atom are generated from glucose.

The return on the energy invested in the biosynthesis of palmitic acid can also be calculated by considering the overall reaction:

$$8\ \text{Acetyl CoA} + 14\ (\text{NADPH} + \text{H}^+) + 7\ \text{ATP} \longrightarrow \textit{palmitate} + 8\ \text{CoA} + 14\ \text{NADP}^+ + 7\ \text{ADP} + 7\ \text{P}_i + 6\ \text{H}_2\text{O}$$

We find 7 ATP plus 14 (NADPH + H^+) consumed for each molecule of palmitate formed. Since each NADPH + H^+ is equivalent to 3 ATP, the cost of producing palmitate from 8 acetyl CoA is 49 ATP. The corresponding degradative reaction for acetyl CoA formation from palmitate (see preceding) yields 33 ATP. In this case the loss is 49 − 33 or 16 ATP per palmitate molecule (close to the 18 ATP value for glucose) and again represents the commission exacted by thermodynamics.

In summary, we have described many of the essential reactions of bioenergetics which fill in the details of the general scheme presented in Fig. 11-1. A large body of metabolic pathways was included to reveal the energy transactions in glucose biosynthesis and oxidation and a comparison with the fatty acid reactions. The mechanism of ATP synthesis in photophosphorylation and oxidative phosphorylation was presented only in bare outline, however. To see just how ATP is generated in the primary processes of metabolism we can now proceed to the chapter on electron flow and ATP synthesis.

PROBLEMS

1 What is the free-energy change expected for the redox reaction of plastoquinone and plastocyanine on the basis of the redox potentials indicated in Fig. 11-4?

2 Glucose is prepared with ^{14}C in the 6 position and metabolized by fermenting yeast. Will the ^{14}C appear in the ethanol produced?

3 If the glucose with ^{14}C in the 6 position is metabolized to pyruvate, which is used to form acetyl CoA, and the acetyl CoA enters the TCA cycle, in which positions of fumarate will the ^{14}C appear?

4 If the acetyl CoA derived from 6-^{14}C-glucose is used to form malonyl CoA, will the ^{14}C be incorporated into fatty acids?

5 How many ATP molecules (or equivalents) are produced in the processing of one acetyl CoA molecule in the TCA cycle?

Chapter 12 Frontispiece *Electron micrograph of a submitochondrial particle. The particle is capable of carrying out oxidative phosphorylation. The knobs have been identified as coupling factor F_1. Magnification 640,000×. Courtesy J. Telford.*

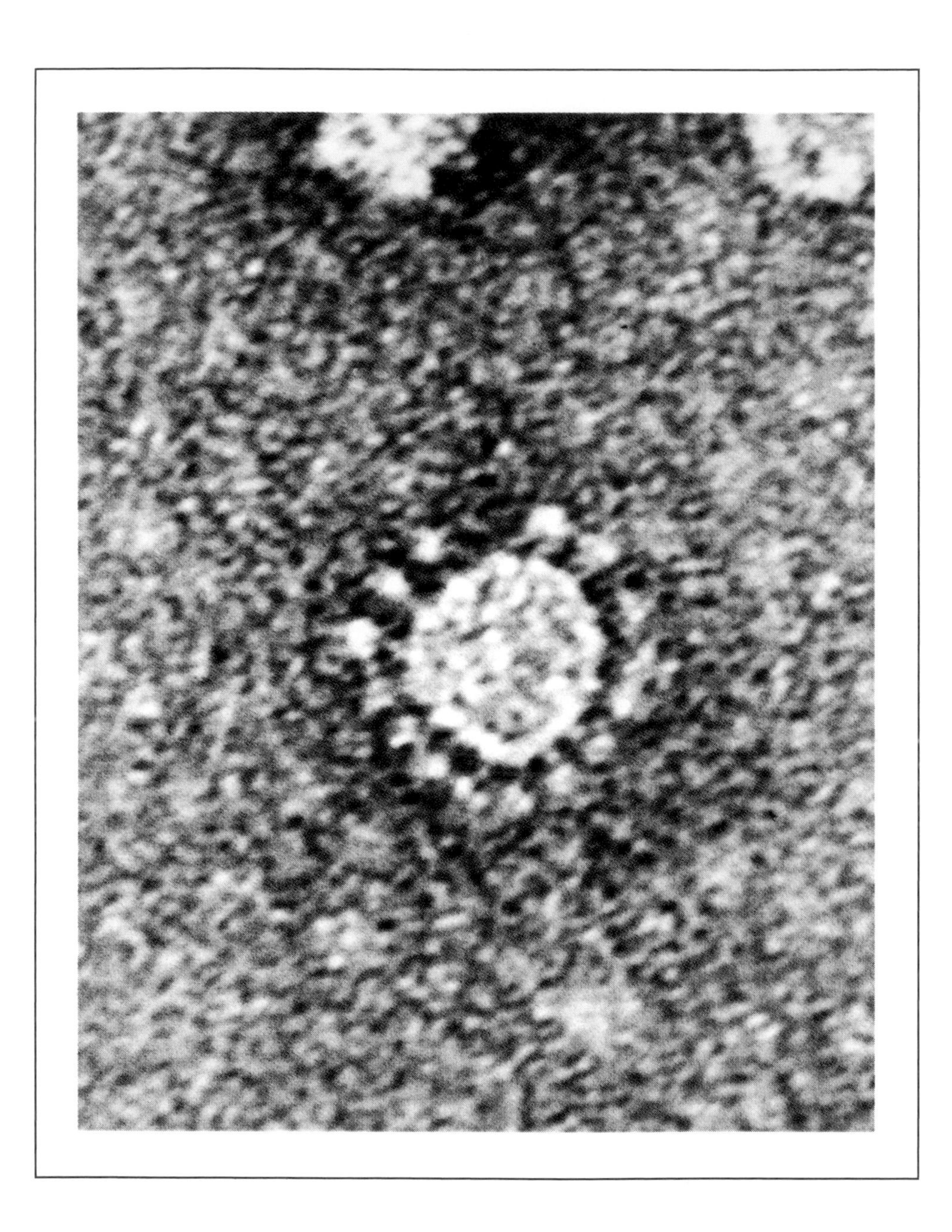

ELECTROCHEMICAL COUPLING

12

In the last chapter we approached metabolism as an accountant would, emphasizing the quantitative aspects of the subject and the numbers of ATP molecules consumed and produced at various stages. We paid little attention to the detailed structural aspects of the machinery that carries out these various reactions. In this chapter we will approach metabolism more as an engineer, emphasizing several of the important structural components of the cell involved in metabolism and some of the important physical-chemical principles involved. This approach leads to a close examination of the cellular organelles known as mitochondria and chloroplasts. In plants, all of the reactions of photosynthesis occur in the chloroplasts. For both plants and animals, mitochondria are responsible for the reactions of oxidative phosphorylation. They also contain the enzymes of the TCA cycle and fatty acid oxidation.

Origin of Mitochondria and Chloroplasts

Some idea of the size and complexity of the organelles of bioenergetics—mitochondria and chloroplasts—is achieved by returning to our first comparison of eucaryotic and procaryotic cells. Eucaryotic cells, we noted in Chapter 1, are very large, complex structures containing a nucleus and many other subcellular structures. We also described the much smaller procaryotic organisms such as bacteria, which are reasonably homogeneous cells composed of a relatively continuous cytoplasmic interior. With this comparison in mind, we can note that mitochondria and chloroplasts are roughly the same size as typical bacterial cells. Dimensions in the range of microns in length are generally encountered. In fact the striking hypothesis has been advanced that mitochondria may represent bacterialike cells that at some early stage in evolution were captured by the "host" cell and incorporated to perform the reactions of oxidative phosphorylation. As evolution proceeded, these organelles included in the host cell became specialized and lost many of their independent functions. In support of this view is the observation that bacteria do resemble mitochondria. Bacteria also carry out oxidative phosphorylation, using enzymes and chemical systems very similar to those of mitochondria. The evolutionary hypothesis also extends to chloroplasts, which resemble photosynthetic bacteria in many of their properties.

This evolutionary analysis has additional support from the observations that chloroplasts and mitochondria actually contain many of their own elements of a genetic and a protein-synthesizing system. Mitochondria and chloroplasts contain DNA, generally in a circular form, as is encountered in bacteria. They also contain ribosomes and other factors for protein synthesis, such as the initiating systems and RNA polymerase. Especially interesting is the observation that the protein-synthesizing systems of the chloroplasts and mitochondria resemble the procaryotic type rather than the type found in the cytoplasm of eucaryotic cells. The initiating system for protein synthesis involves formylmethionine, not methionine as is encountered in eucaryotic cells. Similarities in membrane composition are also found for bacteria and the organelles of bioenergetics. Thus a considerable body of evidence points to the origin of chloroplasts and mitochondria from a line of evolution that includes bacteria.

It should be emphasized that while mitochondria and chloroplasts contain their own DNA and protein-synthesizing systems, they are not truly independent organisms. Indeed, the majority of the proteins found within chloroplasts and mitochondria are apparently coded for by genes in the eucaryotic nuclei. Most, if not all, of these nuclei-coded proteins are synthesized in the cytoplasm and imported into the organelle. While some proteins are synthesized in the organelle, they too are regulated by the host cell. In fact, just how the interactions between endogenous organelle factors and the external components of the cell interact to generate chloroplasts and mitochondria is one of the exciting subjects of contemporary research in biochemistry.

The evolutionary scheme proposed, involving the capture of small bacteria for use in the eucaryotic cells, also poses many fascinating questions. What was the form of the cell that was large enough to incorporate bacteria if the cell itself did not contain the complex machinery needed for oxidative phosphorylation and photosynthesis? Was it a remnant of some earlier phase in cellular evolution, possibly an anaerobic phase in the history of the earth? Questions of this nature are certain to occupy biologists for some time to come.

Membrane Structure

The most striking feature of mitochondria and chloroplasts (Fig. 12-1) is their extensive network of membranes. Not only are many of the components of oxidative phosphorylation and photophosphorylation attached to the membrane, but these membranes have evolved an intricate, convoluted structure to maximize the available surface area. Mitochondria have clearly convoluted surfaces, with invaginations of the inner face of the membrane known as *cristae*. Two distinct membranes can be recognized, an inner and an outer membrane. Chloroplasts contain extensive networks of disklike membranes, stacked together in structures known as *thylakoids*. In both cases the emphasis on extending the area of membrane available suggests a central functional role for the membrane and just what this role may be will be discussed further along in this chapter.

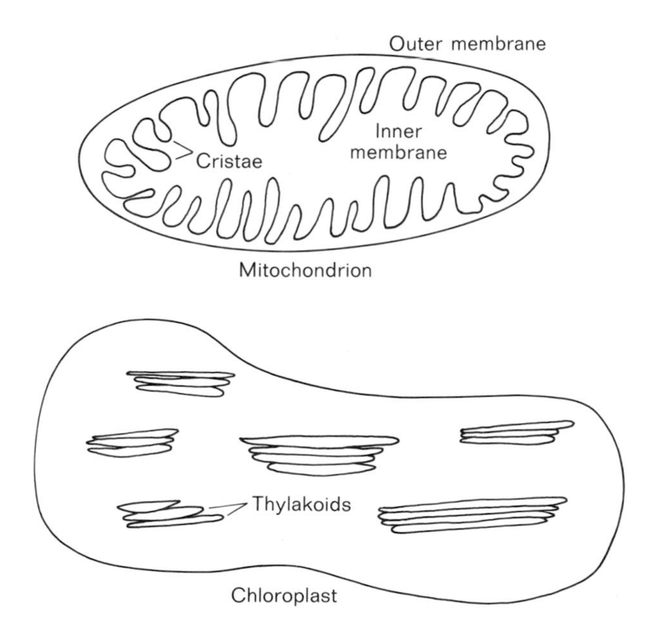

FIG. 12-1 *Mitochondrion and Chloroplast.* The inner volume of the mitochondrion is known as the matrix space. See also the frontispieces of Chapters 11, 12, and 15.

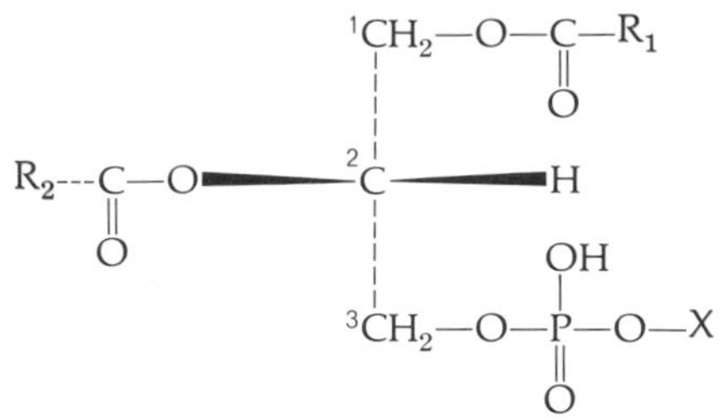

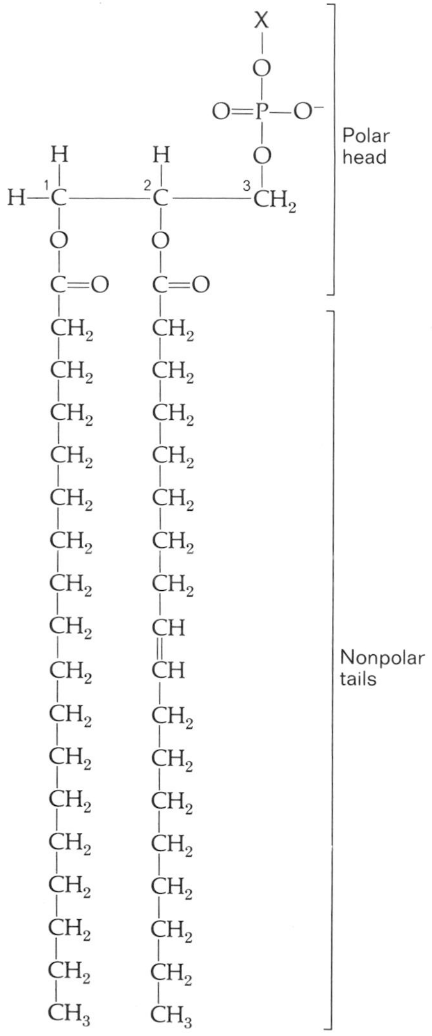

FIG. 12-2 *Phospholipids.* R_1 and R_2 are fatty acids; X is one of a number of alcoholic groups in Table 12-1.

Membranes are an essential part of most biological systems. Their use in bioenergetics is just one example, although perhaps the most thoroughly studied case and the one that will receive most of our attention. However, we shall also note that membranes are important in all cells, simply as a permeability barrier to select those molecules that may enter and leave the cell. Membranes also enjoy a special function in nervous tissue in which their excitability (that is, their ability to change their permeability properties rapidly, particularly toward sodium) can be triggered by a chemical or electrical stimulus. Virtually all these functions, whether in bioenergetics or in nerve-impulse transmission, involve a similar basic membrane structure. One of the central ingredients in the structures is a class of molecules known as *phospholipids.*

Phospholipids are composed of fatty acids linked to a glycerol backbone. Generally two fatty acids are attached to the glycerol with the third position available for one of the specific compounds encountered in phospholipids, such as choline or ethanolamine. These small molecules are frequently bridged to the glycerol through a phosphate linkage. Several typical structures are illustrated in Fig. 12-2 and Table 12-1. Membranes invariably contain appreciable amounts of these phospholipids, particularly phosphatidyl ethanolamine and phosphatidyl choline, as well as a moderate amount of the

TABLE 12-1 Some Alcohol Components of Phospholipids

Phospholipid	Alcohol Component
Phosphatidyl Ethanolamine	$HOCH_2CH_2NH_2$
Phosphatidyl Choline	$HOCH_2CH_2N^+(CH_3)_3$
Phosphatidyl Glycerol	$HOCH_2CHOHCH_2OH$
Phosphatidyl Serine	$HOCH_2CHNH_2COOH$
Phosphatidyl Inositol	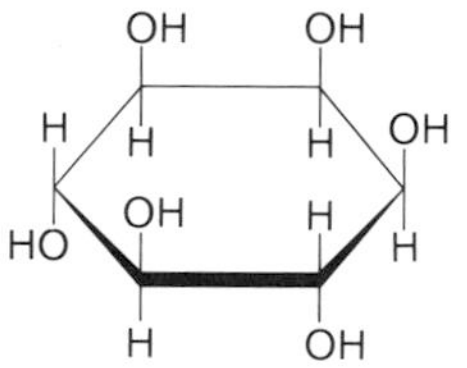 (inositol ring structure: OH, OH, H, H, H, OH, HO, OH, H, H, H, OH)
Cardiolipin	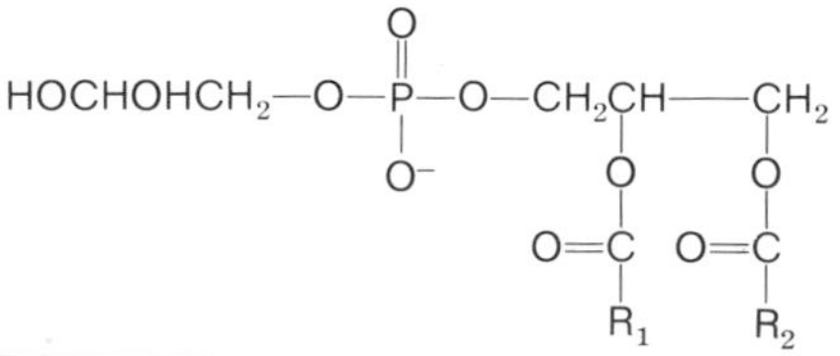 $HOCHOHCH_2$—O—P(=O)(O⁻)—O—CH_2CH—CH_2 with O—C(=O)—R_1 and O—C(=O)—R_2

steroid cholesterol (see Fig. 12-3), a relative of many hormones and vitamins (some are shown in Fig. 12-4). The fascinating sequence of biosynthetic reactions for cholesterol is presented in Chapter 14.

The various lipid molecules, when assembled into a membrane structure, are generally believed to possess a bilayer form (Fig. 12-5). The bilayer is composed of a sandwich of phospholipids with the charged phosphate groups on the outside and the oily fatty acid chains on the inside. This bilayer will form a continuous structure, 100 Å or less in thickness, which may be interpenetrated by the flat cholesterol molecule, endowing a certain element of rigidity. Specific protein molecules may also be included in the bilayer. Two distinct roles can be visualized for the protein: a *structural* role in which a protein contributes to the structural integrity of the membrane itself, and a *functional* role in which the protein performs a catalytic function conveniently localized at the membrane surface. In fact proteins associated with membranes may contribute to an important degree in both areas, structural and functional. While it is difficult to characterize proteins in terms of both of these functions, we will see below that, in the small number of cases that have been studied, proteins associated with a membrane may undergo a modification of function when they are removed from the membrane. Proteins also play an important role in mediating specific permeability of small molecules.

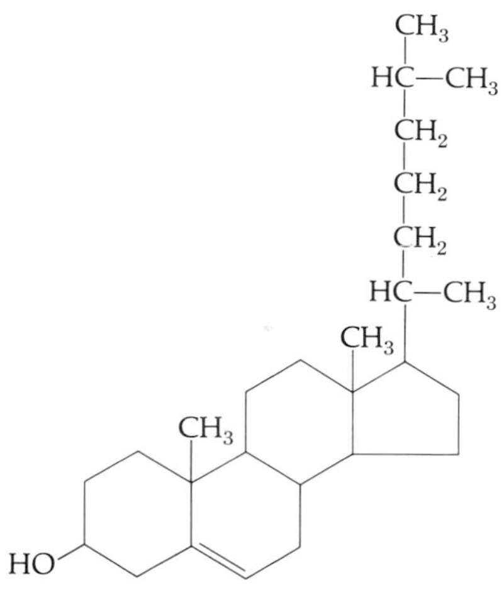

FIG. 12-3 *Cholesterol.*

Membranes and the Electron-transport Chain

In Chapter 11 we discussed the components of oxidative phosphorylation and photophosphorylation in fairly abstract terms. We can point out here that virtually all of the cyto-

FIG. 12-4 *Steroid Derivatives.*

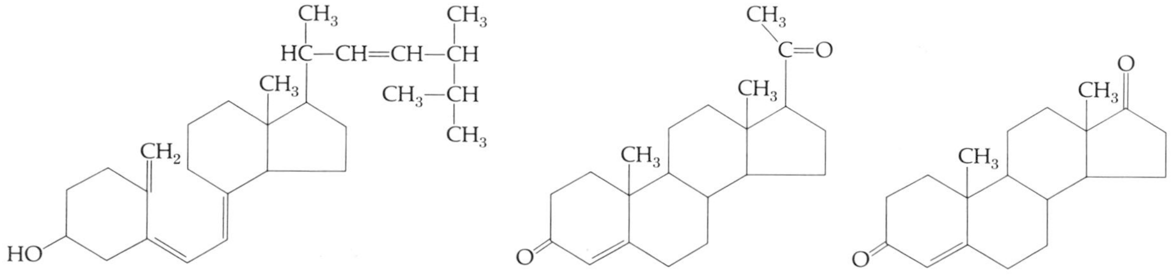

Vitamin D_2 (calciferol)

Progesterone (female sex hormone)

Testosterone (male sex hormone)

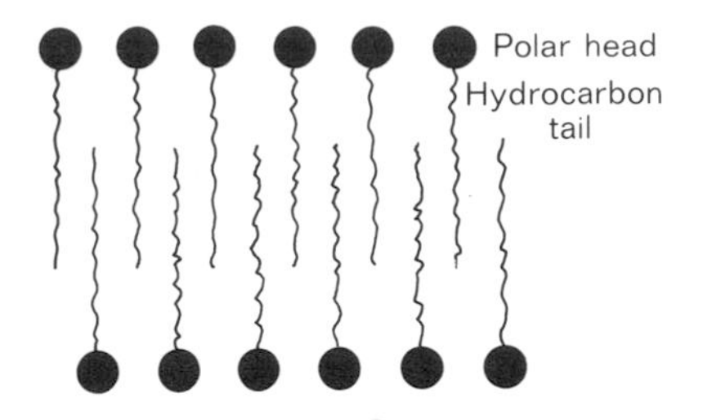

FIG. 12-5 *Postulated Bilayer-membrane Model.*

chromes and other proteins and chromophores involved in these processes are closely associated with their respective chloroplast or mitochondrial membranes. Furthermore, just as we spoke of a similarity in the origin of chloroplasts and mitochondria from photosynthetic and nonphotosynthetic bacteria, we can carry the evolutionary consideration further and point out the similarities between the processes of photophosphorylation and oxidative phosphorylation. Both involve electrons cascading along a redox chain. In oxidative phosphorylation the redox stream is steadily downhill, as highly reduced NADH + H^+ gives its electron to a chain of molecules that ultimately donates the electron to oxygen. In the process three molecules of ATP are formed from ADP and phosphate. A similar stream occurs in photophosphorylation, beginning in this case with the origin of electrons in water. However, this stream is not steadily downhill, and as we noted in Chapter 11, the chain of events involves a number of ups and downs, with electrons raised to highly reducing levels by light in photosystems I and II.

On the basis of what we have noted so far, the two processes are only vaguely similar, involving, for example, heme-containing cytochromes. However, the similarities become more apparent when we consider phosphorylation in greater detail. This consideration involves the essential *coupling factors*. The flow of electrons in mitochondria is coupled to the synthesis of ATP from ADP and phosphate at each of three sites in a reaction mediated by proteins known as coupling factors. One of these factors (F_1) has been isolated from mitochondria and studied extensively in the laboratory of Efraim Racker. A similar factor (CF_1) has been isolated from chloroplasts. For both mitochondria and chloroplasts the appropriate coupling factor restores oxidative phosphorylation or photophosphorylation when returned to the organelle from which the factor had been carefully removed (see Fig. 12-6). Thus a clear-cut requirement for these factors can be demonstrated. Even more striking is the similarity in structure of F_1 and CF_1. Both are large molecules with molecular weights of approximately 280,000 and with strong similarities in subunit structures. For example, both F_1 and CF_1 display a phenomenon known as cold lability. When these proteins are incubated in the cold (at about 4°C), their structures undergo a significant transition, and they dissociate into their component subunits with a loss of coupling activity. At room temperature (25°C) the factors are quite stable (see Fig. 10-6).

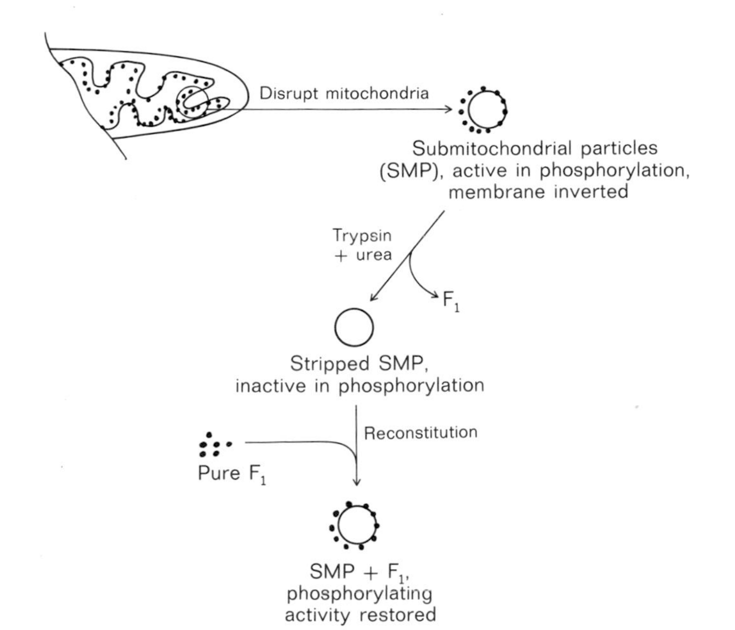

FIG. 12-6 *Mitochondria Coupling-factor Experiment.* Based on data from E. Racker, *Scientific American,* February 1968. An electron micrograph of an actual submitochondrial particle is shown in the Chapter 12 frontispiece.

While both factors F_1 and CF_1 have been implicated in the formation of ATP, they were actually isolated by following an enzymatic activity for the breakdown of ATP. Known as an ATPase, each of these enzymes will catalyze the formation of ADP and phosphate from ATP. While this enzymatic reaction greatly facilitates isolation by providing a convenient assay, the exact role of this enzymatic activity is uncertain when the protein is acting as a coupling factor in phosphorylation while it is affixed to the membrane. Preparations of F_1 usually contain a small molecule known as F_1 inhibitor, which markedly depresses ATPase activity of F_1; this inhibitor may therefore serve to regulate the ATPase activity during phosphorylation. A similar inhibitor is associated with CF_1 from chloroplasts.

Formulation of the Coupling Problem

We now come to the major question of photophosphorylation and oxidative phosphorylation. How is the electron flow down the chain of electron carriers translated into the formation of ATP from ADP and P_i? In other words, how is the electron flow "coded" into a chemical synthesis? And how is this operation performed by the coupling factors? When we considered protein synthesis we sought to discover what arrange-

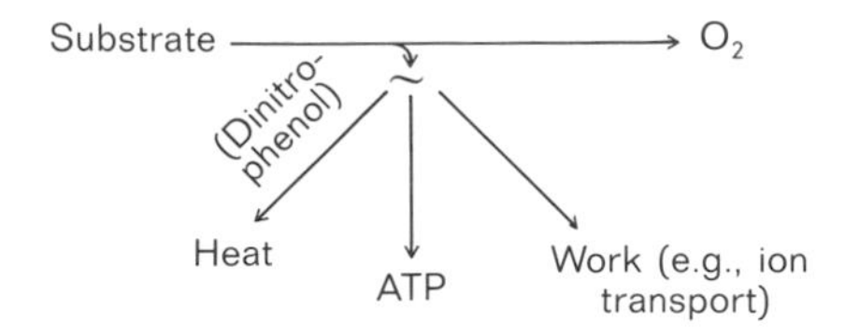

FIG. 12-7 *Energy Coupling.*

ment of nucleotides corresponded to what arrangement of amino acids. Thus we were seeking informational aspects of a code. Now we want to know what form of energy transformation is involved in connecting an electrical flow to a chemical synthesis. Here we are seeking not the information content of the code, but the nature of the energy involved.

It must be emphasized from the start that the final answer is not yet known. Work in this area is still in a speculative stage. However, the mechanisms are so basic and important that we will give attention to the current level of knowledge and possible solutions. A number of clues can be brought to bear on the question and some of these are summarized in Fig. 12-7. If we write the electrical form as a line from substrate to oxygen and the formation of ATP as a perpendicular line from that flow, we can represent this unknown encoded form of energy by a squiggle (~)—the symbol for the high-energy character. Some transitory form of high energy must exist and be captured in the formation of ATP. Just what this form of energy is, how it is generated, how it is coupled to ATP synthesis—these are the points we are seeking to resolve.

We also indicate in the scheme that the transitory high-energy storage can be dissipated as heat through the addition of an *uncoupling agent* such as dinitrophenol (DNP). In the presence of an uncoupling agent, electron transport occurs, but no phosphorylation takes place and P:O ratios drop from 3 to close to 0. The high-energy state can also be expressed directly as work in ion transport. One important element in the coupling mechanism came from experiments on the transport of ions by mitochondria and chloroplasts. It was emphasized by Peter Mitchell that protons are expelled by mitochondria during oxidative phosphorylation. In chloroplasts, protons are actually taken up during photophosphorylation. The fundamental process may be the same, however, since protons flow in during electron transport with fragmented but functional particles of both mitochondria and chloroplasts. The difference for the intact organelles may simply reflect difference in the cristae and thylakoid architecture.

Experiments by Andre Jagendorf and his colleagues took the observations on proton flow one step further and attempted to drive ATP formation with pH changes. Chloroplasts were incubated in the dark at pH 4.5 and then quickly returned to an alkaline solution. As hoped, ATP was formed by the "pH jump." Presumably the acid bath raised the level of protons in the chloroplasts. When they were returned to a neutral or slightly alkaline solution, a proton gradient was formed (more protons on the inside, less on the outside) and ATP synthesis was driven. Observations of this type have led to a tentative explanation of electrochemical coupling that directly involves a flux of protons and also explains the obvious requirement for a membrane in this system. The view has been advocated by Mitchell and is known as the *chemiosmotic hypothesis.*

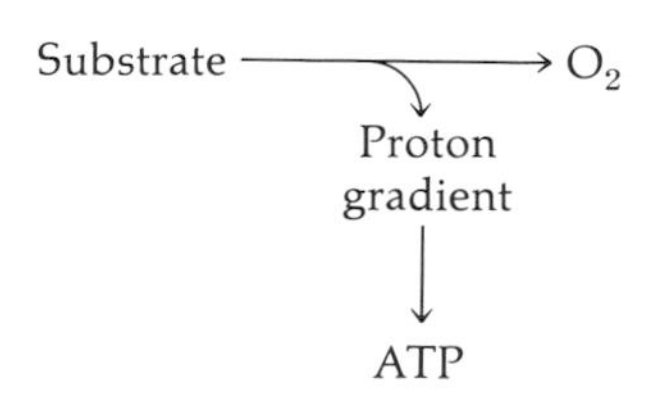

FIG. 12-8 *Intervention of Proton Gradient in Energy Coupling.*

Chemiosmotic Coupling

The missing link in the electrochemical coupling of mitochondria and chloroplasts may be the formation of a proton gradient across the membranes of these organelles. In this view, as summarized in Fig. 12-8, ion transport is not simply one of a number of by-products of electron transport. Rather, the proton gradient is itself the intermediate storage mechanism and is generated directly by the flow of electrons. Thus we can visualize the scheme for oxidative phosphorylation depicted in Fig. 12-9 in which the primary electron donor, some reducing component such as NADH + H^+, donates its electrons and protons to the respiratory chain. The respiratory chain then passes along the electrons while extruding the protons (for mitochondria). The excess protons on the outside thus generate a force called the *proton motive force.* This force tends to equalize the pH difference between the inside and the outside of the mitochondrion and the difference can then be released through some mechanism, evidently involving the coupling factor, which harnesses the reentry of protons to a coupling to the synthesis of ATP. The coupling of proton movement and ATP formation is thus tentatively assigned to coupling factors such as F_1 and CF_1.

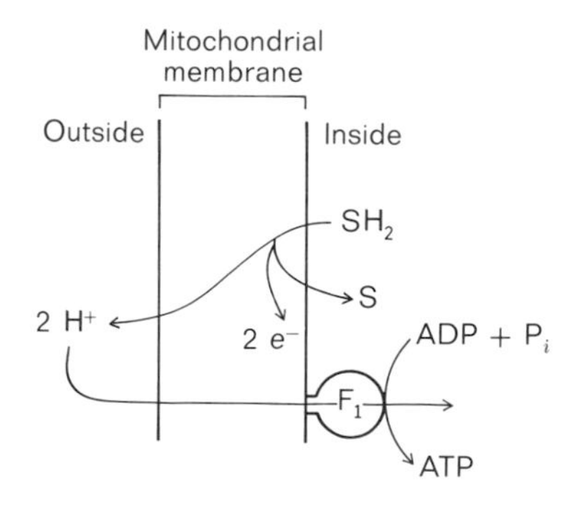

FIG. 12-9 *Principles of Chemiosmotic Coupling.* SH_2 refers to a reduced substrate such as NADH + H^+. For additional details, see P. Mitchell, *Biological Reviews,* **41:** 445 (1965).

Alternative Coupling Hypothesis

Chemiosmotic coupling is an ingenious mechanism for relating oxidative phosphorylation to the presence of the membrane. However, alternative mechanisms involving strictly soluble chemical intermediates cannot be excluded. Such mechanisms could resemble the reaction catalyzed by glyceraldehyde-3-phosphate to yield a high-energy acyl phosphate (Fig. 11-9) with a subsequent transfer to ADP to yield ATP.

This process, known as substrate-level phosphorylation, is discussed in more mechanistic detail in Chapter 14. Such a reaction system could in principle be active in a membrane-free mixture, although efforts to demonstrate oxidative phosphorylation without membrane fractions have been unsuccessful so far. Thus the integrity of the membrane appears to be a strong requirement for oxidative phosphorylation. Recent experiments on the reconstitution of oxidative phosphorylation from purified compounds by Racker and colleagues, for example, display an absolute requirement for the membrane vesicles. The chemiosmotic hypothesis relies on a vectorial construction of the membrane components to translocate protons, and studies of the location of various components in the oxidation chain reveal precise positions for many on one side of the membrane or the other.

For the passage of one pair of electrons down the respiratory chain, beginning with NADH + H^+, three high-energy phosphate bonds, or ATP, are formed. This corresponds to a P:O ratio of 3:1 (three high-energy phosphates P per oxygen O consumed). At the same time, experiments designed to measure the proton flux during electron transport indicate that six protons are transferred for each molecule of NADH oxidized. This observation agrees with the prediction that two protons are translocated for each ATP synthesized. However, some workers have questioned whether the size of the proton gradient is actually large enough to account for the synthesis of three molecules of ATP. Nevertheless, the stoichiometry of proton flux and the requirement for the integrity of the membrane, as well as the location of key factors such as F_1 and CF_1 on the membrane surface, suggest that any comprehensive theory of electrochemical coupling is likely to retain features of the chemiosmotic hypothesis and transitory coding of chemical energy in the form of a proton gradient.

Microsomal Electron Transport

The *microsomes* of liver and certain other eucaryotic cells also contain electron-transport systems. The microsomes are distinct from mitochondria in that they oxidize substrates directly by insertion of molecular oxygen. In some cases (oxygenases) both atoms of O_2 are inserted, while in other cases (hydroxylases or mixed-function oxidases) only one of the two atoms of O_2 is inserted in the substrate. [Examples of both types of reaction are found in the pathway for phenylalanine oxidation described in Chapter 13 on metabolism of nitrogen-containing compounds (see Fig. 13-7).] The microsomes, like mitochondria, are a membranous system, although the microsomes are part of the extensive interior cell network known as the *endoplasmic reticulum* (see Fig. 1-2). The microsomes include electron-transport chains with flavoproteins and cytochromes distinct from those found in mitochondria. Also, the microsome chains are shorter than their mitochondria analogs and are nonphosphorylating. In certain cases a cytochrome, P_{450}, with an absorption peak at 450 nm is a characteristic component. In addition to their role in amino acid degradation, the microsomes play an important part in drug metabolism, detoxification, and steroid metabolism, especially in liver.

Other Forms of Chemical Coupling

In our discussion of oxidative phosphorylation we described how electron flow could lead directly to a manifestation of work in terms of the production of a proton gradient. However, most forms of work in biochemical systems are directly mediated by ATP. The proton gradient represents a more elementary form of energy storage, for the initial production of ATP. Once ATP is available it can be used directly in coupled chemical reactions, as illustrated in the various pathways described in Chapter 11. In addition, one can delineate at least three other examples of biochemical work in which ATP or another high-energy phosphate is employed: transport processes, bioluminescence, and muscle contraction.

Transport of small molecules invariably involves a membrane of the general type described for mitochondria. The membrane per se is impermeable to the small molecule in question, such as a sugar, an amino acid, or, particularly for nerve cells, to an important ion, such as sodium. The first criterion for transport is a specific carrier. Each small molecule must have a receptor associated with the membrane, with the stereospecificity to recognize and bind the small molecule. If only small amounts of the molecule are to be transported, passive diffusion may be enough to drive an adequate number of molecules from a region of high concentration on the outside of the cell to the inside of the cell. However, in many cells active transport is needed, and the small molecule must be concentrated or expelled against a diffusion gradient. In this case energy is required, and generally ATP or another

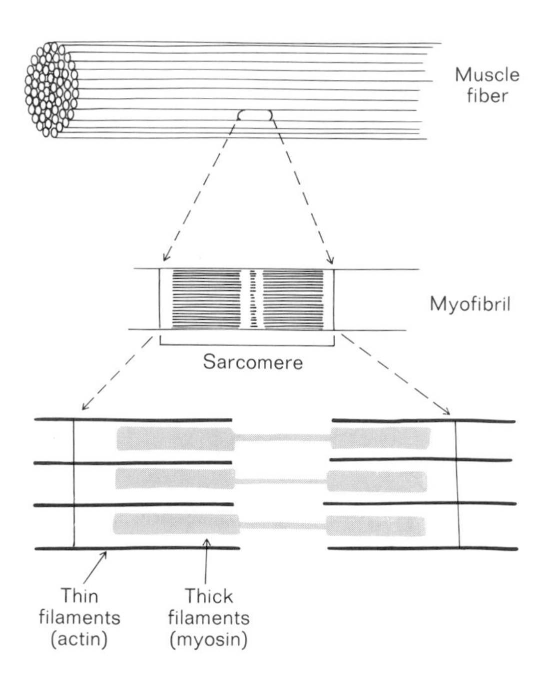

FIG. 12-13 *Muscle Structure.* For additional details, see H. E. Huxley, *Science,* **164:** 1356 (1969).

arranged in a globular "head" region as well as a long ropelike "tail" region. The helical ropelike tail section of myosin provides the major structural component for the thick filament of the muscle. A long helical structure is also found in actin-containing thin filaments (see Fig. 12-15). Here we have examples of proteins serving a structural as well as a catalytic function. Another important structural protein, also with a helical architecture, is the three-stranded collagen molecule found in connective tissue. These structural uses of proteins along with involvement in other complex aggregates such as ribosomes or membranes represent an important aspect of

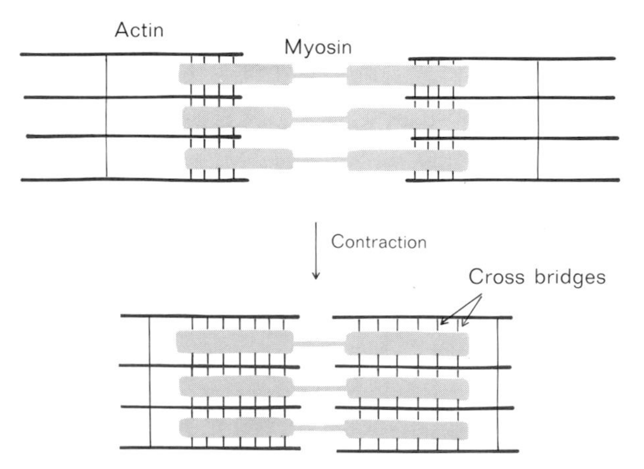

FIG. 12-14 *Sliding Filaments.*

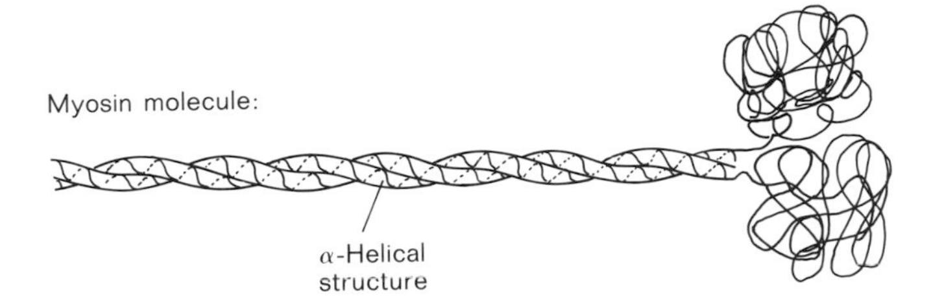

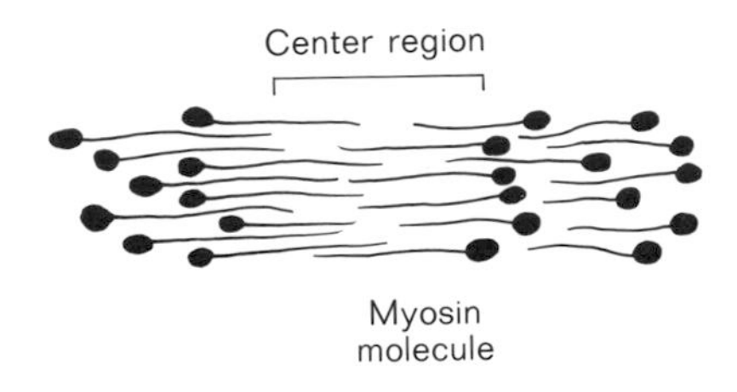

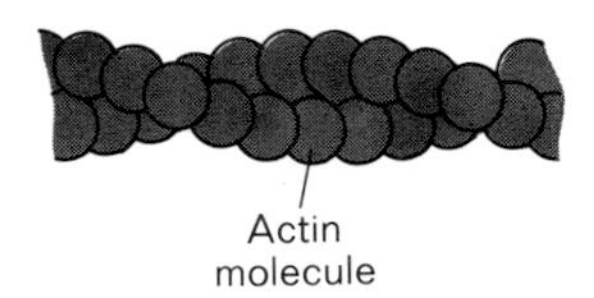

FIG. 12-15 *Muscle Proteins.* Myosin (MW about 0.5 million) contains two main polypeptide chains that form a long helical tail (about 1.5 μ in length) with a double globular head. These myosin molecules associate laterally, with heads pointing away from center, to form thick filaments. Thin filaments are composed of double helical structures made up of globular actin units (MW about 15,000). Cross bridges may be formed by the globular head of myosin, and other, smaller polypeptide chains associated with myosin may be involved.

protein action that supplements the enzymatic and regulatory roles emphasized in Chapters 7 to 10.

PROBLEMS

1 Proton flux may be closely coupled to phosphorylation in mitochondria and chloroplasts. During phosphorylation do protons tend to accumulate on the inside or outside in (a) chloroplasts, (b) mitochondria, (c) submitochondrial particles?

2 Passage of one pair of electrons from NADH + H^+ along the respiratory chain gives phosphorylation with a P:O ratio of 3:1 and translocation of six protons. What P:O ratio and number of translocated protons would be expected for passage of a pair of electrons from $FADH_2$?

Chapter 13 Frontispiece *Electron micrograph of a section of guinea pig pancreas cell showing lysosome (dark vesicular structure). The lysosomes are the sites of the hydrolytic enzymes secreted from the cells in an inactive form as zymogen granules. Magnification 57,000×. Courtesy G. E. Palade.*

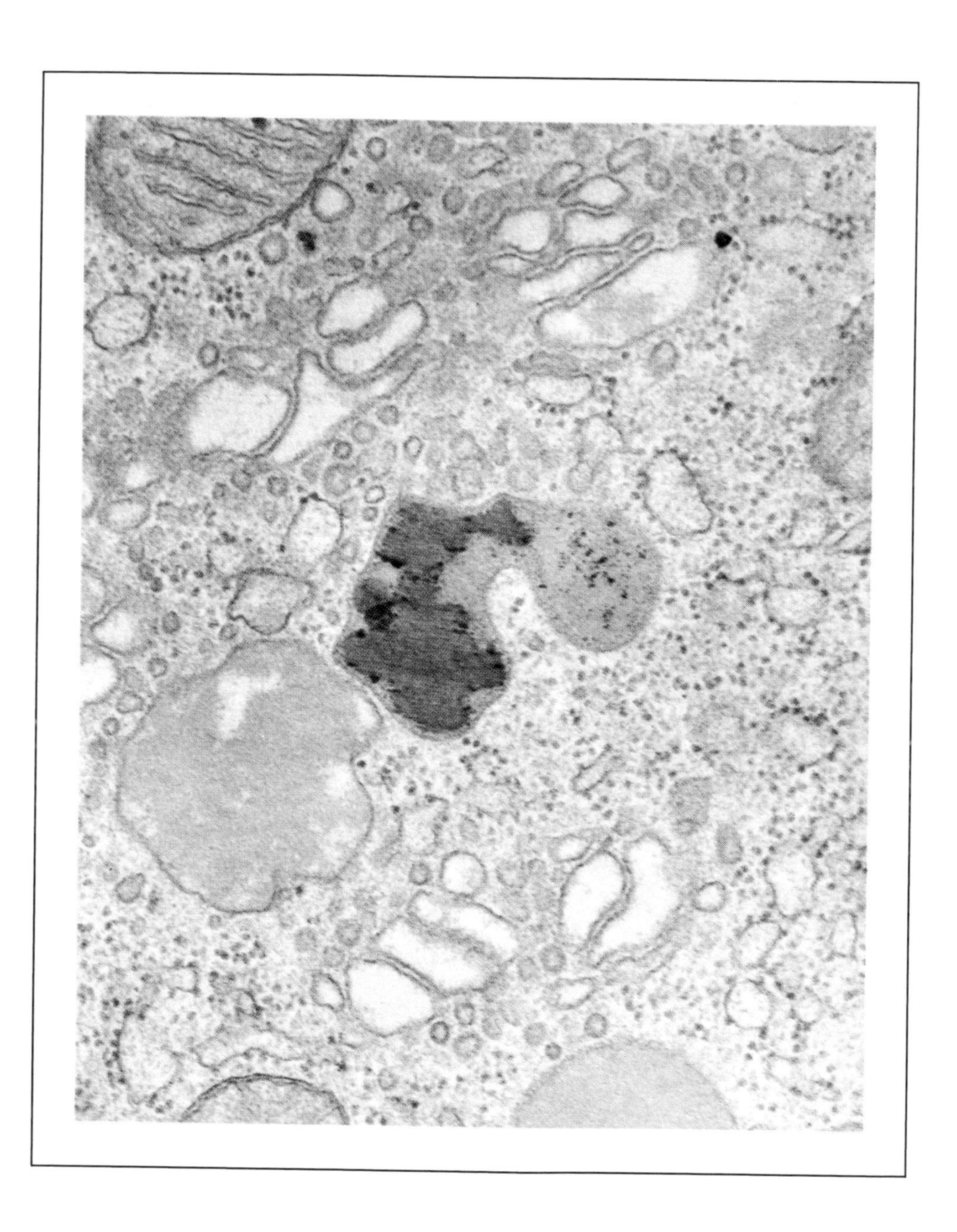

METABOLISM OF NITROGEN-CONTAINING COMPOUNDS

13

To paraphrase an old saying, "Man cannot live by carbohydrates alone." Other types of molecules, particularly nitrogen-containing molecules, play a very essential role in biochemistry. From our earlier discussions of nucleic acids and proteins (Part I) the importance of these nitrogen-containing molecules is evident. In addition many of the coenzymes, which we will consider in the next chapter, contain nitrogen. The widespread use of nitrogen in biochemical substances stems from a number of its interesting chemical properties. Nitrogen readily enters into bonds with partial double-bond character, lending a degree of planarity to biochemical structures. The value of such planarity was evident in the structure of the nucleotide bases of DNA, particularly in their fundamental stacking reaction (Chapter 2). In addition planarity in the peptide bond exerts an important restraint on the conformations accessible to polypeptide structures (Chapter 8). Nitrogen compounds, particularly in the amine form, also possess a high degree of reactivity, essential for the formation of peptide bonds and other covalent linkages in metabolic reactions. Another example of a reaction involving nitrogen is the acid-base catalysis involving the imidazole side chains in the mechanism of ribonuclease (Chapter 9). Thus while biochemistry is often viewed as an area dominated by carbon chemistry, we see that the reactions and structures involving nitrogen also play a crucial role.

The Nitrogen Cycle

Just as carbon passes through a cycle of oxidation reduction—CO_2 to carbohydrates in plants and back to CO_2 in animals—compounds of nitrogen can be reduced to store energy and later oxidized to provide metabolic power. For

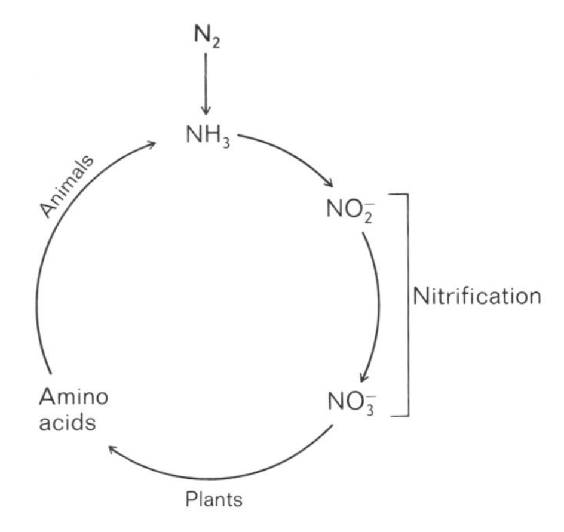

FIG. 13-1 *The Nitrogen Cycle.*

a certain class of living organisms, these oxidation-reduction reactions of nitrogen are in fact the main sustenance of life. In addition, as a by-product of their oxidation-reduction reactions with nitrogen, these organisms provide plants and animals with forms of nitrogen in the source most convenient to them. Some of these various interactions are summarized in Fig. 13-1, a simple view of the global nitrogen cycle on this planet. The cycle begins with the fixation of atmospheric nitrogen into a reduced form, ammonia.

Nitrogen fixation is a complicated process carried out by legumes in symbiosis with certain bacteria in their roots, as well as by other plants and some microbes such as blue-green algae. Nitrogen fixation occurs in the presence of an electron donor and energy provided by ATP. The process involves a highly reduced intermediate, the reduced form of ferredoxin, since molecular nitrogen (N_2) is a stable material that can be reduced only in the presence of a strong reducing agent. Ammonia generally finds its way into the soil, where another large class of organisms, soil bacteria, oxidize the ammonia, first to nitrite (NO_2^-) and then to nitrate (NO_3^-). This oxidation process is an important energy-yielding set of reactions for these microorganisms, just as energy is extracted from the oxidation of carbohydrates and fats by animal cells.

In general plants prefer to assimilate their nitrogen in the form of nitrate. Thus the product of nitrification, NO_3^-, becomes the nitrogen source for the bulk of plant life on this planet. Plants then reduce the nitrate back to nitrite and on to ammonia, which is then incorporated into amino acids. Some of the steps in this incorporation process will be considered shortly. Plant materials are then eaten by animals, and the amino acids from these plant sources are absorbed and incorporated into the proteins of animal tissues. Proteins are eventually broken down as part of a turnover of molecules in an animal organism or as part of the feeding process, and urea or other similar end products of nitrogen metabolism in animals are excreted. These substances generally find their way back into the earth or water of the planet and are passed along the cycle through ammonia, back to the nitrification by bacteria to form nitrates and nitrites, followed by the return to plant life to recycle again and again.

In this chapter we will emphasize how the amino acids are synthesized from ammonia and other intermediates of the carbohydrate pathways we have already described. We will also discuss in some detail how proteins are degraded and digested and how nitrogen is reclaimed from the discarded amino acids. This analysis will bring us into the important urea cycle for excreting nitrogen products, to complete our discussion of amino acid metabolism. We will also briefly describe some of the biosynthetic reactions of nucleotides, which, after amino acids, form the second major class of nitrogen-containing compounds.

Biosynthesis of Amino Acids

In humans, only 12 of the 20 major amino acids can be synthesized by the cells. The other eight, along with vitamins and minerals, are essential nutrients that must be provided in the diet. These essential amino acids are

- Isoleucine
- Leucine
- Lysine
- Methionine
- Phenylalanine
- Threonine
- Tryptophan
- Valine

They are available from animal and plant sources. Particularly rich in amino acids among plant sources are soy beans, nuts, grains, and sprouts, as many vegetarians can testify. Concern for maintaining adequate protein supplies for the world population has also led to exploration of microorganisms as sources of protein. The amino acids are synthesized in *families* with structurally related amino acids sharing common precursors. A description of the synthesis of all 20 amino acids is beyond the scope of this book. However, a few comments on several of the amino acids will be presented here to provide some insight into the general mechanisms involved in amino acid synthesis and the connections to the intermediates of carbohydrate metabolism.

Much of amino acid metabolism centers around the pivotal role of glutamic acid. It is therefore appropriate to consider glutamic acid first. Glutamic acid can be formed from the key TCA cycle intermediate, α-ketoglutaric acid, in a reaction, catalyzed by glutamic acid dehydrogenase involving the addition of ammonia coupled with the oxidation of NADPH + H^+:

$$NH_3 + \alpha\text{-ketoglutaric acid} + NADPH + H^+ \longrightarrow \text{L-glutamic acid} + NADP^+$$

This reaction is of extreme importance in all species, as the principal pathway for the incorporation of ammonia into amino acids. Glutamic acid can then go on to yield the two other amino acids in its family, glutamine and proline. Glutamine is formed by the action of the enzyme, glutamine synthetase, in the reaction

$$NH_3 + \text{glutamic acid} + ATP \longrightarrow \text{glutamine} + ADP + P_i$$

Proline is also formed by rearrangements and reduction of glutamic acid. In this case the α-carboxyl group of glutamic acid is retained as the carboxyl group of proline and the δ-carbon present as the distal carboxylic group of glutamic acid is first bonded in the pyrroline structure and then reduced to give the final product, proline. The structures of this sequence are given below:

Glutamate ⟶ Δ^1-Pyrroline-5 carboxylic acid ⟶ Proline

*δ-Carbon of glutamic acid

Glutamic acid contributes to the formation of many other amino acids by *transamination* reactions. A typical transamination reaction is

$$\text{Glutamic acid} + \text{pyruvic acid} \longrightarrow \alpha\text{-ketoglutaric acid} + \text{alanine}$$

Thus the amine and keto functions of two reactants are exchanged to form the opposite distribution in the products. These reactions proceed with the participation of the coenzyme vitamin B_6 in its derivative form, pyridoxal phosphate (see Fig. 13-2), as deduced by Esmond Snell and his coworkers. The aldehyde function of this coenzyme is generally attached to a lysine ε-amino group of the enzyme that catalyzes the transamination reactions. When the α-amino acid comes onto the surface of the protein, the initial ε-amino group of lysine is displaced, and a Schiff's base with the pyridoxal phosphate aldehyde group and the amino group of the incoming amino acid is formed. A typical structure is shown in Fig. 13-3. The double-bond character of the Schiff's base is partially shared by the adjacent carbon-nitrogen bond, as shown in the third structure in Fig. 13-3. This bond can then be hydrolyzed to yield the α-keto acid and the amino

Pyridoxal phosphate

Pyridoxal phosphate bound to enzyme

Pyridoxamine

FIG. 13-2 *Pyridoxal Phosphate.* The free form is shown as well as the enzyme-bound derivative involving a Schiff's base to an ε-amino group of a lysine residue. When an amino acid donates its amino group, pyridoxamine is formed.

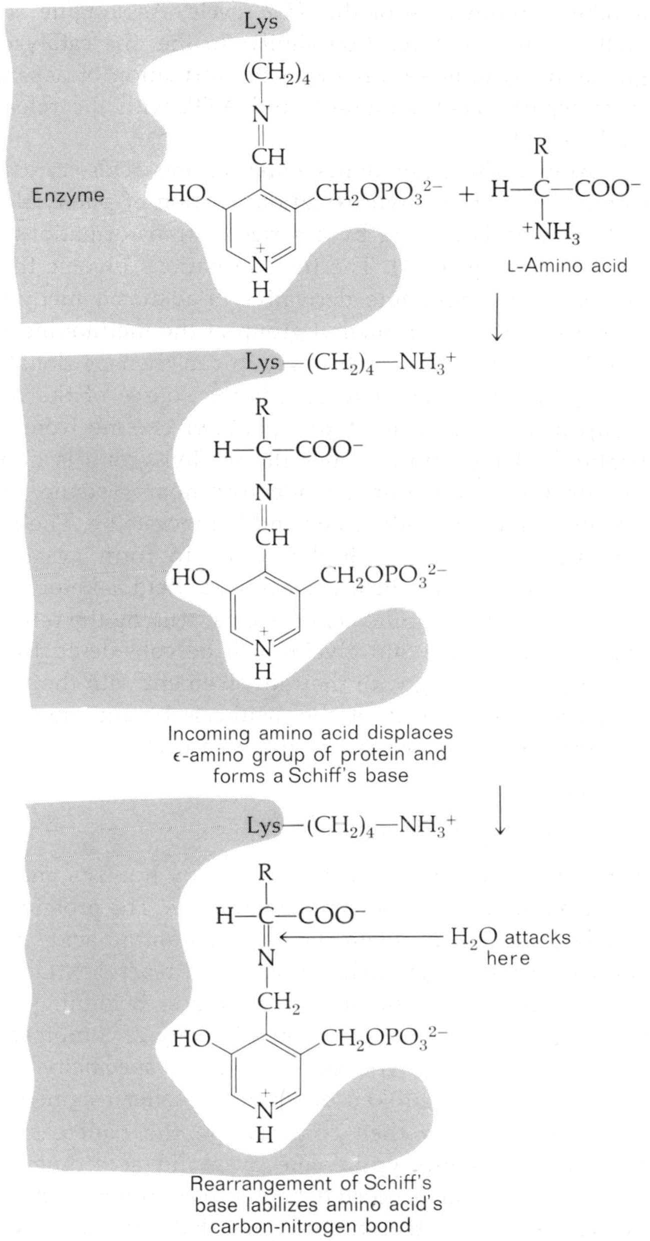

Enzyme

L-Amino acid

Incoming amino acid displaces ε-amino group of protein and forms a Schiff's base

H_2O attacks here

Rearrangement of Schiff's base labilizes amino acid's carbon-nitrogen bond

FIG. 13-3 *Pyridoxal Phosphate: Mechanism of Action.* Incoming amino acid displaces ε-amino group of enzyme to form a Schiff's base. The Schiff's base is rearranged to give a labile carbon-nitrogen bond. The bond can then be hydrolyzed to yield the corresponding α-keto acid and pyridoxamine. The pyridoxamine can then donate the amino group to a different α-keto acid in transamination. In certain other enzyme-catalyzed reactions, the labile compound is decarboxylated to form an amine or racemized to the corresponding D-amino acid. For additional details, see B. M. Guirard and E. E. Snell, in M. Florkin and E. H. Stotz, eds., *Comprehensive Biochemistry,* vol. 15, American Elsevier, New York, 1965.

form of the coenzyme, pyridoxamine phosphate. In this way alanine is formed from the intermediate of glycolysis, pyruvate. Similarly, aspartic acid can be formed in a transamination reaction between glutamic acid and oxaloacetic acid, another intermediate of the TCA cycle. Asparagine is generally formed in a reaction similar to the one catalyzed by glutamine synthetase, involving the formation of asparagine from aspartic acid, ammonia, and ATP, with the release of ADP and P_i.

Among the other nonessential amino acids, tyrosine is formed by a simple hydroxylation reaction of phenylalanine, and cysteine is formed by a series of transformations from methionine (Fig. 13-4). The transformations involve the formation of a methionine derivative, S-adenosyl methionine, in which the terminal methyl group of the methionine molecule is activated. The methyl group can then be donated to an appropriate acceptor (e.g., it is the source of the methyl group in the synthesis of phosphatidyl choline from phosphatidyl ethanolamine). Once the methyl group is donated, the remaining compound, S-adenosyl homocysteine, can be hydrolyzed to yield adenosine and homocysteine. The homocysteine then reacts with the serine to form cystathione, shown in Fig. 13-4, which is cleaved to yield α-ketobutyrate, ammonia, and the amino acid cysteine. Among the remaining amino acids, serine and glycine will be considered together in Chapter 14, along with their involvement with the vitamin folic acid, and aspects of the synthesis of arginine will be encountered below as part of the urea cycle.

Amino Acid Degradation

The bulk of the amino acids taken in by humans and other animals is in the form of complete proteins. The proteins must first be digested to obtain the essential amino acids needed and to permit energy-yielding oxidation reactions. Digestion takes place in the gut in several stages, beginning in the stomach. Here the low pH, approximately 2, combines with the action of a proteolytic enzyme of broad specificity, pepsin, and the protein is broken down into smaller fragments and peptides. Digestion then continues in the pancreas where trypsin and chymotrypsin take over and according to the specificity previously noted (Chapter 8) continue the digestion process. Other proteolytic enzymes that have also been of value in the structural work on proteins, such as carboxypeptidases, contribute to the process.

An interesting aspect of proteolytic enzymes is the strategy evolved to restrict self-digestion. If a cell synthesized a potent proteolytic enzyme, it would be susceptible to immediate digestion of its own proteins by this enzyme. To avoid

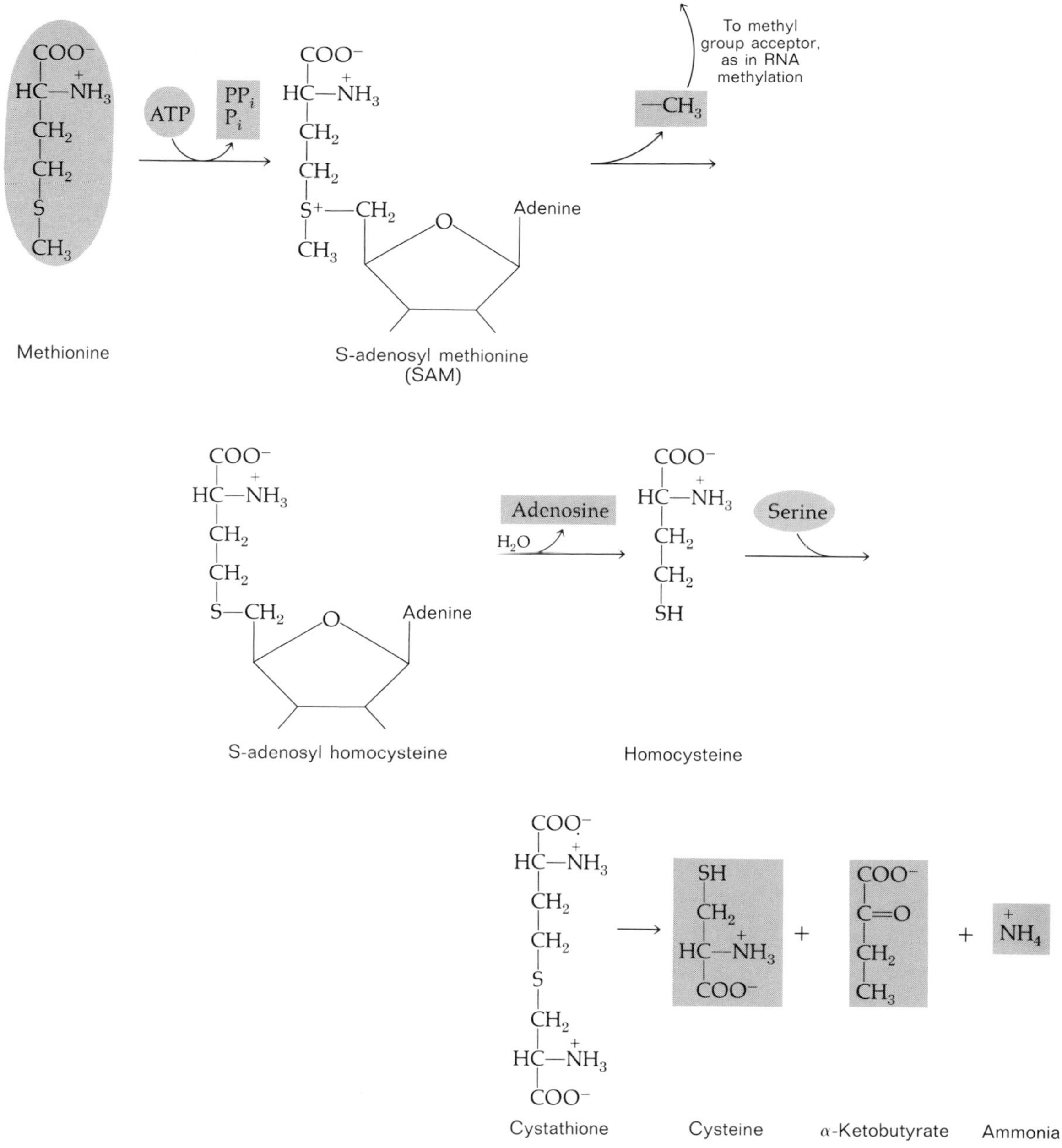

ATP

PP_i, P_i

$-CH_3$

To methyl group acceptor, as in RNA methylation

Methionine

S-adenosyl methionine (SAM)

Adenosine

H_2O

Serine

S-adenosyl homocysteine

Homocysteine

Cystathione

Cysteine

α-Ketobutyrate

Ammonia

FIG. 13-4 *Cysteine Biosynthesis.* Colored circles indicate input, gray rectangles indicate output.

such a predicament, most proteolytic enzymes are synthesized in an inactive form, known as *zymogen*. In this inactive form the particular constellation of amino acids necessary for catalytic peptide-bond splitting is constrained away from the appropriate geometry into some nonfunctional configuration.

Then the proteins are exported, secreted into the gut of the digestive tract. There, activation takes place, either by an autocatalytic self-digestion, digestion by one of the other proteases, or as a result of the new pH medium. Activation involves splitting a small number of labile peptide bonds, sometimes releasing a small peptide fragment. Then the polypeptide snaps into its proper structure for catalytic action. This process of activation has been studied extensively, and some of the steps involved in the activation of one of the typical proteolytic enzymes, chymotrypsin, from inactive chymotrypsinogen are shown in Fig. 13-5. While many chemical events take place, crystallographic studies of the detailed architecture of the molecule reveal that changes at the active site are relatively small. Evidently displacements in the range of only a few angstrom units are adequate to disrupt the proper chemical functioning of the enzyme.

Proteins digested in this way are broken down into their amino acids, and the amino acids can be processed into the general flow of metabolism. Those amino acids that are essential may be directed toward incorporation in newly syn-

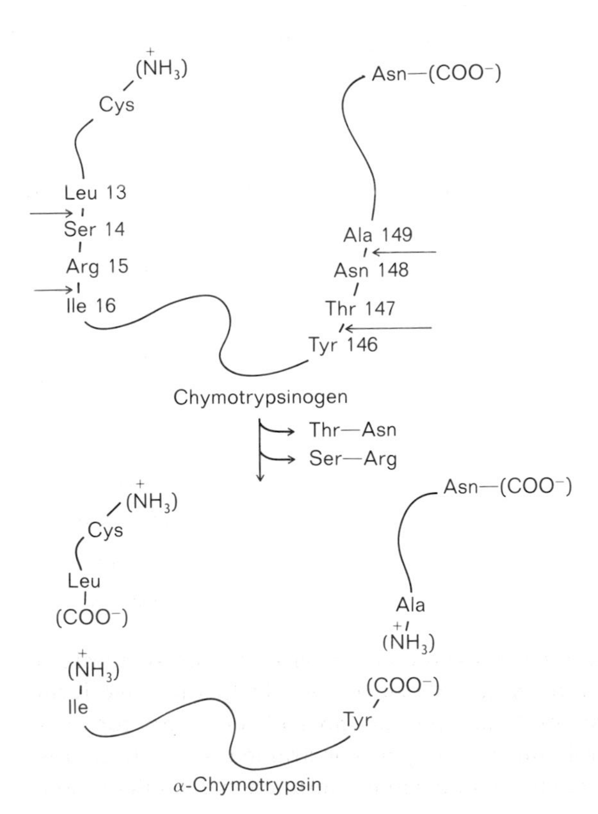

FIG. 13-5 *Activation of Chymotrypsinogen:* colored arrows indicate peptide bonds attached by trypsin and chymotrypsin.

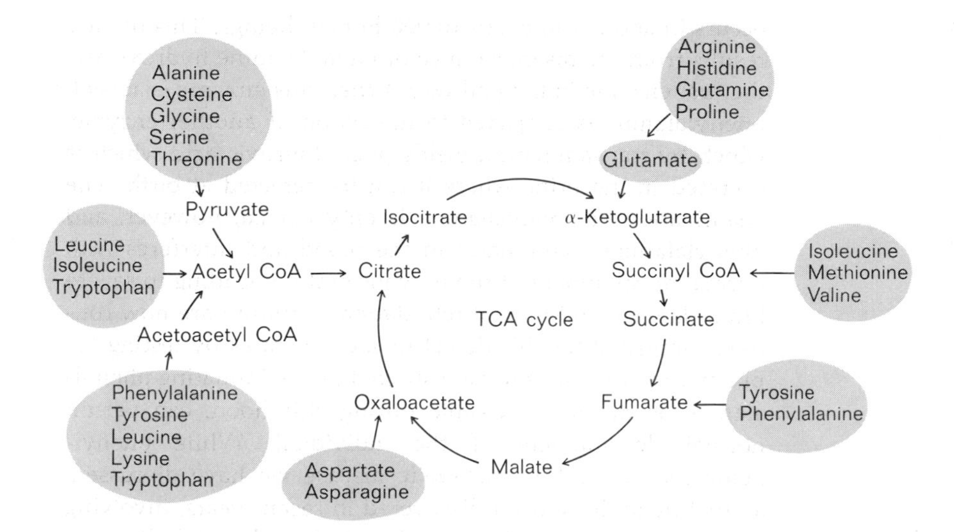

FIG. 13-6 *Input of Amino Acids into Metabolic Oxidations.* For additional details, see A. Meister, *Biochemistry of the Amino Acids,* 2nd ed., Academic Press, New York, 1965.

thesized proteins and smaller peptide hormones in various tissues of the body. Other amino acids may be employed for energy-yielding oxidation reactions. Amino acids to be processed for oxidation are generally first relieved of their amino group by a transamination reaction. The amino acid reacts with α-ketoglutaric acid to yield glutamate and the corresponding keto form of the donor amino acid. The rest of the amino acid structure is then fed into an appropriate point of the metabolic scheme of carbohydrate metabolism. In some cases considerable amounts of rearrangement are required, but eventually oxidation products of each amino acid supply the TCA cycle or one of the earlier stages in metabolism. In broad outline, the various points of entry are summarized in Fig. 13-6. Glutamate, aspartate, and alanine can feed into their respective points simply by a deamination reaction. For the other amino acids, somewhat more complicated transformations are required and each amino acid provides a story of its oxidation all its own. Too many stories are present to provide each one in detail, but we can touch on one of special interest, the oxidation of phenylalanine. The pathway for phenylalanine, shown in Fig. 13-7, first involves the formation of tyrosine by the hydroxylase reaction, and by a series of further transaminations and oxidations eventually yields fumaric acid, for entry into the TCA cycle, and acetoacetic acid, which, when reacted with coenzyme A, becomes an intermediate of fatty acid oxidation.

This particular pathway was singled out because of the medical interest due to the disease phenylketonuria, which

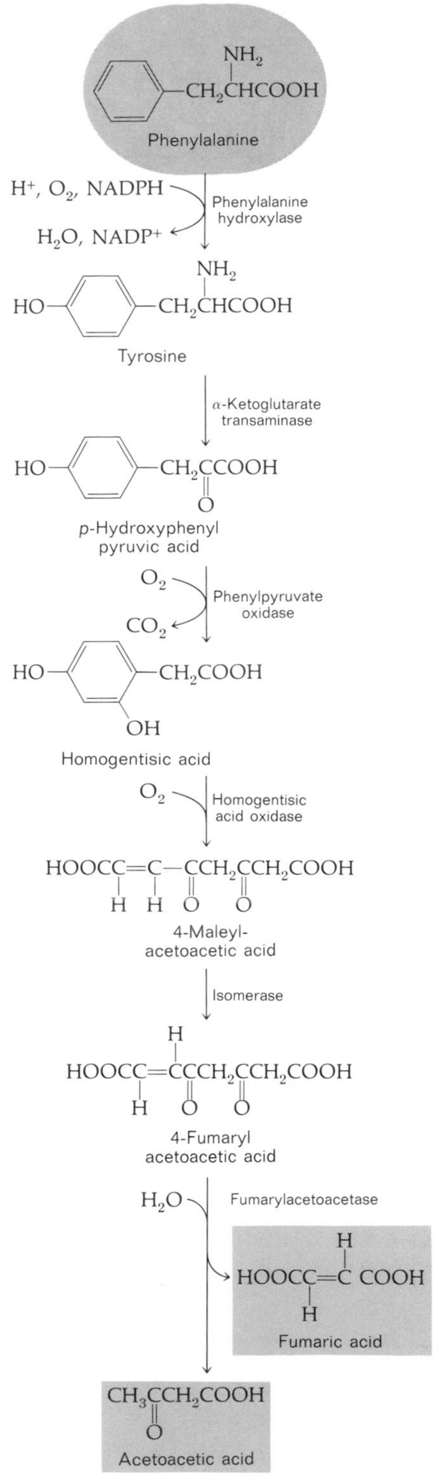

Phenylalanine

H^+, O_2, NADPH → H_2O, $NADP^+$ — Phenylalanine hydroxylase

Tyrosine

α-Ketoglutarate transaminase

p-Hydroxyphenyl pyruvic acid

O_2 → CO_2 — Phenylpyruvate oxidase

Homogentisic acid

O_2 — Homogentisic acid oxidase

4-Maleyl-acetoacetic acid

Isomerase

4-Fumaryl acetoacetic acid

H_2O — Fumarylacetoacetase

Fumaric acid

Acetoacetic acid

FIG. 13-7 *Phenylalanine Oxidation.*

occurs in about 1 in every 10,000 human beings. This disease results from the disappearance of phenylalanine hydroxylase, the first enzyme in the pathway. When this enzyme is absent, phenylalanine is subjected to the action of another enzyme, which by transamination yields phenylpyruvic acid which is excreted in the urine where it can be detected at birth. The deamination of phenylalanine is only partial, however, and phenylalanine accumulates in the blood and interferes with normal development of the nervous system, leading to severe brain damage and mental retardation. Children are now routinely screened for this defect in several states by testing for phenylpyruvic acid, since a restricted diet, eliminating phenylalanine as much as possible during childhood, can permit normal development of the individual. While phenylketonuria is the first metabolic disease we have discussed, a great many have been discovered in recent years, involving carbohydrate metabolism, lipid and steroid metabolism, as well as nucleotide metabolism. Treatment of these diseases and other similar pathologies represents one of the burgeoning areas of medical science.

Three other aspects of the phenylalanine pathway are of general interest. One is the involvement of a coenzyme, dihydrobiopterin (see Fig. 13-8), in the phenylalanine hydroxylase reaction. The reaction involves the donation of one atom of oxygen from O_2 as the hydroxyl group of tyrosine with the donation of the reducing equivalents by tetrahydrobiopterin, acquired from NADPH + H^+. The second point of interest concerns the oxidation of homogentisic acid by the homogentisic acid oxidase. This is a complex reaction, which involves a decarboxylation, oxidation, and migration of a side chain. The enzyme involved contains copper and also requires ascorbic acid, or vitamin C. This reaction is one of several examples of vitamin C involvement in metabolism:

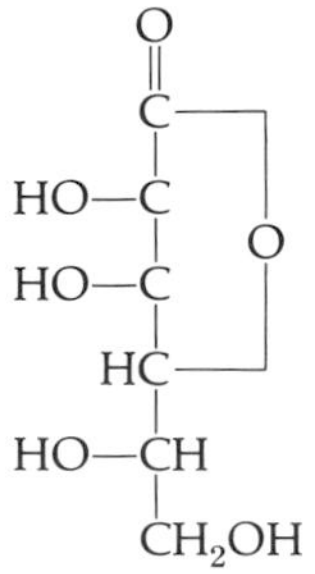

Vitamin C
(ascorbic acid)

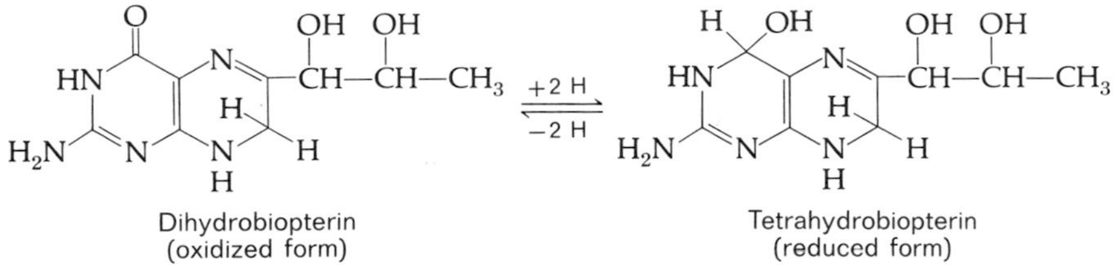

Dihydrobiopterin (oxidized form) $\underset{-2\,H}{\overset{+2\,H}{\rightleftharpoons}}$ Tetrahydrobiopterin (reduced form)

FIG. 13-8 *Structure of Biopterin.*

Finally, the details of the reaction catalyzed by phenylalanine hydroxylase are of interest as an example of a *mixed-function* reaction. Both oxidation and reduction are involved in this one enzymatic process with one atom of oxygen incorporated into the substrate. This reaction may be contrasted to a full *oxidase* reaction, with both atoms of oxygen incorporated, as occurs later in the pathway in the conversion of homogentisic acid to 4-maleylacetoacetic acid.

With the carbon skeletons of the amino acids now accounted for, we can go on to describe the fate of the amine groups. We noted above that one of the first reactions of amino acid degradation is the donation of the amine group to α-ketoglutarate to form glutamic acid. In bacteria and most other organisms, including higher animals, glutamic acid can be metabolized directly to form ammonia and α-ketoglutarate. This reaction, catalyzed by glutamic dehydrogenase, is

$$\text{L-Glutamate} + NAD^+ \longrightarrow \alpha\text{-ketoglutarate} + NH_4^+ + NADH + H^+$$

Other cells can oxidize amino acids directly with the flavoprotein enzyme, L-amino acid oxidase. This enzyme contains FAD (see Chapter 14) as a prosthetic group and the amino acid is oxidized with the release of ammonia and the formation of reduced $FADH_2$. The reduced flavin mononucleotide then reacts with oxygen to form peroxide (H_2O_2), which is broken down by the action of catalase to water and molecular oxygen. Both D- and L-amino acid oxidases are present, and in eucaryotic cells they are generally localized in specialized vesicles known as *peroxisomes.*

Since ammonia is relatively toxic to most terrestrial animals, other pathways have evolved to convert the ammonia into a less toxic form. In reptiles and birds, the main form for exporting amino groups is uric acid, which is represented by the following structure:

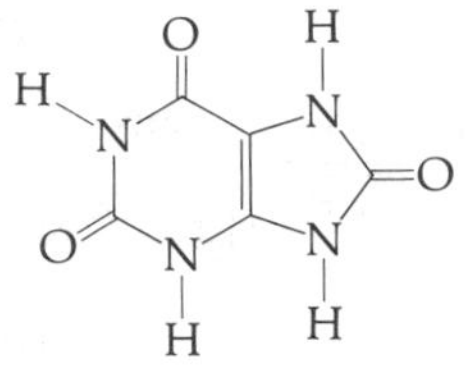

Uric acid

The form of excretion we will consider in detail here is the mechanism found in most mammals, including man, known as the *urea cycle*. The cycle is summarized in Fig. 13-9 and involves many familiar structures, such as the carbohydrate intermediates fumarate, malate, oxaloacetate, and α-ketoglutarate. In addition the amino acids glutamate and aspartate play an important role, along with arginine and ornithine, a homolog of lysine. The overall stoichiometry of the reaction is

$$2\,NH_3^+ + CO_2 + 3\,ATP + 2\,H_2O \longrightarrow \underset{\text{Urea}}{NH_2CONH_2} + AMP + 2\,ADP + 2\,P_i + PP_i + 3\,H^+$$

Since the PP_i produced is hydrolyzed to 2 P_i, four high-energy phosphate bonds are consumed for each molecule of urea produced. However, NADH + H^+ produced in oxidizing fumarate can regenerate 3 ATP in the mitochondria where the reactions of the urea cycle take place. Thus there is a net loss of only one high-energy phosphate per urea produced.

Nucleotide Biosynthesis

While humans and other mammals leave biosynthesis of some of their amino acids to other organisms, synthesis of purines and pyrimidines, the essential components of nucleic acids, are not trusted to anyone else. Virtually all organisms, with the exception of a small number of bacteria, synthesize these molecules themselves, and generally quite similar pathways are involved. Here we will just touch on some of the essential features of the pathways and can begin by noting that both the purine and pyrimidine nuclei are constructed from very simple precursors. The nature of the precursors contributing to these structures is summarized in Fig. 13-10. Formation of the purine compounds begins with the sugar phosphate

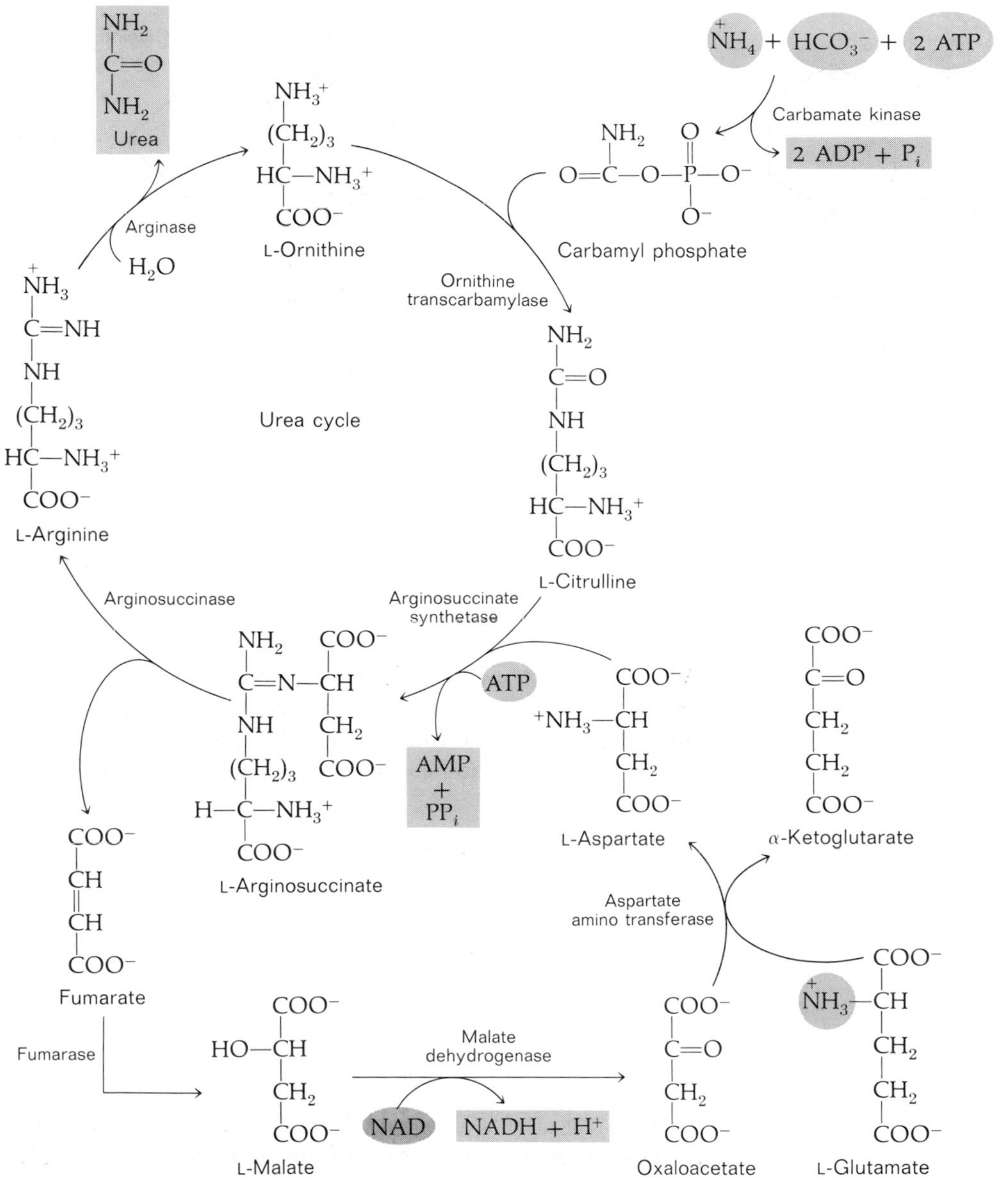

FIG. 13-9 *The Urea Cycle.*

already attached. The first step in the biosynthetic pathway is the amination of phosphoribosyl pyrophosphate (PRPP), shown in Fig. 13-11, by glutamine, to yield phosphoribosyl amine. Then by a series of fairly complicated rearrangements and additions the contributing components shown in Fig.

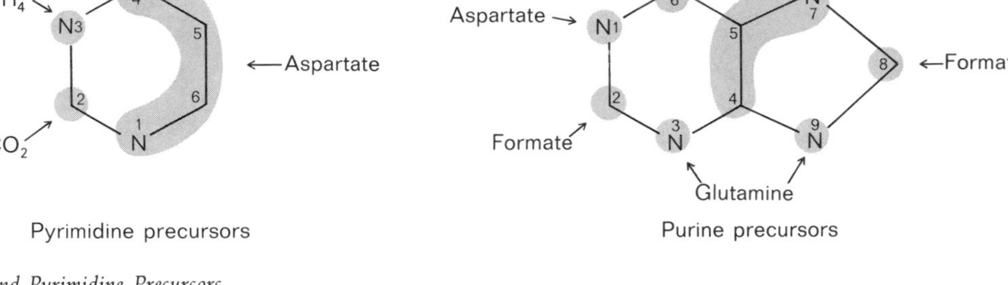

Pyrimidine precursors

Purine precursors

FIG. 13-10 *Purine and Pyrimidine Precursors.*

13-10 are added one at a time to form the first nucleotide product, inosinic acid:

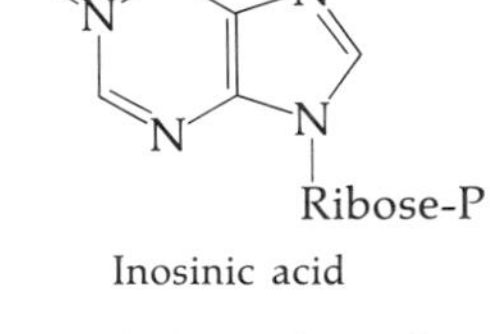

Inosinic acid

Adenylic acid (AMP) is formed from inosinic acid by the addition of aspartic acid to form adenylosuccinate and the subsequent release of fumaric acid, leaving the amino group characteristic of adenine. The structure of adenylosuccinate is shown below:

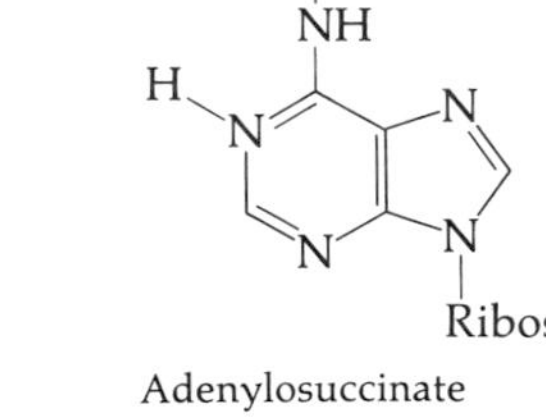

Adenylosuccinate

Guanylic acid (GMP) is formed by oxidation of inosinic acid to xanthylic acid followed by an amination dependent on ATP,

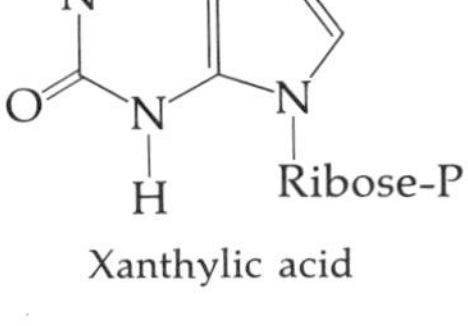

Xanthylic acid

with the amine group donated by glutamine, to yield GMP, glutamate, and AMP plus PP_i. The triphosphates are then formed by two successive phosphokinase steps to form ADP and GDP and then ATP and GTP. Pyrimidine biosynthesis, in contrast, occurs first with the formation of a ring structure, beginning with carbamyl phosphate and aspartate, the enzyme reaction catalyzed by the important archetype of regulatory enzymes, ATCase (see Chapter 10), to yield after a dehydration and a dehydrogenation step, orotic acid. Orotic acid then reacts with phosphoribosyl pyrophosphate to form orotidylic acid (OMP):

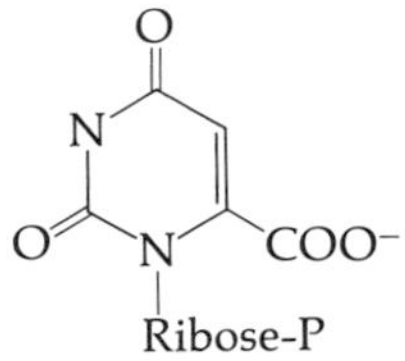

O N O N COO^- Ribose-P

Orotidylic acid

A simple decarboxylation step next yields UMP and in two phosphorylation steps, UTP is formed. Then in contrast to the transformation of I to A and G at the monophosphate level, CTP is formed by a conversion of UTP in an ATP-dependent amination reaction.

Formation of deoxyribonucleotides for incorporation in DNA takes place at the diphosphate level. First a small protein, called thioredoxin, which has two free SH groups, is reduced by NADPH + H^+ in a reaction catalyzed by the enzyme thioreductase. Then in the presence of two other enzyme fractions, reduced thioredoxin reacts with a ribonucleoside-5′-diphosphate to form oxidized thioredoxin (in which the two free SH groups are joined in a disulfide bond) and the deoxyribonucleoside diphosphate. This reaction in *E. coli* has been extensively studied by Reichard and his colleagues. In some organisms, a form of vitamin B_{12}, cyanocobalamine (see Chapter 14), is required to catalyze the for-

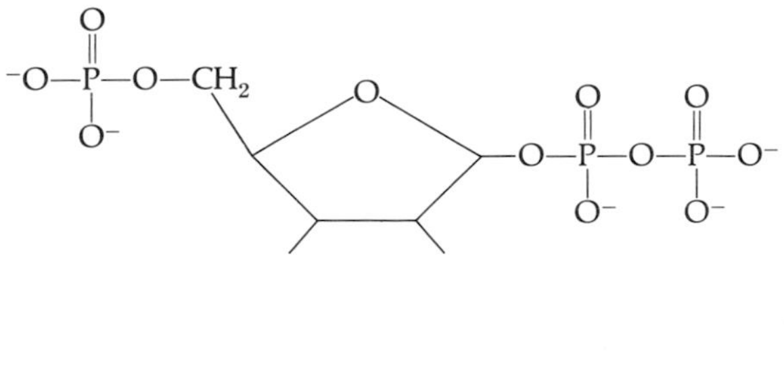

$^-O—P(=O)(O^-)—O—CH_2$ O —O—P(=O)(O^-)—O—P(=O)(O^-)—O^-

FIG. 13-11 *Structure of PRPP.*

mation of deoxyribose. One final point concerning deoxynucleotides is the formation of deoxythymidilic acid (dTMP) for incorporation in DNA. This reaction occurs by the methylation of dUMP with an enzyme system involving a folic acid derivative. Typical methyl-donation reactions involving both cyanocobalamine and folic acid will be described in Chapter 14.

Breakdown of nucleic acids and their constituent nucleotides and bases is also accomplished in digestion and metabolism by a variety of enzymes. In addition, biosynthesis of hemes and many coenzymes has also been described and occurs by a number of intriguing pathways. However, we will stop at this point in the description of the metabolism of nitrogenous compounds and go on in the next chapter to some of the finer details of some of the reactions already described.

PROBLEMS

1. Methionine is a precursor of which atoms in cysteine? What is the source of the remaining atoms in cysteine?
2. What is the common feature of oxidation of isoleucine, methionine, and valine?
3. Is the urea cycle an energy-yielding process?
4. What are the sources of the nitrogen atoms in (a) purines (b) pyrimidines
5. Does deoxyribose formation from ribose occur at the level of free ribose?

Chapter 14 Frontispiece *Electron micrographs of E. coli pyruvate dehydrogenase complex and its component enzymes negatively stained with phosphotungstate. Magnification 230,000×. Upper left, pyruvate dehydrogenase complex; upper right, dihydrolipoyl transacetylase; lower left, pyruvate dehydrogenase; lower right, dihydrolipoyl dehydrogenase. Courtesy L. J. Reed. For additional details see Archives of Biochemistry and Biophysics* **152:** *655 (1972).*

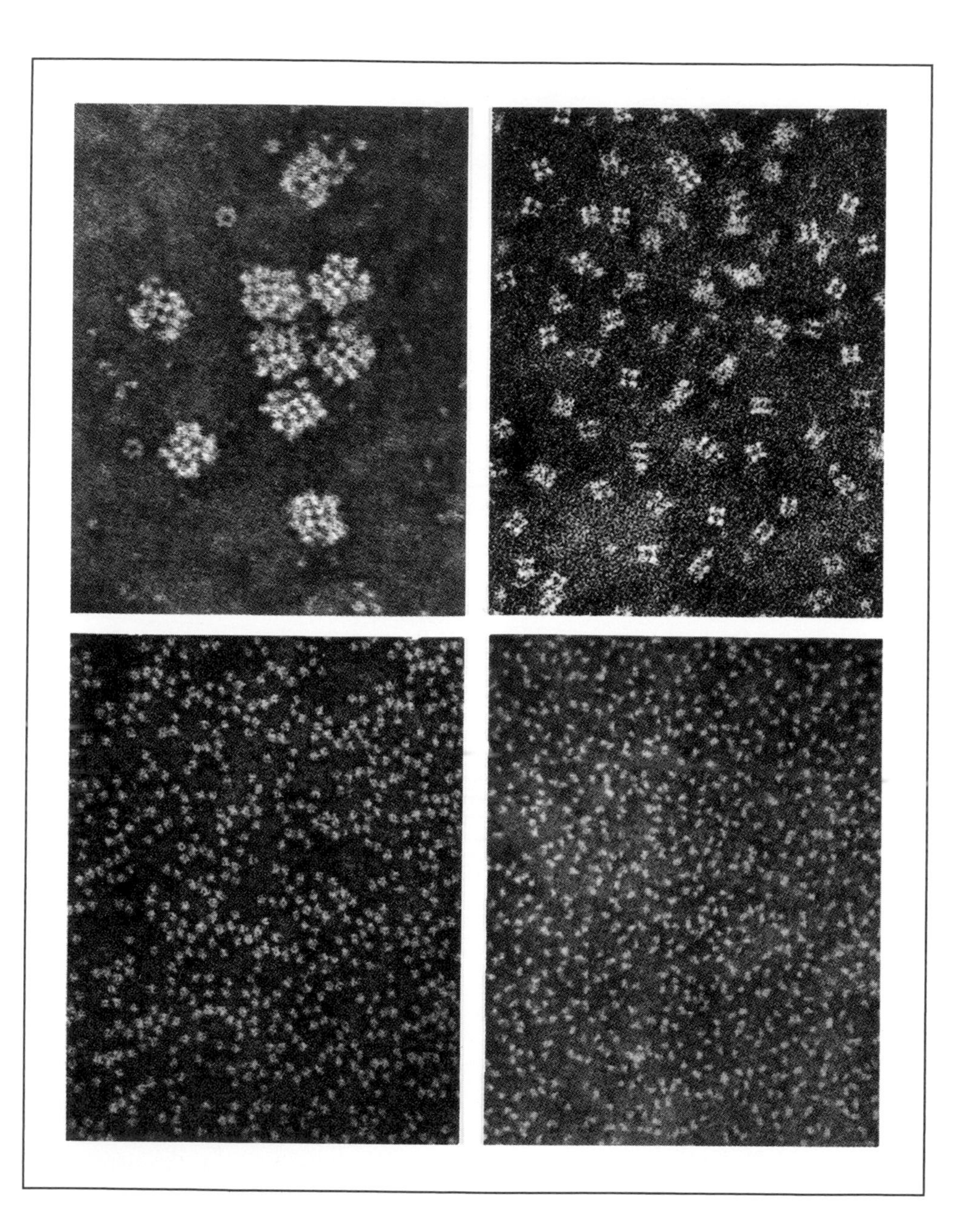

14 COENZYMES AND REACTION MECHANISMS—BETTER CHEMISTRY THROUGH LIVING

As we have emphasized through our discussion of biochemistry, proteins—the workers in the biochemical world—are extremely versatile molecules. Ribonuclease can split RNAs specifically after uracil and cytosine bases by imitating base pairing with hydrogen-bond-forming amino acids (Chapter 9). ATCase mediates the cooperative binding of its substrates by a subtle interaction between its subunits and regulatory molecules such as CTP (Chapter 10). Proteins even assume varied shapes, such as the lollipop structure of myosin (Chapter 12) with its unique role in both the structure and contraction mechanism of muscle. Thus proteins have demonstrated an extraordinary dexterity in biochemical reactions, but they still at times require help. Certain chemical transformations are simply beyond the scope of the chemical repertoire of amino acid side chains. Amino acids cannot mediate oxidation-reduction reactions, for example. Also reactions involving certain kinds of carbon-carbon bond formation are impossible for proteins alone. For these and other reactions, specialized molecules must be called on and these biochemical collaborators are known as *coenzymes*. When very tightly bound to a particular protein, a coenzyme is designated as a prosthetic group.

We have already encountered coenzymes such as NAD^+ and $NADP^+$ in our discussion of bioenergetics in Chapter 11. The spectral property of NAD^+ in mediating oxidation-reduction reactions was clearly demonstrated. A number of other coenzymes were described in Chapter 13. In this chapter we will go through a selection of reactions more systematically, emphasizing the special role of coenzymes, or in a few other cases an especially interesting reaction mechanism that has evolved. We will begin with an example that does not involve a new coenzyme, but rather illustrates an important biochemical principle, substrate-level phosphorylation.

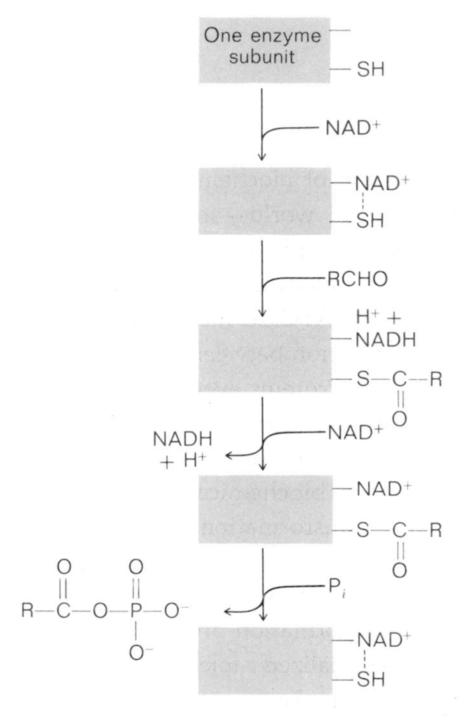

FIG. 14-1 *Mechanism of Glyceraldehyde-3-phosphate Dehydrogenase.* The enzyme contains four subunits. NAD^+ binds to each subunit near an SH group. The aldehyde (glyceraldehyde-3-phosphate or a close analog) reacts with the enzyme-NAD complex with reduction of the NAD^+ and formation of a covalent acyl-enzyme intermediate. A molecule of NAD^+ is then exchanged for $NADH + H^+$ leaving an acyl-enzyme–NAD complex. The acyl-enzyme is finally transferred to inorganic phosphate to yield the acyl phosphate product (1,3-diphosphoglycerate) and regenerate the NAD-enzyme complex for another round of catalysis. For additional details, see E. Racker, *Mechanisms in Bioenergetics*, Academic Press, New York, 1965.

Substrate-level Phosphorylation

While the bulk of ATP generated by the cell is derived from the reactions of oxidative phosphorylation in mitochondria, we have described the ability of glycolytic reactions to furnish a modest amount of ATP (Chapter 11). One of the principal steps involved in substrate-level phosphorylation involves the enzyme glyceraldehyde-3-phosphate dehydrogenase. This enzyme, in a complex sequence of steps, couples the oxidation of glyceraldehyde with the reduction of NAD^+ and the introduction of a high-energy phosphate bond. The high-energy phosphate bond is then delivered to ADP to form ATP in the next step in the glycolytic sequence, the reaction catalyzed by phosphoglycerate kinase.

The steps in the reaction catalyzed by glyceraldehyde-3-phosphate dehydrogenase are summarized in Fig. 14-1. The reaction involves formation of a covalent bond between the aldehyde and a sulfhydryl group on the protein, with removal of the hydrogen from the substrate to NAD^+. As first proposed by Racker, this reaction produces a thioester between the enzyme sulfhydryl group and the carbonyl group of the substrate. This bond has a relatively high energy and the acyl group can be transferred to inorganic phosphate to yield the product of the reaction, 1,3-diphosphoglycerate.

The overall reaction can be dissected in a thermodynamic sense into two individual parts. The first reaction is the oxidation of the aldehyde substrate to an acid. This reaction depends on the presence of NAD^+ as the hydrogen acceptor and is a very favorable reaction. The standard free-energy change is about -10 kcal. The second part of the reaction can be viewed as the phosphorylation of the carboxylic acid. This is an unfavorable reaction since the product is a relatively high-energy compound. A standard free-energy change of 12 kcal can be assigned to the second reaction. However, if we combine the reactions and consider the sum

$$\text{Aldehyde} + P_i + NAD^+ \longrightarrow \text{acylphosphate} + NADH + H^+$$

the free-energy change is the sum of the free-energy changes of the two half-reactions and is a small positive number. This number is sufficiently close to 0 that by removal of the products through subsequent reactions the overall reaction can proceed to good yield. In this way a favorable reaction is *coupled* to an unfavorable reaction, permitting formation of moderate quantities of a product that would otherwise be very hard to synthesize. The reaction of glyceraldehyde-3-phosphate dehydrogenase thus illustrates the important principle of coupling, which is found in many biochemical reactions, several of which occur in the pathways summarized in Chapter 11.

Substrate-level phosphorylation illustrates an important concept in bioenergetics. ATP can be formed by simple chemical reactions that do not involve the complex machinery of mitochondria or chloroplasts. Since the glyceraldehyde-3-phosphate dehydrogenase reaction involves only a soluble enzyme, it is not surprising that early workers in the field of bioenergetics sought a similar soluble component or components for the generation of ATP in mitochondria and chloroplasts. However, from our further considerations of oxidative phosphorylation and the intricate involvement of the membrane components and proton gradients (Chapter 12), we see that nature has evolved a mechanism for generating the bulk of this ATP that is more complex and evidently more efficient than substrate-level phosphorylation. Hence the reaction of glyceraldehyde-3-phosphate dehydrogenase may represent a living biochemical fossil, reflecting a primitive anaerobic time when ATP was synthesized in this more simple way. The reaction also serves to illustrate the use of a thiol group from cysteine in an enzyme-reaction mechanism. Cysteine is one of the reactive amino acids and, like serine and histidine, is found at the active sites of a number of enzymes. The formation of an acyl cysteine intermediate with cysteine may be contrasted to the acyl serine formed by the proteolytic enzymes such as chymotrypsin, described in Chapter 7.

Pyruvate Dehydrogenase Complex

Pyruvate dehydrogenase complex carries out one of the essential reactions of intermediate metabolism, the preparation of pyruvate for its entry into the TCA cycle. This one reaction, formation of acetyl CoA from pyruvate, actually involves five different coenzymes. Therefore by examining this one reaction in detail, we will achieve a good understanding of the working of several major coenzymes. The enzyme complex itself is one of the most striking examples of versatility and complexity among metabolic enzymes. The complex has a total molecular weight of about 4 million and contains three distinct enzymatic activities. Each enzymatic activity arises from a distinct polypeptide chain and numerous copies of each type of chain are present in the complex. Elegant studies

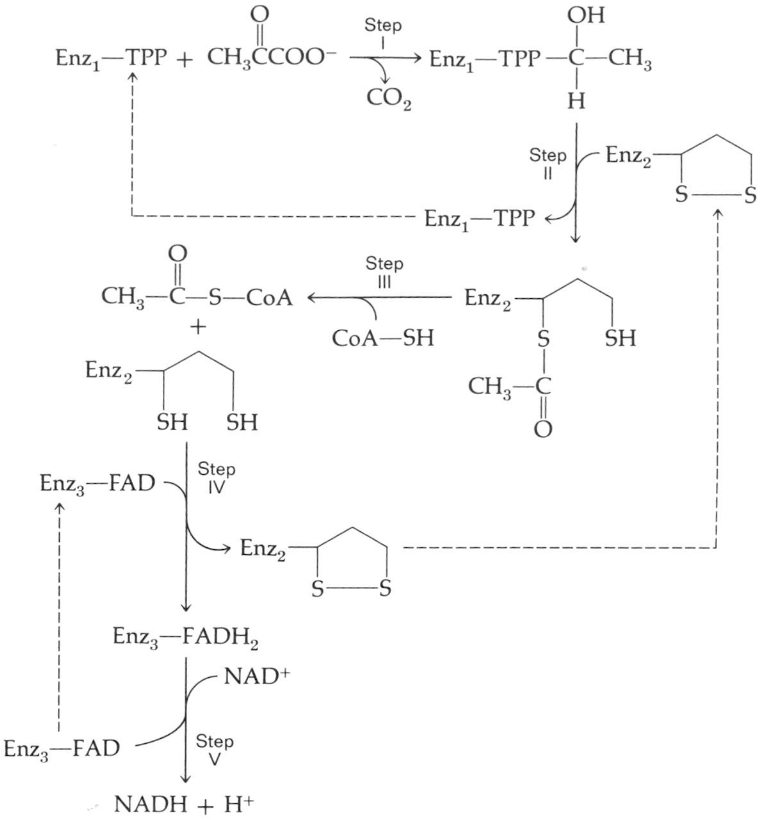

$$\text{Enz}_1\text{—TPP} + \text{CH}_3\overset{\text{O}}{\overset{\|}{\text{C}}}\text{COO}^- \xrightarrow[\text{CO}_2]{\text{Step I}} \text{Enz}_1\text{—TPP—}\underset{\text{H}}{\overset{\text{OH}}{\text{C}}}\text{—CH}_3$$

Step II: $\text{Enz}_1\text{—TPP—C(OH)(H)—CH}_3 + \text{Enz}_2(\text{S—S}) \rightarrow \text{Enz}_1\text{—TPP} + \text{Enz}_2(\text{S—C(O)—CH}_3)(\text{SH})$

Step III: $\text{Enz}_2(\text{S—C(O)—CH}_3)(\text{SH}) + \text{CoA—SH} \rightarrow \text{CH}_3\text{—C(O)—S—CoA} + \text{Enz}_2(\text{SH})(\text{SH})$

Step IV: $\text{Enz}_2(\text{SH})(\text{SH}) + \text{Enz}_3\text{—FAD} \rightarrow \text{Enz}_2(\text{S—S}) + \text{Enz}_3\text{—FADH}_2$

Step V: $\text{Enz}_3\text{—FADH}_2 + \text{NAD}^+ \rightarrow \text{Enz}_3\text{—FAD} + \text{NADH} + \text{H}^+$

FIG. 14-2 *Steps in the Oxidation of Pyruvate to Acetyl CoA.* Structure of the enzyme complex in Chapter 14 frontispiece. For additional details, see L. J. Reed, in P. D. Boyer, ed., *The Enzymes,* 3rd ed., vol. 1, Academic Press, New York, 1970, p. 213.

of the entire complex have been carried out by Reed and his coworkers revealing the symmetry and organization of these protein components to form one tightly integrated unit. Here we will be concerned principally with the chemical details of the reactions involved. The various steps in the reaction sequence are summarized in Fig. 14-2. The first step, involving enzyme one, pyruvate dehydrogenase, also includes the cofactor thiamine pyrophosphate (TPP). The structure of TPP is shown in Fig. 14-3. It is usually isolated in the unphosphorylated form, known as vitamin B_1, the anti-beriberi fac-

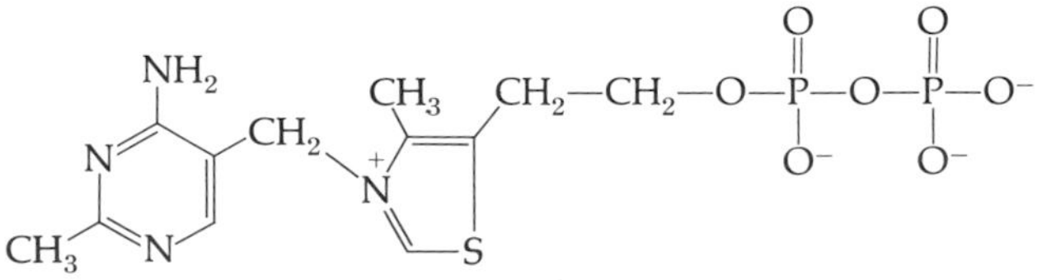

FIG. 14-3 *Structure of Thiamine Pyrophosphate (TPP).*

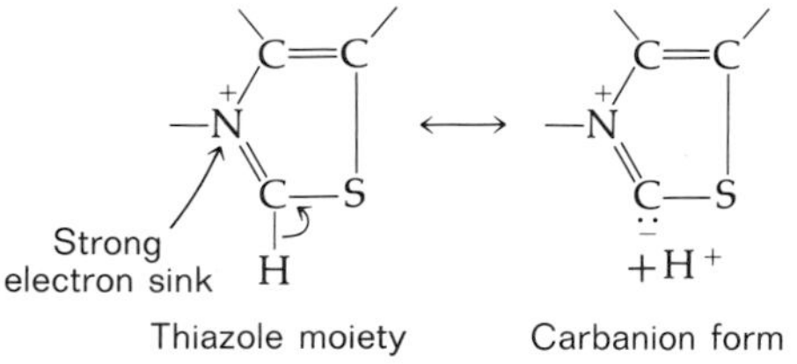

Strong electron sink

Thiazole moiety — Carbanion form

FIG. 14-4 *Chemistry of TPP.*

tor. TPP plays an important role in decarboxylation reactions by forming a stabilized carbanion, as shown in Fig. 14-4. Figure 14-5 indicates some of the electron movements postulated for the mechanism of decarboxylation.

The second step on the way to acetyl CoA involves the transfer of the hydroxyethyl group on TPP to another coenzyme, lipoic acid, bound to the second enzyme in the sequence, dihydrolipoyl transacetylase. The structure of lipoic acid is shown in Fig. 14-6 and includes the oxidized and reduced forms, as well as the structure generally present in enzyme reactions: the coenzyme in an amide linkage between its terminal carboxyl and the ε-amino group of a lysine residue on the protein. The transfer to the lipoic acid in the second step is accompanied by an oxidation-reduction reaction on the hydroxyethyl group bound to the TPP, oxidizing it to an acetyl group by the loss of hydrogen atoms to the disulfide bond of lipoic acid. The acetyl group can then be

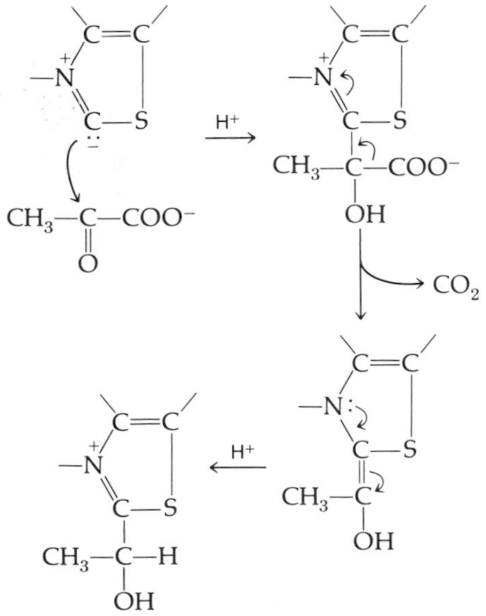

FIG. 14-5 *The Decarboxylation Reaction.*

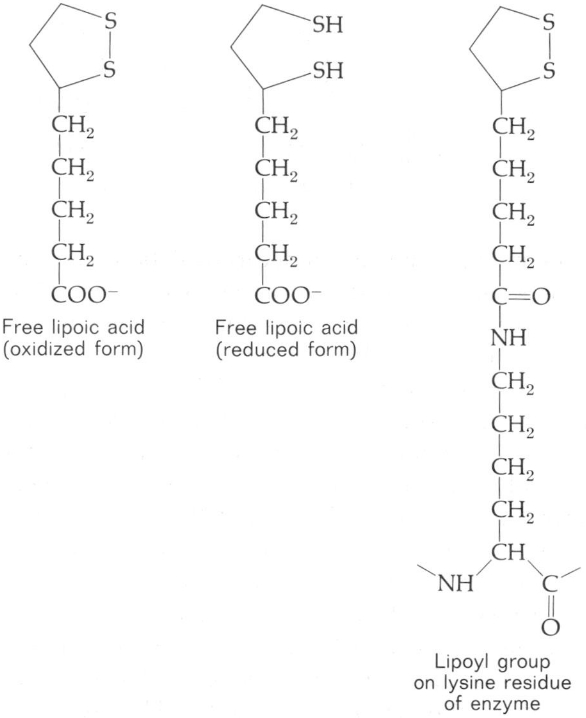

S S $(CH_2)_4$ COO^-
Free lipoic acid (oxidized form)

SH SH $(CH_2)_4$ COO^-
Free lipoic acid (reduced form)

S S $(CH_2)_4$ C=O NH $(CH_2)_4$ CH NH C=O
Lipoyl group on lysine residue of enzyme

FIG. 14-6 *Lipoic Acid.*

transferred to coenzyme A to yield, in the third step, acetyl CoA and reduced lipoic acid (dihydrolipoic acid) attached to enzyme two. In order for lipoic acid to function catalytically, it must be reoxidized in the fourth step and made ready for the next cycle. Reoxidation is accomplished by the third enzyme of the complex, dihydrolipoyl dehydrogenase, which contains another cofactor attached to it, flavin adenine dinucleotide (FAD), encountered earlier in oxidative phosphorylation (Chapter 11).

Thus in the third and fourth steps, two new cofactors, acetyl CoA and FAD, are encountered which we have mentioned earlier but not discussed in detail. Acetyl CoA is shown in Fig. 14-7 and terminates in a free SH group. The special virtue of coenzyme A is the transformation of the relatively inert acetyl groups into more reactive species when bound to the coenzyme as thioesters. This quality results from the property of sulfur next to the carbonyl group holding tightly to its electrons and not participating in the partial double-bond formation typical of most esters (see Fig. 14-8). Rather, the sulfur atom tends to stabilize a partial positive charge on

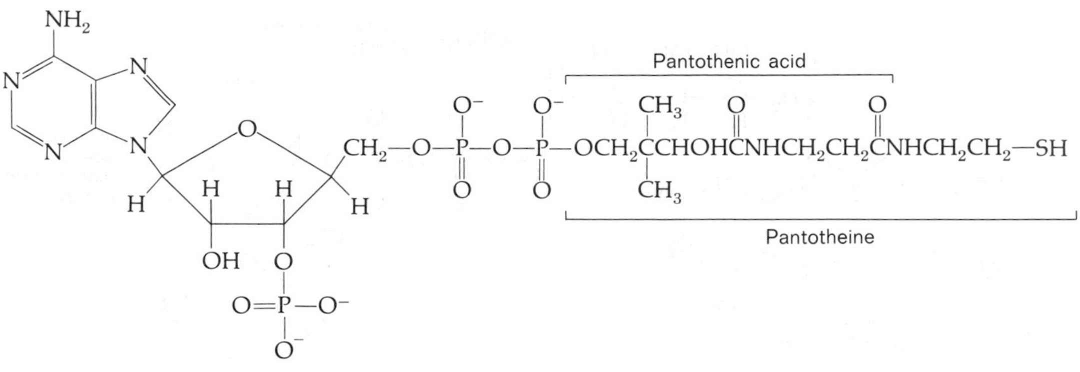

Pantothenic acid

Pantotheine

FIG. 14-7 *Structure of CoA.*

the adjacent carbon atom and a partial negative charge on the carbonyl oxygen. This arrangement in turn withdraws protons from the carbon in the next position, endowing it with a partial negative charge. Structures with partial positive character are especially susceptible to nucleophilic attack—that is, to attack by such compounds as amines, water, or other thiol compounds that are electron rich and attracted to the positive charge. The adjacent atom, which has a partial negative charge, is susceptible to attack by electrophilic groups, such as CO_2, as will be described when we consider the reactions involving biotin.

The structure of FAD is shown in Fig. 14-9. It is involved with reactions somewhat similar to those of NAD^+, although the reduced-oxidized transformation involves a much different active unit, flavin, also shown in Fig. 14-9. Two forms of the coenzyme are generally encountered, FAD (flavin adenine dinucleotide) and FMN (flavin mononucleotide). The FMN form was described while discussing the reaction of the enzyme amino acid oxidase in Chapter 13.

The oxidation of pyruvate to acetyl CoA is completed in the fifth step when the reduced FAD bound to the third enzyme is oxidized by NAD^+ to yield FAD and reduced NAD^+. The oxidation-reduction properties and structure of the coenzyme NAD^+ and the closely related $NADP^+$ were described in Chapter 11.

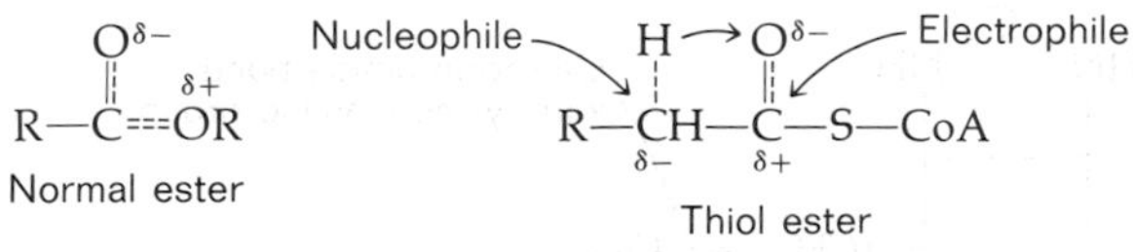

Nucleophile

Electrophile

Normal ester

Thiol ester

FIG. 14-8 *Chemistry of Acetyl CoA.*

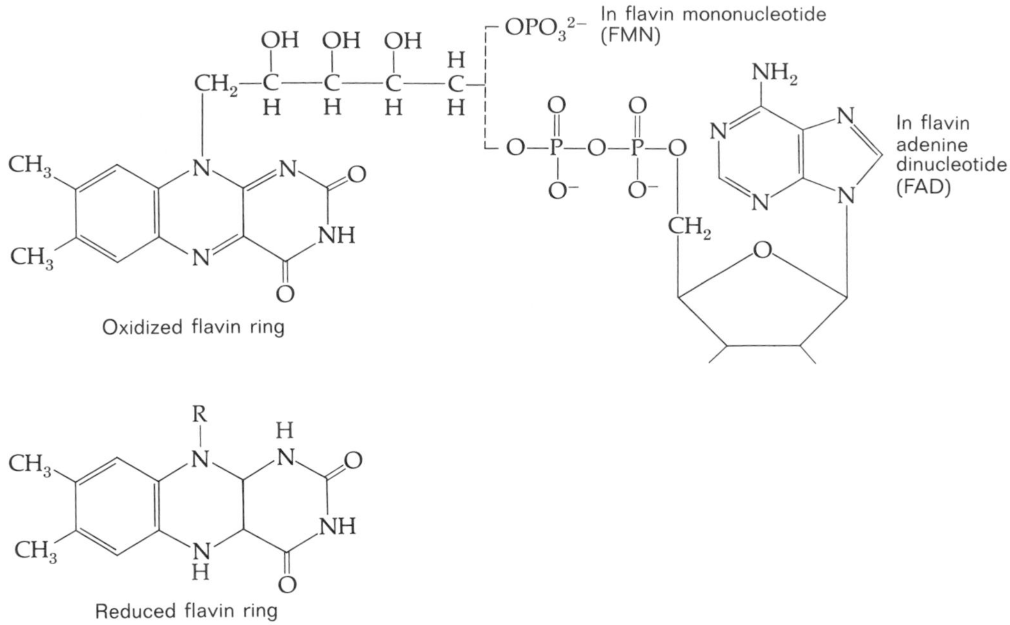

Oxidized flavin ring

Reduced flavin ring

In flavin mononucleotide (FMN)

In flavin adenine dinucleotide (FAD)

FIG. 14-9 *Flavin Structures.*

Acetyl CoA Carboxylase

In discussing the decarboxylation reaction included in the pyruvate dehydrogenase complex we encountered several important coenzymes. Reactions involving carboxylation and decarboxylation are among the most taxing for biochemical systems. While TPP is endowed with special properties for decarboxylation reactions, we will now consider a coenzyme, biotin, which facilitates carboxylation reactions. The structure of biotin is shown in Fig. 14-10, and the reaction we will consider here is the ATP-dependent formation of malonyl CoA catalyzed by acetyl CoA carboxylase. This reaction is the first step in the synthesis of fatty acids described in Chapter 11. The reaction proceeds in two steps (see Fig. 14-11): formation of a carboxybiotin intermediate with the hydrolysis of ATP followed by donation of the carboxyl group to the α-carbon of the acceptor. As in the case of lipoic acid,

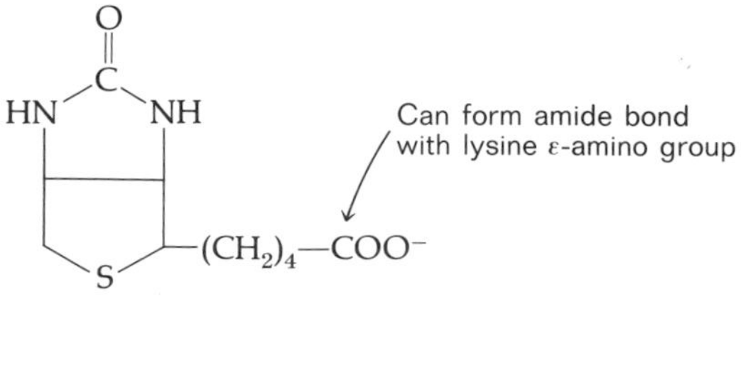

Can form amide bond with lysine ε-amino group

$-(CH_2)_4-COO^-$

FIG. 14-10 *Biotin.*

Two stages in the reaction:

CO_2 + ATP + Biotin—Enz $\rightleftharpoons$ Carboxybiotin—Enz + ADP + P_i

Carboxybiotin—Enz + Acetyl CoA $\rightleftharpoons$ Malonyl CoA + Biotin—Enz

Possible structure of carboxybiotin:

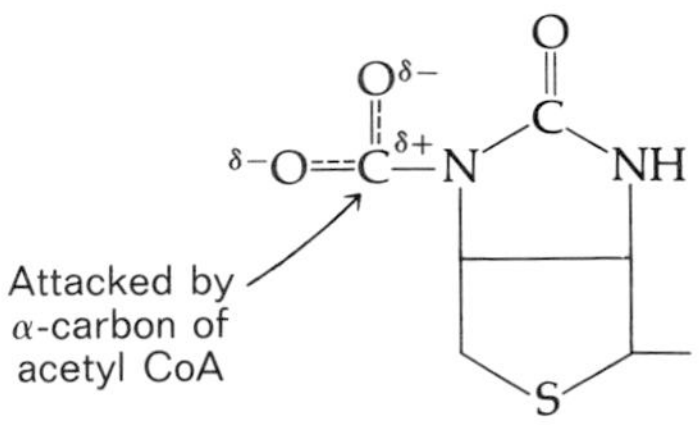

Attacked by α-carbon of acetyl CoA

FIG. 14-11 *Role of Biotin in Carboxylation.*

biotin is generally associated with proteins through a peptide bond formed between its carboxyl function and an ε-amino group of an enzyme lysine. However, in contrast to lipoic acid and TPP, which are also involved in CO_2 transfer, biotin is believed to react directly with free CO_2 and then to assist in donating the CO_2 to acetyl CoA or other possible acceptors. One interesting aside concerning the acetyl CoA carboxylase is the structure of this enzyme. It exists in an unusual form, as an extended filament. The filament is composed of smaller units of molecular weight about 400,000, each containing 4 subunits that aggregate into long, platelike filaments. These properties are in marked contrast to the simple globular nature of the vast majority of other soluble enzymes of metabolism.

One-carbon Transfers

Following our discussion on the incorporation and removal of CO_2, it should now be of interest to pursue the properties of transfer reactions involving other one-carbon units. The coenzyme involved in such reactions is tetrahydrofolic acid, derived from the vitamin folic acid. The structure is given in Fig. 14-12 and permits us to return to the topic of glycine synthesis from serine, deferred from Chapter 13. The reaction of serine to form glycine is shown in Fig. 14-13 and illustrates the role of tetrahydrofolate in one-carbon transfers. Viewing the reaction in this way, a carbon unit is donated from serine to the coenzyme, leaving glycine. For the tetrahydrofolate to be regenerated from the methylene tetrahydrofolate derivative, a carbon atom must then be transferred to another donor.

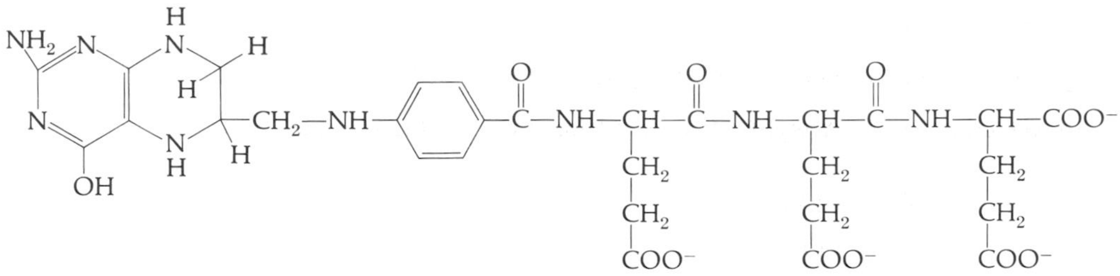

FIG. 14-12 *Structure of Tetrahydrofolic Acid.*

We have already encountered one such reaction in Chapter 13, involving the synthesis of thymidine from uracil. Other acceptors are provided at several points in the reactions of amino acid and nucleotide biosynthesis.

One final consideration of one-carbon transfers concerns the coenzyme derived from vitamin B_{12}, cyanocobalamine, one of the more recently discovered coenzymes (found by Barker and his coworkers in 1958). The structure of coenzyme B_{12}, shown in Fig. 14-14, serves to illustrate the high degree of complexity found in coenzyme structures. This particular coenzyme involves not only nucleoside units, in the form shown with 5′-deoxyadenosyl bound to the cobalt atom, but an extended ring system resembling the porphyrin structure, known as the corrin ring system. In this case two of the pyrrole rings are joined directly, rather than through methene bridges. Coenzyme B_{12} has been implicated in reactions involving the isomerization of dicarboxylic acid, the methylation of homocysteine to form methionine, and the conversion of dihydroxy compounds to the deoxy or monohydroxy grouping. An especially interesting reaction of the second type involves the formation of deoxynucleotides from ribonucleotides in the bacteria *Lactobacillus leichmanii*. This particular pathway is distinct from the more common, non-B_{12}-requiring route described for deoxy synthesis in Chapter 13.

Biosynthesis of Steroids

One additional reaction sequence, the formation of a steroid, must be included for the exotic intermediates involved and the complexity of the product formed. The biosynthetic pathway of cholesterol will be described, as solved by the elegant chemical studies of Konrad Bloch and his coinvestigators. Beginning with simple acetyl CoA and acetoacetyl CoA, condensation and reduction yields mevalonic acid, as shown in Fig. 14-15. This compound is then transformed by several further reactions to the major intermediate in cholesterol

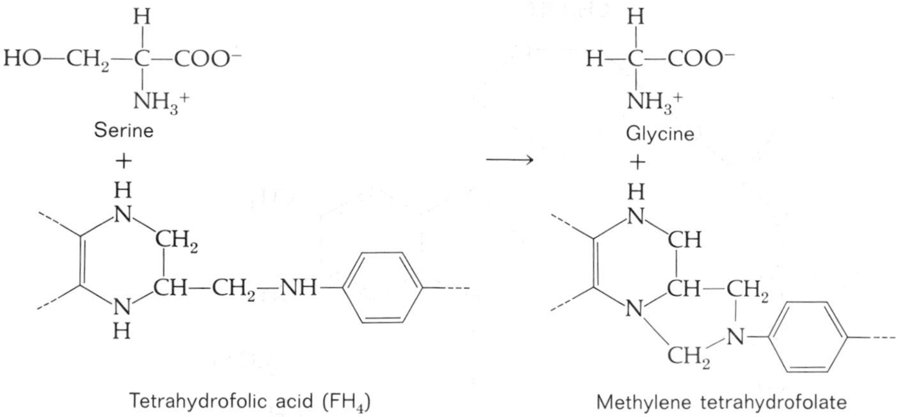

$HO{-}CH_2{-}CH(NH_3^+){-}COO^-$
Serine
+
Tetrahydrofolic acid (FH_4)
⟶
$H{-}CH(NH_3^+){-}COO^-$
Glycine
+
Methylene tetrahydrofolate

FIG. 14-13 *Reactions of Tetrahydrofolic Acid.*

synthesis, 3-isopentenyl pyrophosphate, also shown in Fig. 14-15. The isopentenyl pyrophosphate is the key intermediate in the synthesis of steroids, all of which can be visualized as consisting of these isoprene building blocks. Six isopentenyl pyrophosphate units are polymerized to give the linear, unsaturated squalene. Two units, each containing three isoprenoid residues, are attached head to head, as shown in Fig. 14-16. The final stage involves a cyclization of squalene and some additional transformations involving removal of three methyl groups, migration of a double bond, and introduction of an alcoholic OH to yield cholesterol. Cholesterol plays an important role in membrane formation (Chapter 12), and many other compounds based on the steroid structure act as hormones.

Summary of the Coenzymes

In this chapter and in the selected examples elsewhere we have described most of the major coenzymes in biochemical systems. As noted, many of the coenzymes are derivatives of vitamins. Several of the vitamins were neglected in this discussion, particularly vitamin A, involved in vision (briefly noted in Fig. 12-12); vitamin D, concerned with calcium absorption; vitamin E, which may be involved in reproductive physiology; and vitamin K, which appears to play a role in blood coagulation. In addition we have paid no special attention to metals, even though metals when involved in metalloenzyme complexes play an important function somewhat analogous to that of the coenzymes. In other sections, we have noted the involvement of iron in hemes and cytochromes and the role of copper in a number of oxidases. Zinc is found

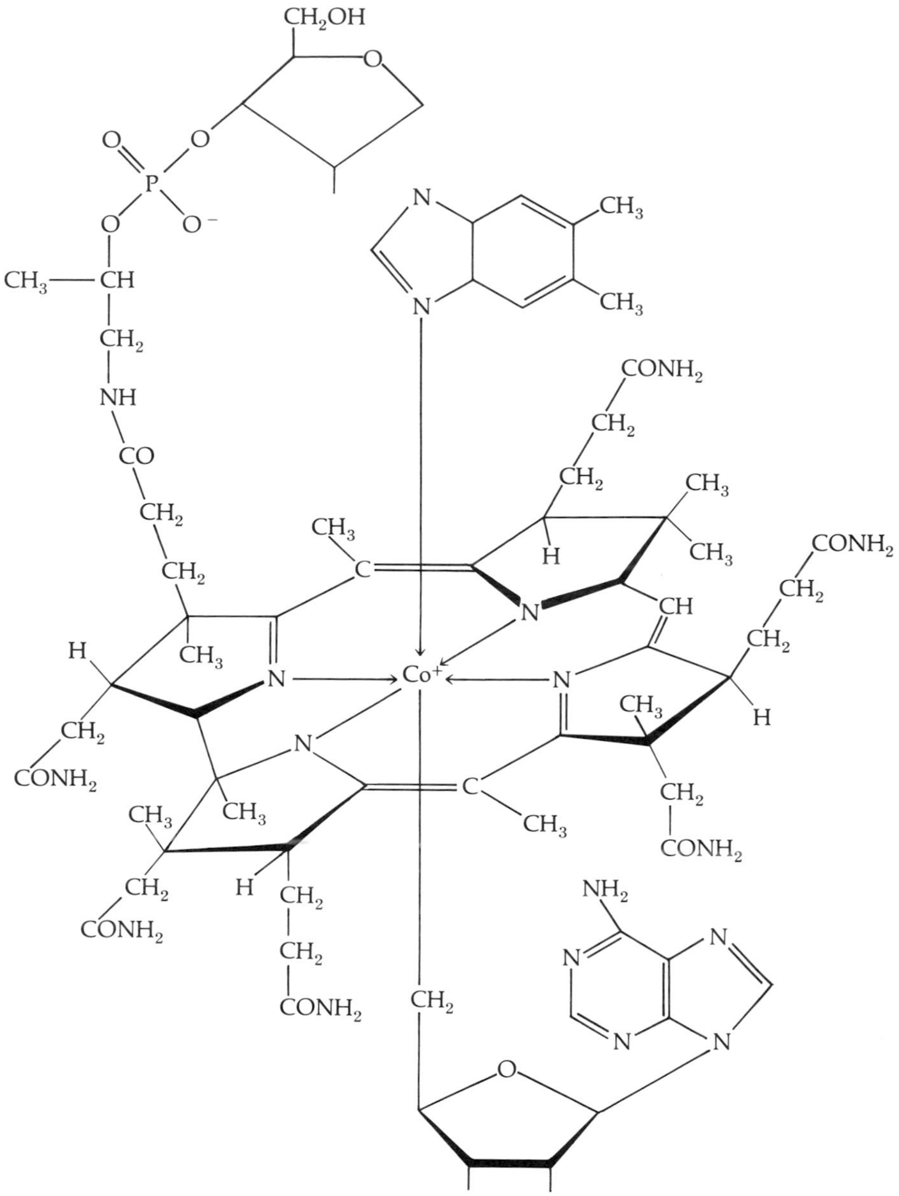

FIG. 14-14 *Coenzyme* B_{12}.

in peptidases and other proteins, such as aspartate transcarbamylase. Magnesium is commonly found in phosphatases and kinases, and manganese has a widespread occurrence in kinases as well. Molybdenum plays an important role in nitrate reductase, and cobalt was described at the central position in vitamin B_{12}. Sodium and potassium are also in-

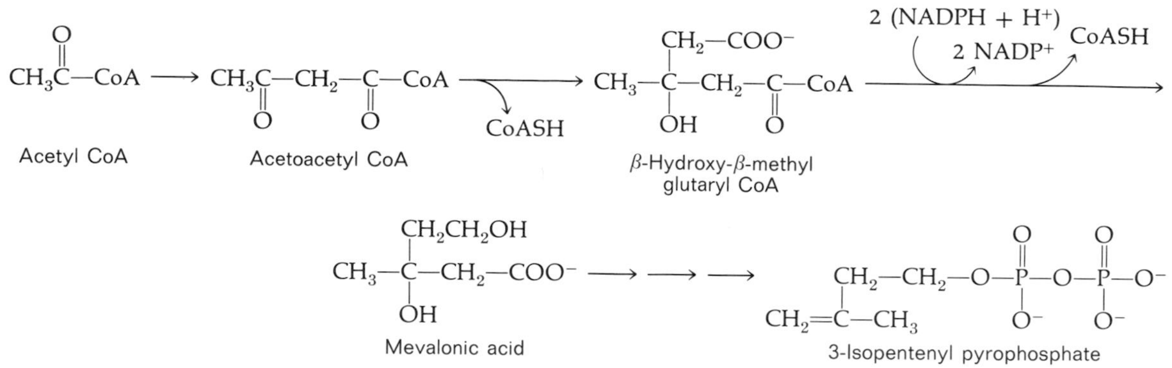

$CH_3C(=O){-}CoA \longrightarrow CH_3C(=O){-}CH_2{-}C(=O){-}CoA \longrightarrow CH_3{-}C(OH)(CH_2{-}COO^-){-}CH_2{-}C(=O){-}CoA$ (CoASH released) $\xrightarrow{2\,(NADPH + H^+) \rightarrow 2\,NADP^+;\ CoASH}$

Acetyl CoA — Acetoacetyl CoA — β-Hydroxy-β-methyl glutaryl CoA

$CH_3{-}C(OH)(CH_2CH_2OH){-}CH_2{-}COO^- \longrightarrow \longrightarrow \longrightarrow CH_2{=}C(CH_3){-}CH_2{-}CH_2{-}O{-}P(=O)(O^-){-}O{-}P(=O)(O^-){-}O^-$

Mevalonic acid — 3-Isopentenyl pyrophosphate

FIG. 14-15 *Cholesterol Biosynthesis: Initial Reactions.* For additional details, see K. Bloch, *Science,* **150:** 19 (1965).

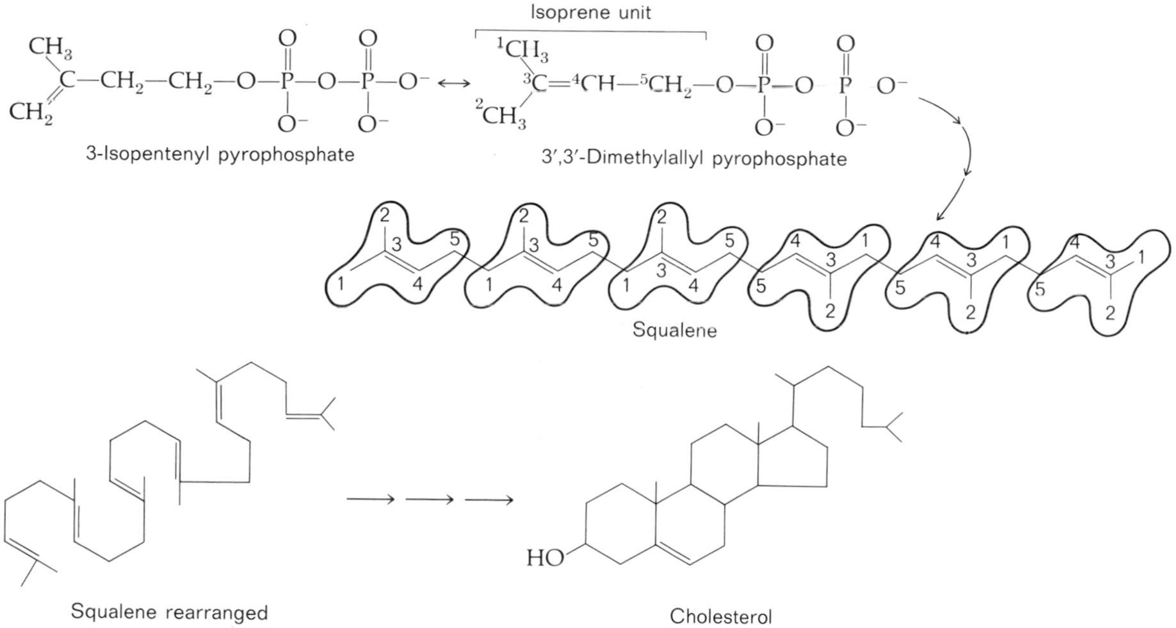

$CH_2{=}C(CH_3){-}CH_2{-}CH_2{-}O{-}P(=O)(O^-){-}O{-}P(=O)(O^-){-}O^- \longleftrightarrow {}^1CH_3({}^2CH_3){}^3C{=}{}^4CH{-}{}^5CH_2{-}O{-}P(=O)(O^-){-}O\ P(=O)(O^-)\ O^-$

3-Isopentenyl pyrophosphate — Isoprene unit — 3′,3′-Dimethylallyl pyrophosphate

Squalene

Squalene rearranged $\longrightarrow \longrightarrow \longrightarrow$ Cholesterol

FIG. 14-16 *Cholesterol Biosynthesis: Final Reactions.*

volved in a number of enzyme systems, as well as in nerve excitability. Thus, along with the vitamins and eight critical amino acids, metals are among the essential materials that must be provided in the human diet.

Chapter 15 Frontispiece *Electron micrograph of the fracture faces of a freeze-etched chloroplast internal membrane. Magnification 90,000×. Courtesy D. Branton and R. B. Park.*

INTEGRATION OF METABOLISM

15

In the past few chapters we have attempted to survey the important reactions of metabolism and some of the detailed mechanisms that have evolved to perform these reactions. In this chapter we will describe a number of the ways in which individual pathways are integrated and regulated in the general flow of metabolic events. The emphasis will be on mechanisms of *feedback regulation*. Feedback regulation is a process whereby the level of a key metabolite influences the rate of catalysis for an enzyme whose substrates and products bear no direct stereochemical resemblance to the regulatory effector. A detailed example of such a feedback mechanism, which we discussed in Chapter 10, is the regulation of the first enzyme of pyrimidine biosynthesis by the end product of the reaction pathway, CTP. This feedback loop will be encountered again at the end of this chapter with our closing comments on the regulation of nucleotide and amino acid biosynthesis. Before we reach that topic, we will describe some of the key regulatory loops in carbohydrate metabolism and the ways in which carbohydrate and lipid metabolism are interrelated. We will touch on a number of interesting mechanisms in this analysis, including the strategies evolved for integrating the metabolic reactions of the mitochondria with those in the outer cytoplasm. The reactions and interactions we consider here all concern effects at the level of enzyme catalysis.

In the next chapter we will go on to describe some of the ways in which the amounts of key metabolites can influence the quantities of the enzymes present by acting at the level of protein synthesis.

Controls in Carbohydrate Metabolism

In Chapter 11 we presented two distinct mechanisms for the metabolism of carbohydrates and the formation of ATP: glycolysis, a process that may take place in the absence of oxygen, and oxidative phosphorylation, a process that depends on oxygen for acceptance of electrons at the end of the respiratory chain. Many cells can exist both in the presence and in the absence of oxygen. These cells are known as *facultative anaerobes* and when oxygen is absent they metabolize glucose to the stage of lactic acid to generate ATP. (Some cells can live only in the absence of oxygen and are termed strict anaerobes.) Since the amount of energy provided by glycolysis is relatively small, as we have noted, large amounts of glucose must be metabolized for cells existing anaerobically.

Louis Pasteur, the brilliant nineteenth-century French biologist, first observed that facultative anaerobes, grown in the absence of oxygen, immediately curtail glucose consumption when exposed to oxygen. In addition accumulation of lactic acid is halted. These very old observations have been explained in more recent years and provide an interesting point of view for the way in which carbohydrate metabolism is regulated. We will pursue control in carbohydrate metabolism by discussing the regulatory aspects implied by the Pasteur effect and then go on to a discussion of the role of glycogen and other storage polymers in carbohydrate metabolism.

The two important aspects of the Pasteur effect, arrest of lactate accumulation and the fall of glucose consumption, are due to two relatively independent mechanisms. The lactate effect is the more straightforward of the two phenomena and is related to the relative affinities of mitochondria and the enzyme lactate dehydrogenase for NADH. When oxygen is present and oxidative phosphorylation occurs, NADH is rapidly consumed by mitochondria with an apparent affinity for NADH much higher than found with lactic dehydrogenase. Thus virtually all of the reduced NAD available passes to the mitochondria by a shuttle system (to be described later in this chapter) and lactate accumulation stops.

The rapid alteration in glucose consumption is a somewhat more complicated process and one that was only fully appreciated with the discovery and characterization of allosteric enzymes in the last decade. A key observation in this

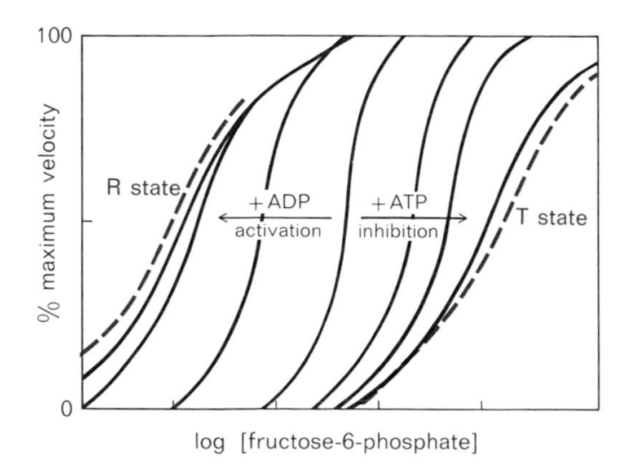

FIG. 15-1 *Properties of Phosphofructokinase.* From D. Blangy, H. Buc, and J. Monod, *Journal of Molecular Biology,* **31:** 13 (1968).

area was the finding that uncouplers of oxidative phosphorylation, such as dinitrophenol (see Chapter 12), eliminated the ability of oxygen to lower glucose consumption. It is now evident that the presence of ATP is an essential factor in executing the effect of oxygen. When mitochondria are uncoupled, even in the presence of oxygen, levels of ATP are low. When mitochondria function normally, ATP levels are high and a specific effect on glycolysis is triggered, principally through the enzyme phosphofructokinase.

Phosphofructokinase is an allosteric enzyme specifically activated by ADP and phosphate. In contrast, ATP and citrate are negative effectors that lower the activity of the enzyme. The profile of reaction velocities as a function of substrate concentration in the presence of various effectors is shown in Fig. 15-1. The typical transitions for an allosteric enzyme are evident: positive effectors shift the curve to the left and negative effectors to the right. We also see that the cooperativity of the curves diminishes at extremes of inhibitor or activator concentration, indicating that the molecular population is pulled almost entirely to the R state at the left or the T state at the right. (In the illustration presented here all curves are S-shaped because substrate concentration is given on a logarithmic scale.) Similar patterns were discussed for hemoglobin in Chapter 10.

Thus we see that the level of *energy charge* or amount of ATP, the critical currency of metabolism, has a striking influence on the conduct of metabolism. The fact that ATP exerts this influence on phosphofructokinase can also be demonstrated by the observation that the earlier substrates in the reaction sequence, fructose-6-phosphate and glucose-6-

phosphate, pile up. We also note that citrate contributes to the inhibition; therefore, when levels of mitochondrial-TCA intermediates are high, glycolysis is further depressed. Hexokinase also participates in the effect by an inhibition from its own reaction product, glucose-6-phosphate. A further example of the role of energy charge on carbohydrate metabolism is seen for the mitochondrial enzyme isocitric dehydrogenase. In this case as well, where ATP levels are high, the enzyme is inhibited. In contrast, ADP is a positive activator while NADH also inhibits this enzyme. Thus, through combination of these factors, the rates of enzyme reaction are delicately poised to operate at the optimal levels.

The Interplay of Glucose and Glycogen

In our earlier discussion of carbohydrate metabolism we emphasized the reactions proceeding from glucose. However, in most animals glucose is stored as the long carbohydrate polymer, glycogen (see Appendix III). A similar product, starch, is found in plants. The digestion of glycogen polymers to glucose is controlled by an important enzyme and one that has been studied for many years, glycogen phosphorylase. This enzyme catalyzes the phosphorolytic cleavage of glycogen polymers as described by the reaction

$$(\text{Glucose})_n + P_i \longrightarrow (\text{glucose})_{n-1} + \text{glucose-1-phosphate}$$

Glucose-1-phosphate is then readily converted into glucose-6-phosphate by another enzyme, phosphoglucomutase. The phosphorylase exists in two dimeric forms, an active, phosphorylated structure, and a less active, nonphosphory-

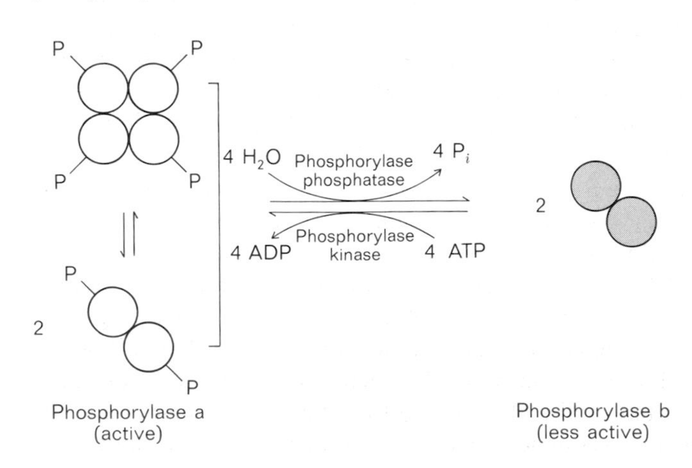

FIG. 15-2 *Glycogen Phosphorylase.*

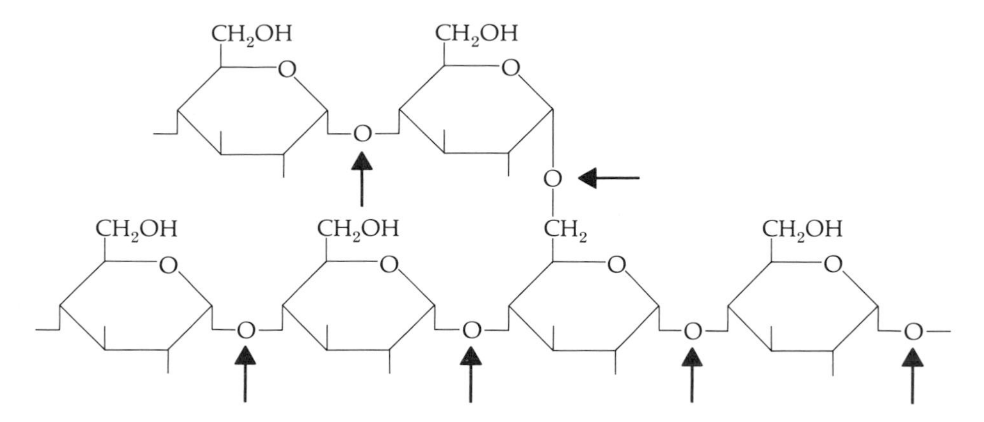

FIG. 15-3 *Glycogen,* composed of glucose units joined in α-(1–4) linkages (colored arrows) with occasional branch points involving α-(1–6) linkages (black arrow).

lated structure. The phosphorylated form also tends to associate to form tetramers at high concentration. These various structural forms are summarized in Fig. 15-2. When the reaction is permitted to go to completion, glycogen, which exists as a complex branched polymer (see Fig. 15-3), is digested to the branch points. A further enzymatic rearrangement by another enzyme is necessary to cleave at the branch points.

Resynthesis of glycogen in animal cells involves a totally different enzyme system centered around the enzyme glycogen synthetase. Glycogen synthesis is not catalyzed by a reversal of the phosphorylase reaction, as was once thought. The biosynthesis starts with an activation of the sugars through coupling to UDP. Thus the first reaction is

$$\text{UTP} + \text{glucose-1-phosphate} \longrightarrow \text{UDP-glucose} + P_i$$

The UDP-glucose substrate is then acted on by the principal enzyme glycogen synthetase, in the reaction

$$\text{UDP-glucose} + (\text{glucose})_{n-1} \longrightarrow \text{UDP} + (\text{glucose})_n$$

Just as phosphorylase action is limited at the branches, so glycogen synthetase is only capable of adding on to already existing glycogen units. In addition it does not catalyze branching; another specific enzyme is needed for that process.

Glycogen synthetase also exists in two discrete forms, an active form known as synthetase A and a less active form, synthetase D. Further striking parallels with phosphorylase exist, for example, the presence of the D form as a phosphorylated derivative. As described in Fig. 15-4, the phosphorylation and dephosphorylation reactions for both enzymes involve kinases that phosphorylate the enzyme with

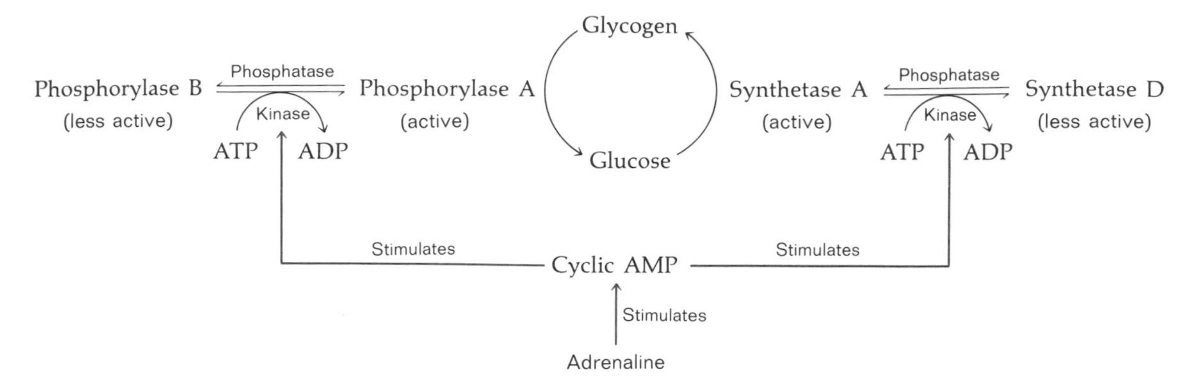

FIG. 15-4 *Control of Glucose-Glycogen Interconversions.* Adrenaline acts by activating adenyl cyclase, an enzyme associated with cell membranes, which forms cyclic AMP from ATP with the release of pyrophosphate. In a "cascade" effect of activators, cyclic AMP stimulates phosphorylase kinase by activating an enzyme called phosphorylase kinase kinase, which phosphorylates phosphorylase kinase, enhancing its activity.

ATP as the phosphate donor and phosphatases that hydrolyze the phosphate from the enzyme. The kinase activities for both phosphorylase and glycogen synthetase are under further control. They are stimulated by a key metabolic effector, cyclic AMP. This compound, 3′,5′-cyclic AMP, will be discussed in somewhat more detail in the next chapter, in terms of its role in the control of protein synthesis. Here we will note that cyclic AMP is formed in high levels in cells exposed to the hormone adrenaline.

Adrenaline is a general stimulant, and by raising cyclic AMP levels, it stimulates the kinase for phosphorylase to produce phosphorylase A, the active form of phosphorylase which will accelerate the breakdown of glycogen to glucose, needed for energy-requiring processes. Concurrently, cyclic AMP stimulates the kinase for the glycogen synthetase. This effect drives the glycogen synthetase into the phosphorylated D form, which in this case is less active in glycogen synthesis. Therefore the glucose formed is not short-circuited back to glycogen but rather held as the free carbohydrate for operations in carbohydrate metabolism. These various interconversions are summarized in Fig. 15-4, and illustrate one of the ways that hormones interact with other biochemical systems.

Gluconeogenesis

While the reactions of glucogen synthetase permit excess glucose to be placed in reserve as glycogen polymers, we can go one step further to describe how transformations of other intermediates in carbohydrate metabolism can be used to produce glucose. Reactions of this type are especially important under fasting conditions in higher animals, when amino acids are broken down to form the glucose needed for the maintenance of the brain. The process of new synthesis of glucose is known as *gluconeogenesis* and involves several special reactions. A simple reversal of the glycolytic scheme, shown in Chapter 11, is not possible. Three barriers are present:

1. The formation of phosphoenol pyruvate from pyruvate cannot be achieved simply by a reversal of pyruvate kinase. The reaction lies too far in the favor of pyruvate for reversal to pass meaningful amounts of material. Therefore a special mechanism, the malate shuttle, has evolved, which involves passage of pyruvate through the oxaloacetate and malate present in the mitochondria. The carbon units are returned to the cytoplasm in the form of malate where further transformations provide phosphoenol pyruvate. The various reactions and enzymes of the malate shuttle are summarized in Fig. 15-5.
2. The second barrier concerns the reaction of phosphofructokinase. Here, as in the pyruvate kinase reaction, the equilibrium lies too far to the right for reversal to provide meaningful amounts of fructose-6-phosphate. Therefore a new route has evolved involving the enzyme fructose diphosphatase. This pathway is also summarized in Fig. 15-5.
3. The third and last barrier is the reaction catalyzed by hexokinase. As gluconeogenesis yields glucose-6-phosphate, the reaction catalyzed by hexokinase lies too far to the right to reverse and yield free glucose. Therefore a new enzyme is called into play, glucose-6-phosphatase, and glucose is produced. In some cases gluconeogenesis will stop at the glucose-6-phosphate level and branch into the reactions of the glycogen synthetase pathway via the mutase reaction to convert glucose-6-phosphate to glucose-1-phosphate, as is also shown in Fig. 15-5.

Coordination of Carbohydrate and Fatty Acid Metabolism

Carbohydrates and fats provide the two great reduced commodities of metabolism. In animals glucose is readily converted to fats for storage via the intermediary acetyl CoA. Acetyl CoA synthesized for possible use in the TCA cycle can be diverted to the reactions of fatty acid synthesis to be stored for later use rather than metabolized for energy at that time. The opposite reaction from fats to glucose is somewhat

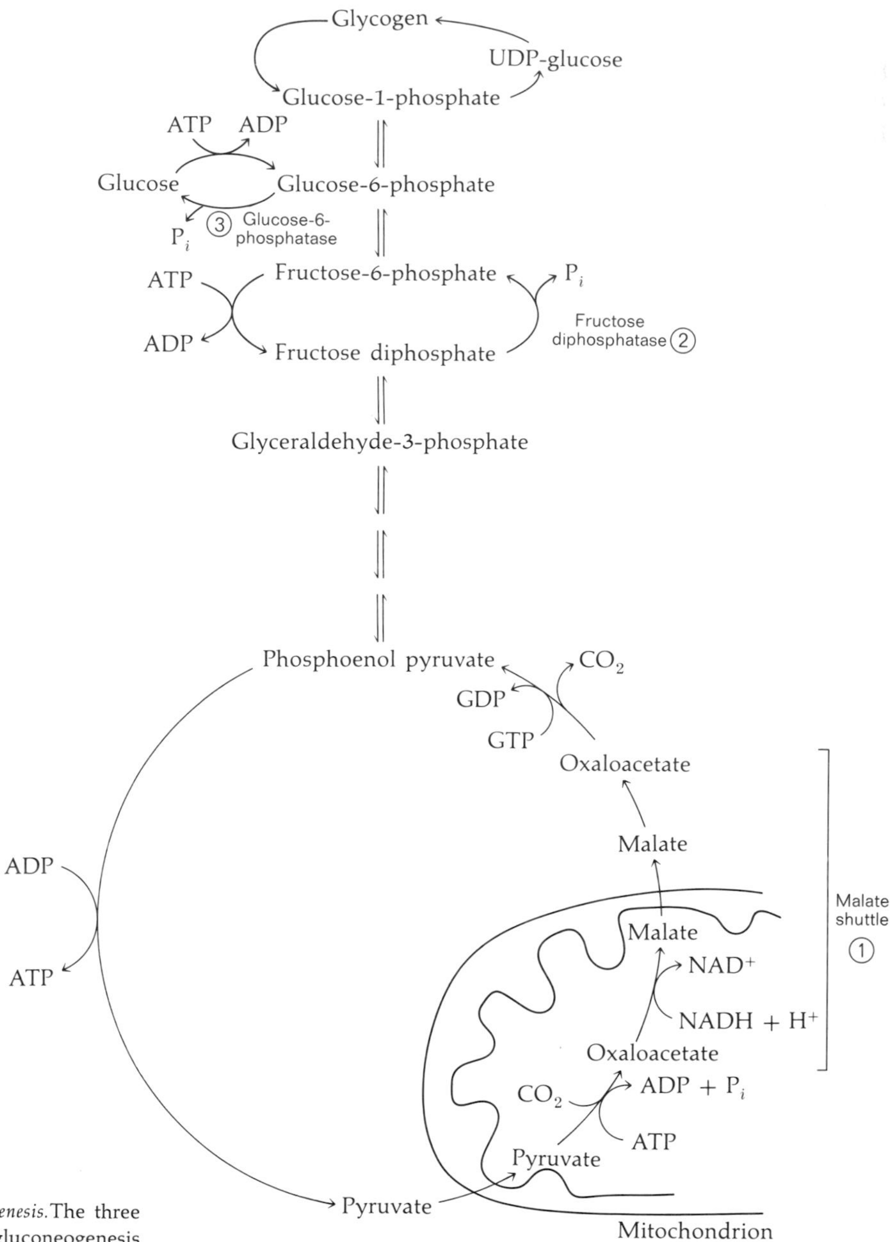

FIG. 15-5 *Steps in Gluconeogenesis.* The three points indicated are where gluconeogenesis requires alternate routes to reversal of glycolysis. Modified from A. L. Lehninger, *Biochemistry*, Worth Publishers, New York, 1970.

less direct. In mammals metabolism of fatty acids cannot lead to a net synthesis of glucose. Even though molecules of acetyl CoA, derived from fatty acids in fatty acid oxidation, will be incorporated into the TCA cycle, equivalent amounts of CO_2

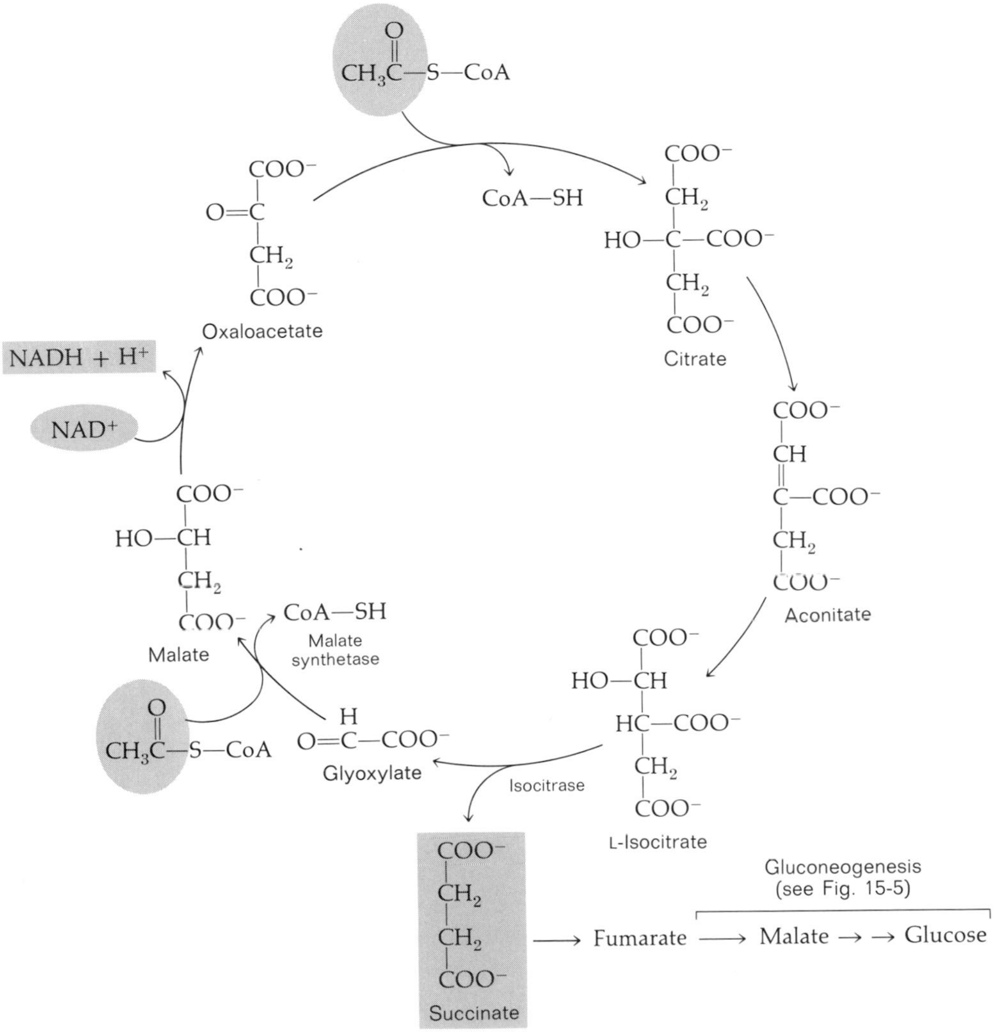

FIG. 15-6 *The Glyoxylate Cycle.* The malate synthetase and isocitrase are special enzymes of this cycle.

are displaced at subsequent steps in the cycle so that no *net* incorporation of carbon atoms takes place (see Fig. 11-10). In plants, however, net synthesis of glucose is accomplished from fatty acids, and is an important aspect of several metabolic processes including germination. In this case two molecules of acetyl CoA are converted into a succinate through the action of the glyoxylate cycle (Fig. 15-6). The formation of the key intermediate, glyoxylate, occurs from the breakdown of isocitrate to glyoxylate and succinate. The glyoxylate can condense with a molecule of acetyl CoA to form malate.

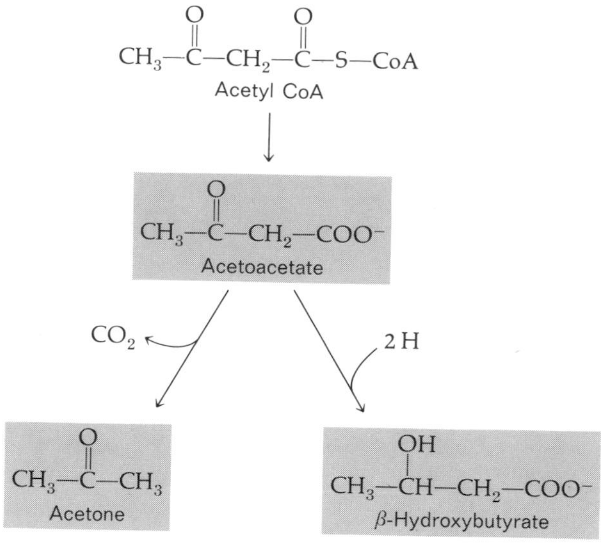

$$CH_3{-}\overset{\overset{\displaystyle O}{\|}}{C}{-}CH_2{-}\overset{\overset{\displaystyle O}{\|}}{C}{-}S{-}CoA$$

Acetyl CoA

↓

$$CH_3{-}\overset{\overset{\displaystyle O}{\|}}{C}{-}CH_2{-}COO^-$$

Acetoacetate

CO_2 ← (to Acetone) 2 H → (to β-Hydroxybutyrate)

$$CH_3{-}\overset{\overset{\displaystyle O}{\|}}{C}{-}CH_3$$

Acetone

$$CH_3{-}\overset{\overset{\displaystyle OH}{|}}{CH}{-}CH_2{-}COO^-$$

β-Hydroxybutyrate

FIG. 15-7 *Formation of Ketone Bodies* (gray rectangles).

Malate goes on to oxaloacetate, which can combine with another molecule of acetyl CoA to re-form citrate. The succinate produced can then go on to form glucose, as indicated, or one of a number of other compounds for which succinate is an intermediate. In addition one molecule of NAD is reduced and can provide electrons for oxidative phosphorylation.

An interesting aside in the reactions of acetyl CoA concerns the role of insulin. Insulin is one of the class of protein hormones composed entirely of amino acids and is produced in the pancreas. Individuals with a deficiency in insulin production, known as *diabetes,* develop a number of dramatic symptoms related to the inability of their cells to utilize glucose. Glucose accumulates in the blood and is excreted in the urine. The organism responds in a number of ways, many leading to the accumulation of acetyl CoA. The acetyl CoA condenses to form acetoacetyl CoA, familiar in the mechanism of fatty acid synthesis, which goes on to the free acetoacetic acid. This molecule can undergo decarboxylation to form acetone or reduction to yield β-hydroxybutyrate. These substances are known as ketone bodies (see Fig. 15-7). In acute conditions the individual may actually give off an odor of acetone from accumulation of these bodies. Formation of these molecules also occurs during prolonged fasting or during starvation.

Crossing the Mitochondrial Barrier

Most of the transformations we have described so far occur in the inner compartment of the mitochondria or the outer cytoplasm of cells. We have noted a few instances where transfers take place in which small molecules are passed across the mitochondrial membrane. This transport process is very important in aerobic glycolysis where $NADH + H^+$ must find its way from the cytoplasm to the electron-transport chain, and also in the malate-shuttle system for the formation of phosphoenol pyruvate from pyruvate in gluconeogenesis, discussed above. The principal pathway for moving molecules across the mitochondrial border, or metabolic smuggling, as one biochemist has put it, involves the glycerol phosphate shuttle. The coenzyme $NADH + H^+$ itself cannot permeate across the mitochondrial inner membrane. However, it reacts with glycerol phosphate dehydrogenase to catalyze the transformation of dihydroxyacetone phosphate into glycerol phosphate in the scheme shown in Fig. 15-8. The glycerol phosphate readily passes through the membrane and reacts with another enzyme system to regenerate dihydroxyacetone phosphate and to donate its electrons to the reduction of a flavin dehydrogenase. The electrons are then passed along the

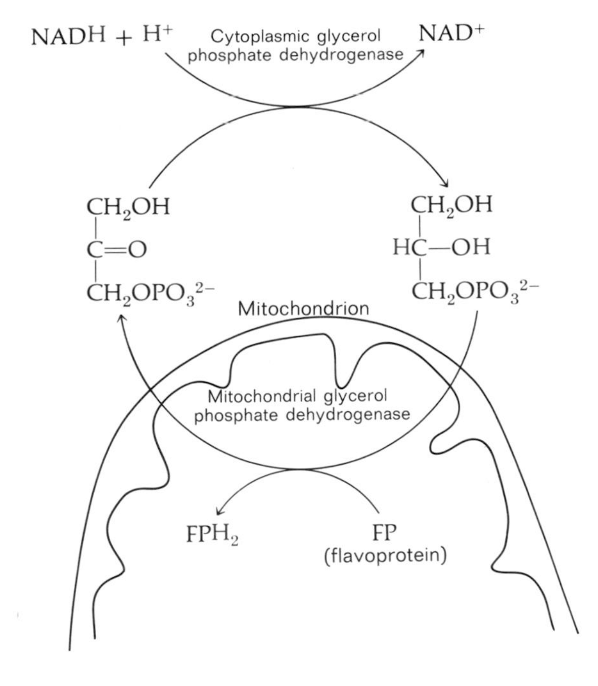

FIG. 15-8 *Glycerol Phosphate Shuttle.*

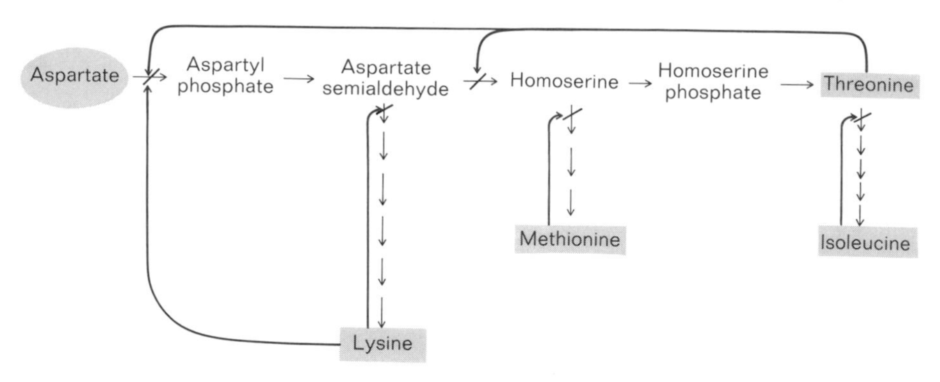

FIG. 15-9 *Control in a Branched Pathway.* Colored arrows represent feedback inhibition loops, in which end-product amino acids inhibit enzymes at reactions indicated. For a general discussion of control patterns, see E. R. Stadtman, in P. D. Boyer, ed., *The Enzymes,* 3rd ed., vol. 1, Academic Press, New York, 1970, p. 397.

oxidative phosphorylation chain. However, they have entered somewhat farther along the chain than electrons from intramitochondrial NADH + H^+, so that only two molecules of ATP are formed from each NADH + H^+ metabolized in this way. This decreased output of ATP from NADH was noted in our bookkeeping of glycolysis in Chapter 11.

The glycerol phosphate shuttle is unidirectional: it can only bring reducing equivalents into the mitochondria. However, the other shuttle we discussed, the malate shuttle (Fig. 15-5), is a reversible shuttle. Therefore a reversal of the scheme shown in Fig. 15-5 can also be employed to bring reducing equivalents into the mitochondria in those cells where the shuttle is present. In this case the reducing equivalents are transferred to the electron-transport chain to yield a full three ATP molecules.

Feedback Control in Amino Acid and Nucleotide Biosynthetic Pathways

The final topic in regulation at the metabolic level concerns the important precursors such as amino acids and nucleotides. All of the biosynthetic pathways for amino acid synthesis are under control by systematic feedback inhibition in a very regular way. The amino acid, the end product of its respective synthetic pathway, inhibits the first enzyme in the pathway. Somewhat more complex patterns are observed in pathways involving branch points leading to the synthesis of more than one amino acid. An example of this type of pathway, with its multiple levels of feedback control, is summarized in Fig. 15-9. Here we see that the principal pathway is the synthesis of L-threonine, beginning with L-aspartate. (Some of the cofactors and intermediates are left out for simplicity.) Threonine goes on to yield isoleucine, and other intermediates in the pathway for threonine give rise to lysine and methionine. All the amino acids formed feed back to the first enzyme in the pathway to regulate their own synthesis, and in the case of lysine, a feedback loop to the initial enzyme in the entire scheme is also present.

Control in the purine biosynthetic pathways involves the action of either ATP or GTP as the regulatory feedback inhibitor on the first enzyme of purine biosynthesis, the amidotransferase for the formation of 5-phosphoribosyl amine from PRPP. The general scheme of this pathway and its inhibition are shown in Fig. 15-10. The feedback loops are fairly complicated and can involve adenosine and guanosine at various levels of phosphorylation. In addition there is positive control for the formation of GMP when high levels of ATP are present and the reciprocal formation of AMP when high levels of GTP are present. The enzymes for the synthesis of ATP and GTP diverging from inosinic acid are also subject to feedback control. Feedback inhibition in the pyrimidine biosynthetic pathway has already been described in Chapter 10, with reference to the first enzyme in the pathway, aspartate transcarbamylase, and its inhibition by CTP and its activation by ATP.

In the next chapter we will go on to the final level of regulation, control of protein synthesis, and show how the action of intermediates in the metabolic pathways can feed back to the level of DNA itself and influence the synthesis of proteins.

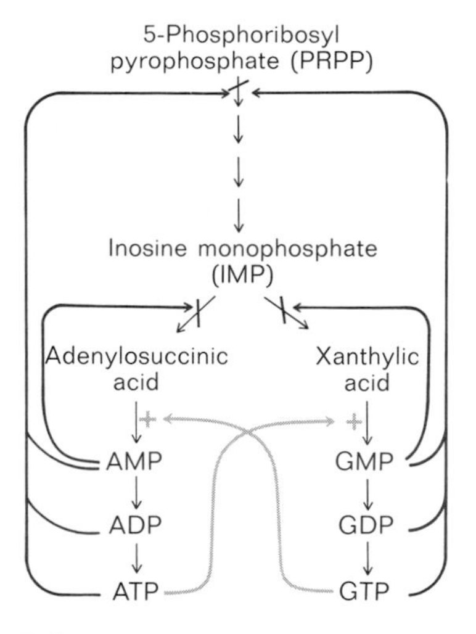

FIG. 15-10 *Regulation of Purine Biosynthesis.* Colored arrows represent feedback inhibition loop, in which products inhibit enzymes catalyzing the reactions indicated. Gray arrows represent positive feedback loops, in which the products of the pathways raise the activity of the enzymes catalyzing the reactions indicated.

PROBLEMS

1. What is the effect of ATP on the allosteric equilibrium of phosphofructokinase, and how does this effect relate to the overall metabolic pattern of the cell?
2. What are the responses of glycogen metabolism to an increase in adrenaline levels?
3. At which points does gluconeogenesis differ from a simple reversal of glycolysis?

Chapter 16 Frontispiece *Electron micrograph of isolated lac operon. Courtesy J. Beckwith.*

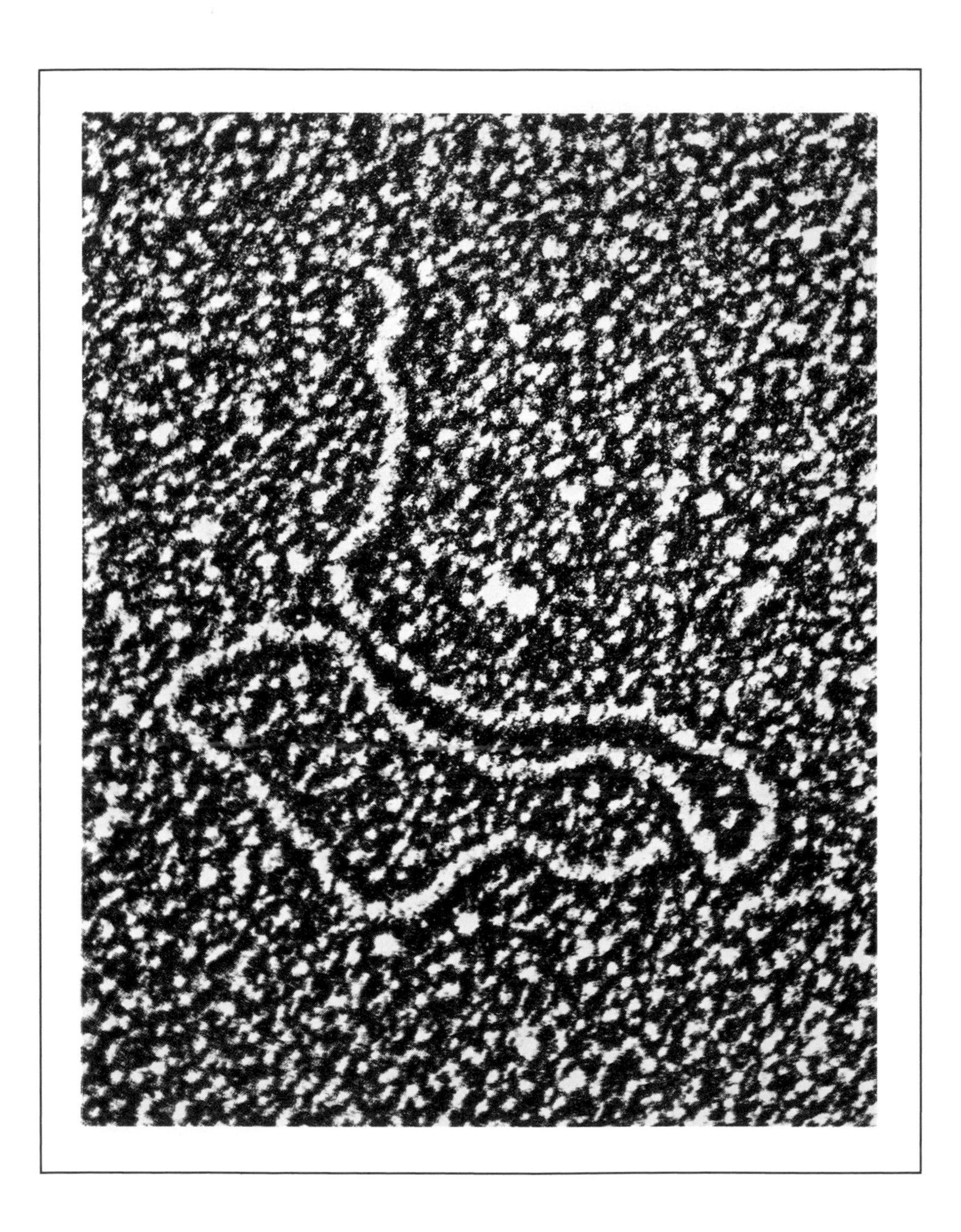

CONTROL OF PROTEIN SYNTHESIS

16

Through our discussion of nucleic acids and proteins we have emphasized the concept of *adaptation* or change. We began our consideration of biochemistry with a discussion of evolution, a most powerful form of biochemical change, and a very slow one. Natural mutation rates generally are less than one in a million per gene duplication, and evolution takes a long time, as we can see by recalling the billions of years required to develop human beings. Here we are, so we must give every credit to evolution, but we can note that it is slow and largely irreversible.

A second type of adaptation is the allosteric control exhibited by many proteins, particularly the enzymes at key metabolic junctions (Chapter 15) and illustrated with aspartate transcarbamylase (Chapter 10). In contrast to evolution, allosteric control is extremely rapid, but it is rather limited. We can only change the conformation of a protein already synthesized from one state to another, at most from an active state to an inactive state.

What is missing in our discussion of adaptation is the intermediate type of control—something between allosteric control and evolution. This level is found in the control systems of protein synthesis—control at the level of nucleic acids. Here control is intermediate in speed between evolution and allostery, yet extremely valuable, especially in

1. *Cell economy,* shutting off the synthesis of proteins when they are not needed and keeping all levels at peak efficiency
2. *Differentiation,* the remarkable process whereby a single fertilized egg can multiply, enlarge, invaginate, and ultimately elaborate distinct cells that comprise the tissues and organs of a tremendously complex organism. One

> cell becomes a neuron of the brain; another serves as a muscle of the arm; a third secretes mucus in the intestine. All contain the same complement of DNA, yet they possess strikingly different characteristics. How do these differences arise?

Unfortunately we can say very little about the way in which protein synthesis is controlled in the cellular differentiation that occurs in the embryological development of the higher forms of life, although some effort will be made in the next chapter. Here we will consider a few well-studied but much simpler examples in the control of protein synthesis, examples that come mainly from the familiar procaryotic organisms. These examples will provide a basis for the exploratory and speculative discussion of differentiation in Chapter 17.

If we consider a cell such as *E. coli,* we recognize that the amounts of individual proteins, such as DNA polymerase and β-galactosidase, needed to sustain life may be very different. Only a few molecules of DNA polymerase might be enough to act on the single macromolecule of DNA in the cell. However, many molecules of β-galactosidase could be usefully employed if the cell were growing in a medium rich in lactose. A further complication arises from the possible changes in nutrient source. While high levels of β-galactosidase would be important in a lactose medium, if the cell switched to glucose the high levels of the β-galactosidase would be a handicap. Enzymes are expensive to synthesize. The amino acids must be synthesized and activated, and the RNA synthesized and transcribed. Therefore the cell is challenged to regulate the levels of synthesis of proteins to take into account those that are needed in different amounts and, moreover, the changes in amounts that may be required at different times. A relatively simple system that exhibits such responses is the lactose operon of *E. coli.*

The Lactose Operon

The regulatory system that *E. coli* has evolved for β-galactosidase is known as the *lactose* (*lac*) *operon* (Fig. 16-1). The *lac* operon contains in close proximity on the DNA the *z gene* for β-galactosidase, the *y gene* for permease (a protein molecule that mediates the transport of galactosides across the cell membrane), and the *a gene* for another protein, acetylase, of

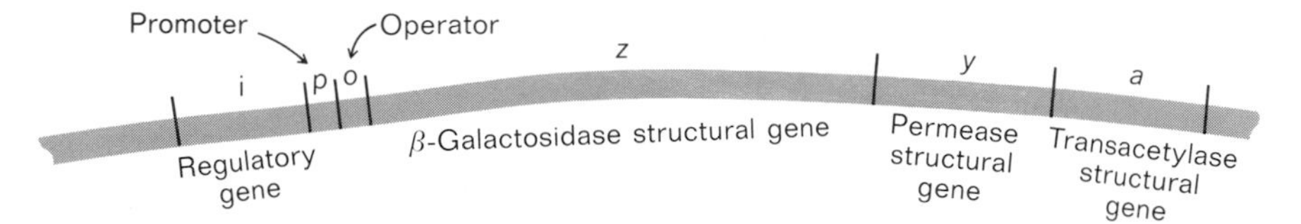

FIG. 16-1 *The Lactose Operon*. The cistrons of the lactose (*lac*) operon are drawn approximately to the scale they occupy on the chromosome. For additional details, see D. Zipser and J. Beckwith, *The lac Operon*, Cold Spring Harbor Laboratory of Quantitative Biology, New York, 1970.

unknown function. In addition the *lac* operon contains a regulatory gene known as the *i gene,* which controls the levels of synthesis of the proteins in the operon.

What happens when *E. coli* is transferred from a medium containing glycerol as the principal nutrient to a medium containing lactose? We can see in Fig. 16-2 that the amount of enzyme present in the cell increases dramatically in the presence of the lactose. In this case lactose is called an *inducer* since it induces the synthesis of the enzyme β-galactosidase. Levels of permease and acetylase are also markedly stimulated in the presence of the inducer, to nearly the same degree as β-galactosidase. Thus the three proteins are *coordinately* induced and constitute an *operon*. The coordination of the levels of the three proteins derives from the translation of the three genes from a *single* messenger RNA molecule.

One of the questions first raised was whether induction is a control at the level of DNA, messenger RNA, or possibly at an allosteric level. Lactose could induce de novo synthesis of the β-galactosidase or simply provoke a conformational change from an inactive to an active form. Experiments were designed to test this question by collecting cells grown with radioactive amino acids in the absence of an inducer. The cells were then collected on a filter, washed, and resuspended in a medium containing inducer and unlabeled amino acids. The β-galactosidase appeared quickly and was then isolated and examined for radioactivity. If the enzyme had been present initially in a masked form, it should contain radioactively labeled amino acids. But if the enzyme were newly synthesized after transfer to the medium with inducer and unlabeled amino acids, little or no radioactivity should be detected in the enzyme. The results, summarized in Fig. 16-3, clearly indicate that the enzyme is newly formed. Therefore the process of induction must be involved with the *synthesis* of the protein. These experiments were performed with the "gratuitous" inducer, isopropyl-β-thiogalactoside (IPTG), shown in Fig. 16-4, which participates in the control process of the lactose operon but is not actually degraded by β-galactosidase. Therefore factors relating to inducer level

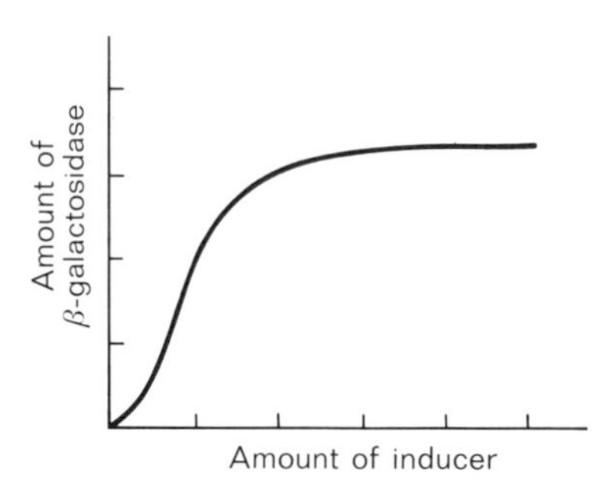

FIG. 16-2 *Induction of β-Galactosidase.* In the absence of inducer, very little β-galactosidase is produced. As inducer is added, enzyme levels increase to a maximal plateau value at high levels of inducer.

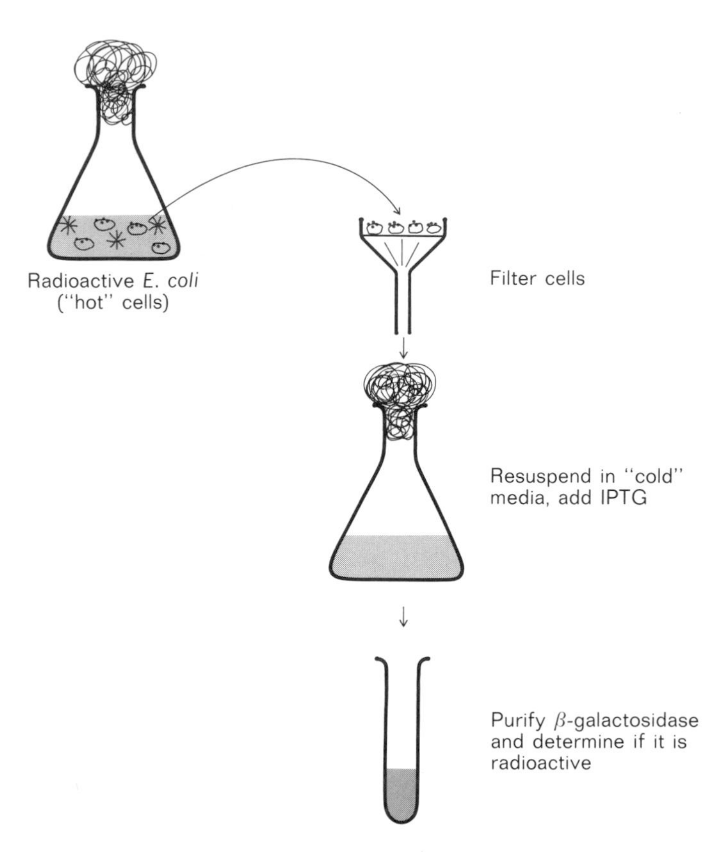

FIG. 16-3 *Synthesis of β-Galactosidase.* Radioactive ("hot") cells of *E. coli* are filtered and resuspended in nonradioactive ("cold") media with inducer (IPTG) added. The β-galactoside induced is isolated and examined for radioactivity. It is found to be nonradioactive, indicating that it was newly synthesized from amino acids in the nonradioactive media in response to inducer.

were easily separated from losses of inducer when it was metabolized by enzymes.

Having established that β-galactosidase is synthesized from amino acids in the presence of inducer, we can now go on to consider just how this induction process is accomplished. Characterization of the induction system was achieved by Jacob and Monod at the Pasteur Institute in Paris and illustrates an area in which biochemistry and genetics have joined in a very fruitful collaboration. Mutants of *E. coli* were obtained that could be localized as single-base changes in the *i* gene, and that rendered the cells noninducible. These mutations, known as i^- mutations, produced a high level of β-galactosidase (as well as permease and acetylase), regardless of whether or not the inducer was present. This behavior of noncontrollable high-level synthesis is known as *constitutive* synthesis.

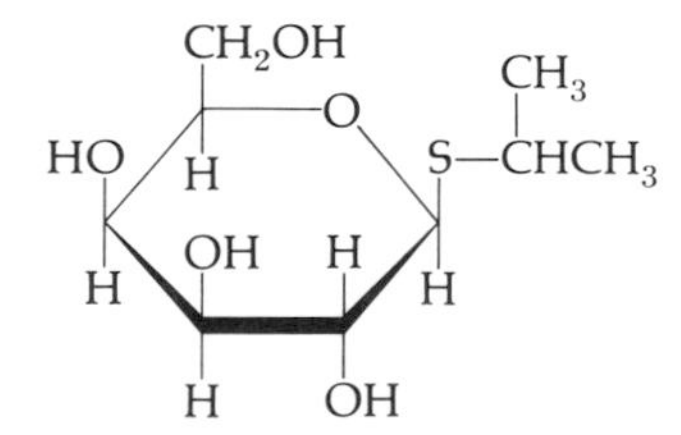

FIG. 16-4 *Structure of Isopropyl-β-D-thiogalactoside (IPTG).*

Another class of mutants was obtained and mapped at the beginning of the *z* gene. These mutants were also constitutive mutants and are abbreviated o^c. The behavior of these constitutive mutants can be comprehended by referring back to the scheme for the *lac* operon given in Fig. 16-1. Here we recall that in addition to the three structural genes *z*, *y*, and *a*, a regulatory gene known as the *i* gene is also present. If we assume that the *i* gene produces a protein called the *repressor*, which binds to DNA at an *operator* site at the beginning of the *z* gene, we can see that the repressor could block messenger RNA synthesis and effectively shut off the lactose genes. The further postulate that lactose, an inducer, stabilizes a conformation of the repressor protein that renders it unable to bind to the operator site would explain the induction process. Such a conformational change in the protein could be easily understood in terms of the conformational changes we have already considered in allosteric proteins (Chapter 10). For example, as we show in Fig. 16-5, a two-state allosteric-type model for the repressor protein would accommodate this view. The T state would have a high affinity for the operator and a low affinity for the inducer. However, when the inducer is added to the system, the conformation called R, which has a high affinity for the inducer and a low affinity for the operator, would be favored, leaving the free operator.

With this simple view of the repressor, we can also explain the constitutive mutants. An alternation in the *i* gene, which results in an inactive or modified repressor molecule, would yield a protein unable to bind to the operator. Therefore constitutive synthesis would occur. Similarly, an altera-

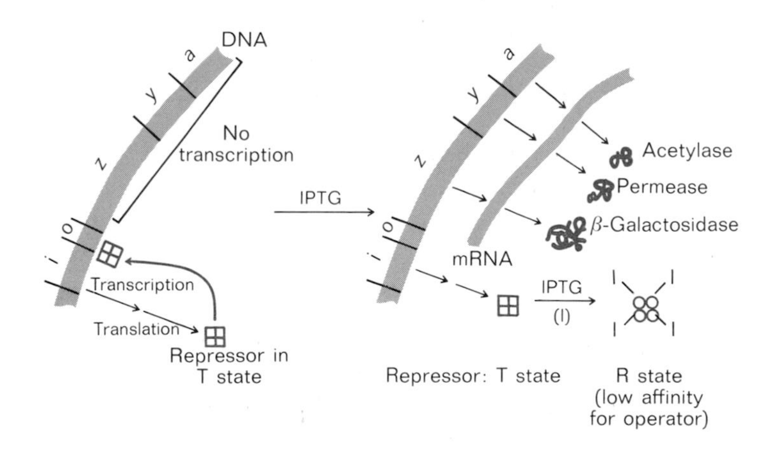

FIG. 16-5 *Induction.*

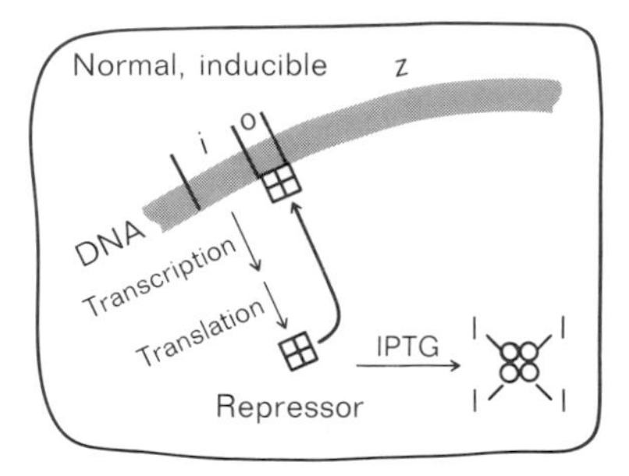

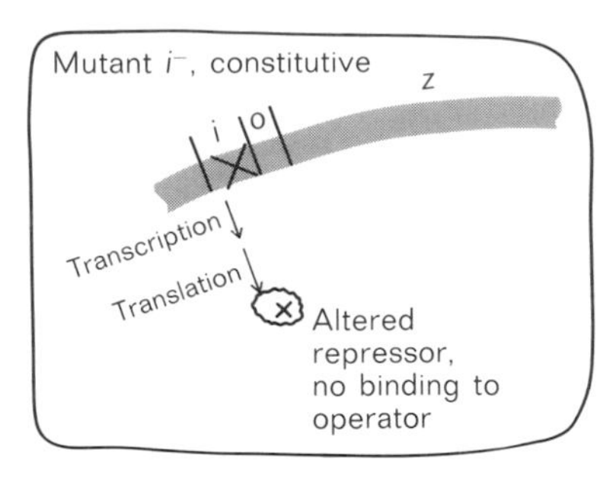

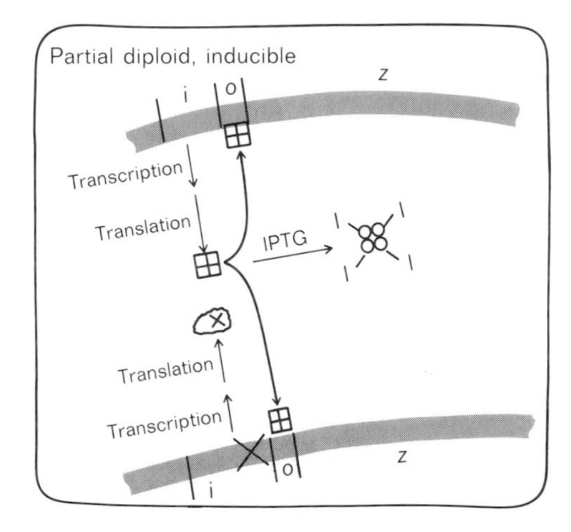

FIG. 16-6 *Mutations in the i Gene.* The basis of the operon model is presented in F. Jacob and J. Monod, *Journal of Molecular Biology,* **3:** 318 (1961).

tion in the operator of the *z* gene itself may render the site unrecognizable to the repressor, and with the inability of the repressor to bind, constitutive synthesis takes place. Support for just such a view came from experiments with a partial diploid strain of *E. coli.* Although procaryotic organisms generally contain a single copy of the genetic message, and thus are distinct from the diploid character of eucaryotic organisms, it is possible to make partial diploids. In this case a small region of the *E. coli* genome is present in two copies, one on the genetic message and the second on a short independent piece of DNA known as an *episome.* By proper genetic manipulation it is possible to have a wild-type copy of a gene on the chromosome and a mutated form of the same gene on the episome, or vice versa. When strains of this type were constructed with the constitutive *lac*-operon mutations of *E. coli,* an interesting pattern of behavior was found. When one copy of the *i* gene is present in the wild-type form, even though the second copy is nonfunctional, the cell is rendered

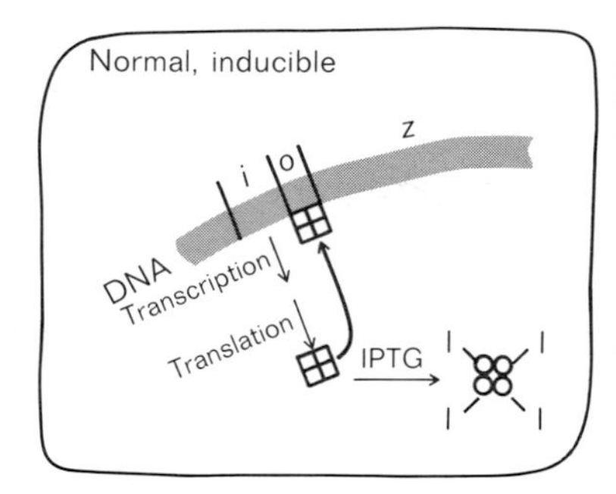

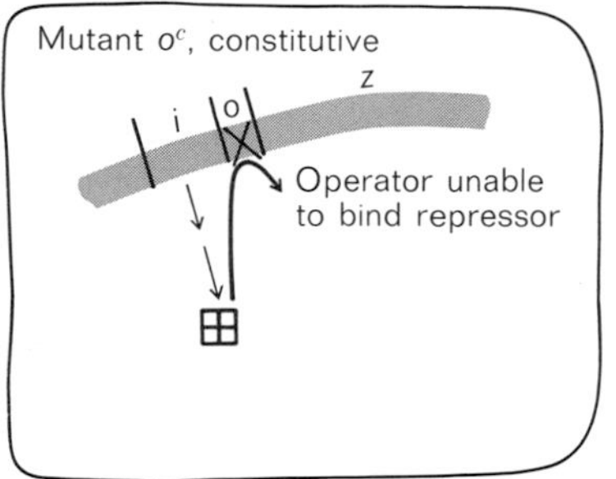

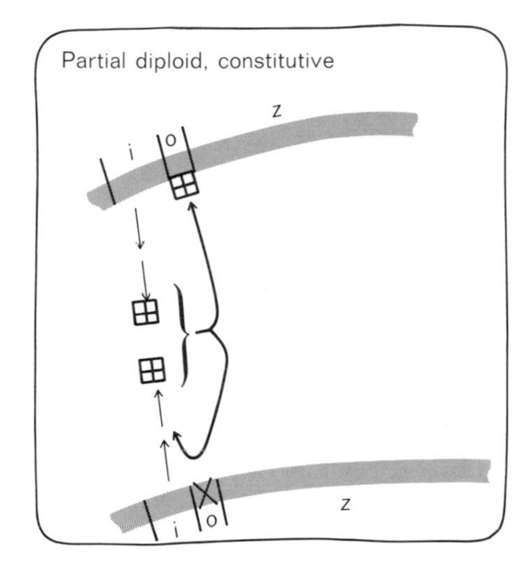

FIG. 16-7 *Mutations in the Operator.*

inducible. Constitutive synthesis no longer occurs, and lactose or IPTG must be added to stimulate the appearance of enzyme. Thus the repressor formed from the "good" *i* gene can act on the operators of both the chromosome and episome *z* genes. As long as one good *i* gene is present, enough repressor is made to control synthesis. This behavior is summarized in Fig. 16-6 and provides strong suggestive evidence for the existence of a diffusible repressor molecule.

When similar experiments are performed with a mutation in the operator—the o^c mutation located on the episome and the wild-type operator on the normal chromosome—constitutive synthesis still occurs. This pattern is summarized in Fig. 16-7 and implies that even in the presence of a competent repressor the operator constitutive mutant is unable to provide an adequate binding site for the repressor, and constitutive synthesis is maintained.

Thus the analysis of the *lac* operon reveals a simple yet highly efficient method for regulating the levels of an enzyme. The description just formulated is the basic outline of the model and was for some time thought to be just about all that one needed to know concerning the control of protein synthesis. Simplicity reigned at the level of DNA. Complex interactions such as differentiation were viewed as elaborations and variations of the basic theme. However, as we will now see, a number of recently discovered factors render the picture considerably more complex.

Positive and Negative Control

Continuing our description of regulation of protein synthesis, we can make one interesting distinction between negative and positive control. The type of control illustrated with the *lac* operon of β-galactosidase is a negative type of control. The presence of the control protein, the regulator protein, is negative: it shuts off control of the structural gene. Another type of negative control is also found in regulatory systems, as illustrated by the work of Bruce Ames and others with the histidine (*his*) operon. Histidine is synthesized by an elaborate series of 10 enzymatic reactions. The general pattern is summarized in Fig. 16-8. Here we see that 10 steps simply abbreviated by arrows are necessary for generation of histidine. As might be expected for a sensible pathway, histidine also inhibits the first enzyme in the sequence by allosteric feedback inhibition. Interestingly, the genes for all of these enzymes lie adjacent to one another on the DNA chromosome. In fact the genes are all incorporated on a single messenger RNA molecule of great length and thus constitute an operon. What distinguishes the *his* operon is that the presence of a

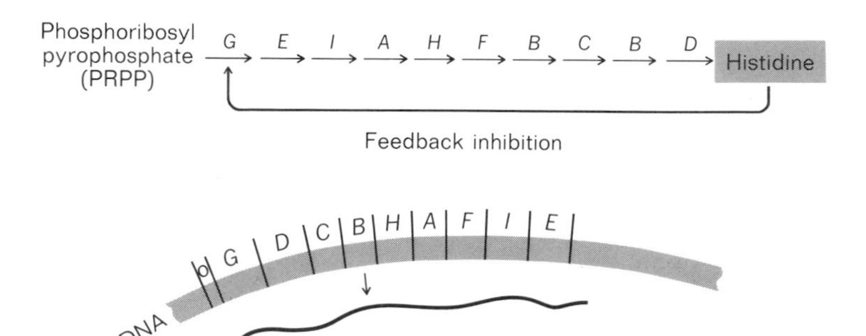

FIG. 16-8 *The Histidine (his) Operon.* For additional details, see M. Brenner and B. N. Ames, in H. J. Vogel, ed., *Metabolic Pathways,* 3rd ed., vol. 5, Academic Press, New York, 1971. Regulation of histidine biosynthesis in higher organisms is discussed in an article by G. R. Fink in the same volume.

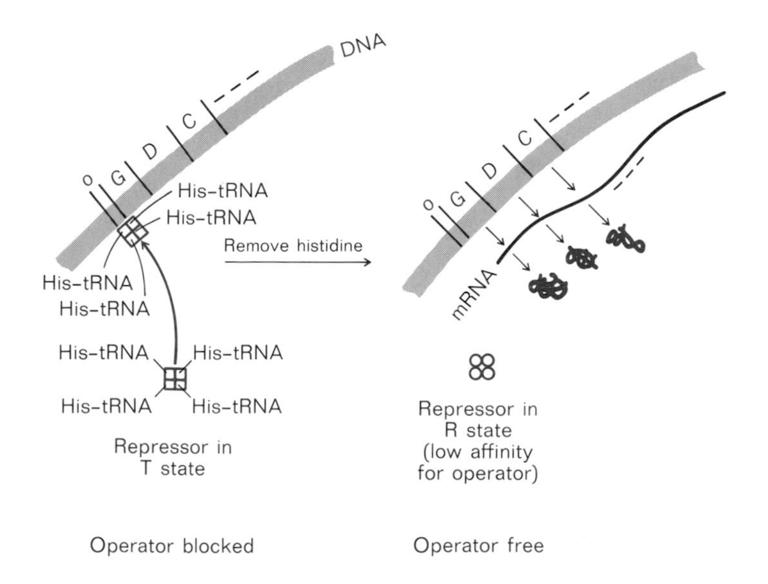

FIG. 16-9 *Repression.* Histidine acts as a corepressor and stabilizes the repressor in the conformation that binds to the operator. Histidine in acyl linkage to its transfer RNA is the agent involved in repression. Representation of the repressor as a tetramer is hypothetical.

small molecule, histidine, acts not as an inducer, but as a repressor. When histidine is present in the growth medium, the many genes required to synthesize it are no longer needed and can be retired. To accomplish this "switching off," a repressor is invoked that, in contrast to the repressor of the *i* gene, binds to the operator more tightly in the presence of the small molecule of histidine. This behavior is illustrated in Fig. 16-9 and results in the disappearance of all the histidine biosynthesis, as summarized in Fig. 16-10. Here we see that negative control systems, that is, systems in which the control protein sits on the operator and shuts it off, can be useful in controlling genes concerned either with *biosynthetic* processes, such as the production of histidine, or *catabolic* process, such as the metabolism of lactose. In one case genes are shut off in the presence of a small molecule; in the other they are induced in the presence of a small molecule. Where genes are repressed as in the case of histidine, the histidine is called not an inducer but a *corepressor*.

Having defined negative control systems, it will be valuable to now contrast them with *positive control systems*. Positive control systems represent an interesting variant to the control process and also introduce us to features present in the *lac* operon that represent some of the complicating factors we referred to previously. Positive control systems refer to protein regulators required for expression of the gene. An example of a positive control system is the arabinose (*ara*) operon

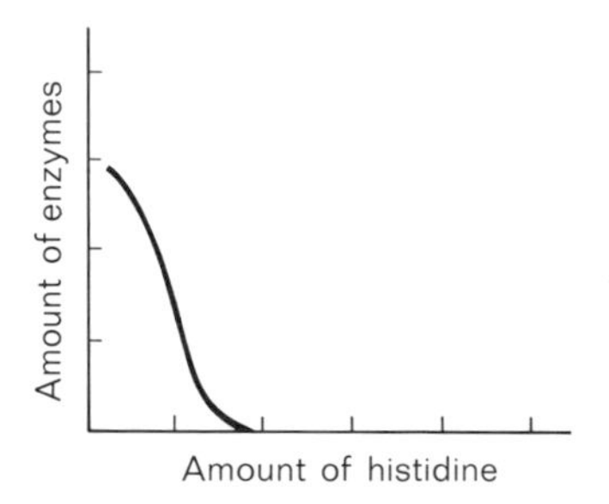

FIG. 16-10 *Repression of Histidine Biosynthetic Enzymes.* Addition of histidine to the growth medium results in the disappearance of the histidine biosynthetic enzymes.

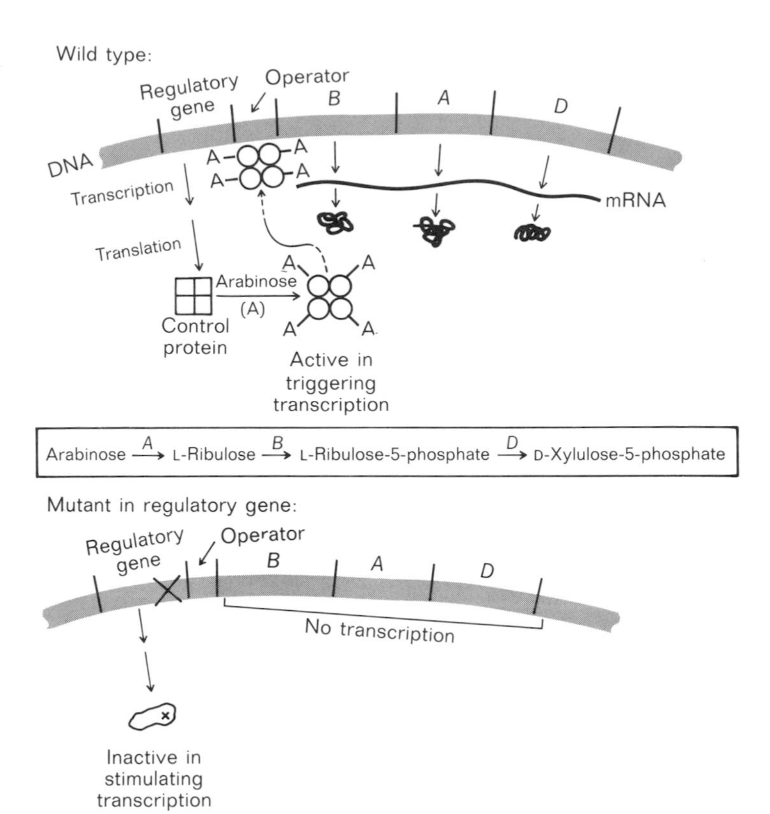

FIG. 16-11 *The Arabinose (ara) Operon: Positive Control.* In a positive control system, mutations of the regulatory gene abolish transcription of the operon. For additional details, see E. Englesberg, in H. J. Vogel, ed., *Metabolic Pathways,* 3rd ed., vol. 5, Academic Press, New York, 1971.

of *E. coli* that produces enzymes for the conversion of arabinose into a pentose shunt intermediate (see Chapter 11). In this case the control protein C is needed for induction. Therefore mutants in C are noninducible and nonconstitutive, in contrast to the i^- mutants of the *lac* operon, which are constitutive. The properties of this positive control system are summarized in Fig. 16-11.

We can further characterize the difference between positive and negative control systems by considering the following analogy. If we want to protect our castle from intruders, we can construct a moat with a drawbridge or a gate to block the passage. If we find that after a number of years our drawbridge breaks down and we cannot find another to replace it, our condition would be analogous to the positive type of control system with no C protein. Without the drawbridge present, we cannot pass and no movement occurs. In the cell no transcription can take place. In contrast if we had originally installed a gate at the door that had worn out in time and could not be replaced, our status would be similar to a negative control system. Without the gate, free passage would be possible. In the cell, transcription proceeds unhampered, in the constitutive mode. Therefore we can easily see the distinction between positive and negative control systems and are also left with an interesting moral: Build your castles in the sky—no intruders.

Positive Control Features of the Lactose Operon

In Fig. 16-1 describing the *lac* operon we included a site labeled *o* for the operator, the point of attachment of repressor, and also indicated the presence of a nearby site, the promoter *p*. The *p* site is the postulated site of attachment of RNA polymerase. RNA polymerase catalyzes the first step in protein synthesis, the transcription of DNA to messenger RNA (Chapter 4). Messenger RNA molecules must be produced for a definite region of the DNA molecule and at definite levels. Therefore it is not surprising that one aspect of the RNA polymerase molecule involves a positive control system for recognizing promoters. This feature of the RNA polymerase molecule is embodied in one of its subunits known as σ. The *E. coli* RNA polymerase functional unit or core protein contains four subunits, as summarized in Fig. 16-12. In addition the subunit σ attaches to the RNA polymerase molecule and mediates the attachment to the promoter site for the initiation of transcription. As will be noted in the following chapter, σ factors specific for different promoters can drastically alter the character of protein synthesis in a cell. One additional factor may be mentioned here, the ρ factor with a molecular weight of about 80,000, which plays a role in release of RNA polymerase from DNA at the termination of messenger RNA synthesis. Not all RNA polymerase molecules operate with a dissociable σ factor, however. In fact the RNA polymerase produced by bacteriophage T7 contains only a single polypeptide chain in the fully functional unit.

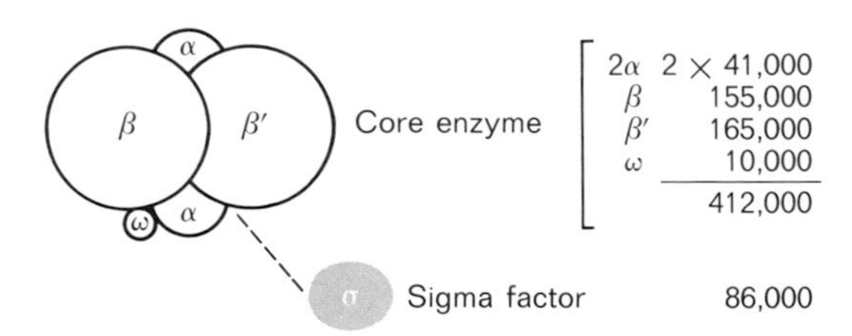

FIG. 16-12 *RNA Polymerase.* For additional details, see R. R. Burgess, *Annual Review of Biochemistry*, **40:** 711 (1971).

A second positive element in the regulation of the *lac* operon concerns the *glucose effect*. As *E. coli* are grown in medium that contains both glucose and lactose, the bacterium selects the glucose and metabolizes it completely, with the enzymes for lactose repressed until the glucose supply is exhausted. At that time the enzymes used for lactose will appear. The emerging picture of this process concerns another protein factor required in a positive sense for transcription, just as the C protein of the *ara* operon is required for transcription, and this protein factor only operates successfully in the presence of 3′,5′-cyclic AMP. Evidently glucose or one of its transformed catabolite products interacts with the enzymes that form cyclic AMP or degrade it to lower levels of the molecule and to quantities below those needed to promote transcription successfully. This pattern of action is summarized in Fig. 16-13. The details of this mechanism have been worked out in very elegant studies on cell-free systems on β-galactosidase synthesis. In these systems DNA incubated with the proper RNA polymerase forms messenger RNA needed to generate synthesis, and in addition the reaction mixture contains all the ribosomes and factors and amino acids needed for protein synthesis itself. Therefore the actual appearance of enzymatic activity can be monitored under various conditions of repressor addition or by changes in the positive factors such as σ and the cyclic AMP–protein factor, which always plays a role in transcription.

Also known as *catabolite repression*, the glucose effect actu-

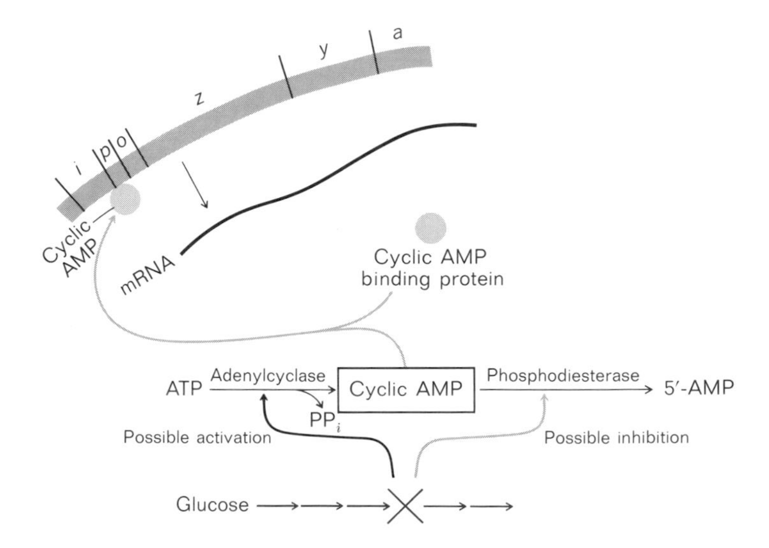

FIG. 16-13 *Catabolite Repression and Cyclic AMP.* A protein–cyclic AMP complex is a positive control element needed for transcription. It probably acts in the region of the promoter. Glucose represses transcription by lowering levels of cyclic AMP. For additional details, see R. R. Arditti et al., Cold Spring Harbor Symposium on Quantitative Biology, **35:** 437 (1970).

ally restrains protein synthesis at a number of operons, presumably through the action of 3′,5′-cyclic AMP in each case. Thus cyclic AMP plays a pivotal role in *E. coli*. In fact cyclic AMP, discovered by Earl Sutherland in 1957, plays an important role in many other cells. While its action on the *lac* operon is one of the most studied examples of its involvement in regulation, other phenomena influenced by cyclic AMP include mediation of hormone effects in a wide variety of cells, triggering of a clumping of amoeba in what resembles a primitive form of differentiation, and alterations of metabolism in certain cancerous lines of connective-tissue cells. The response of cyclic AMP to hormone messages is so widespread that, for a great many hormones, arrival at the target cell accomplishes the sole effect of stimulating adenyl cyclase with a concurrent generation of cyclic AMP as the "second messenger" inside the cell.

Studies with Isolated Repressor

The indirect genetic analysis that led to the model for the *lac* operon was compelling, but final proof was delayed until the individual components could actually be isolated and found to exist chemically. Confirmation of the model from such experiments was somewhat long in coming due to a number of technical difficulties, but a repressor molecule was eventually isolated and shown to possess many of the properties predicted. The difficulties encountered in attempting to isolate the repressor stem from its occurrence at extremely low levels in the cell. Since a cell contains generally only one chromosome or one genome and the repressor is required to bind only at one site on the genome, a very small number of molecules can effectively shut off synthesis for the entire cell. Estimates of the amount of repressor in the cell indicate that it is only about 0.002% of the proteins. Since it has no enzymatic function, which could be monitored in an assay, repressor could only be studied by direct binding measurements. This type of measurement is conveniently performed by equilibrium dialysis (Fig. 16-14). A small molecule, such as IPTG radioactively labeled, is put into the buffer and protein is put inside the dialysis sac. Since the IPTG can pass freely through the membrane but the proteins cannot, the IPTG would be concentrated inside the dialysis bag and therefore would provide an assay for the repressor. However, the 0.002% of the protein present as repressor is too low a

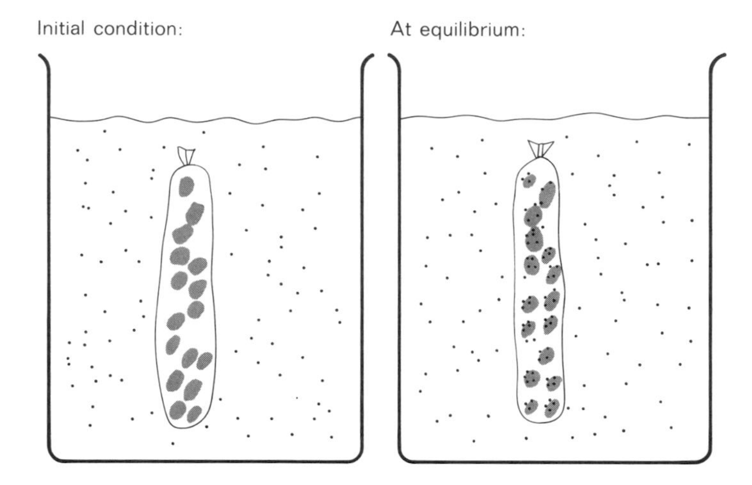

FIG. 16-14 *Equilibrium Dialysis.* A protein solution is placed in dialysis tubing and put into a beaker containing a radioactive small molecule. The protein is too large to pass through the dialysis tubing but the small molecule passes freely. After several hours the small molecule reaches an equilibrium condition. If the protein binds the small molecule, its concentration inside the bag will be higher than the concentration outside the bag.

quantity to detect by equilibrium dialysis, which is not nearly so sensitive as enzymatic assays. Therefore Gilbert and his colleagues, who defied the odds against them and succeeded in purifying the repressor, were forced to purify it blindly, concentrating various fractions obtained by usual methods of protein purification until one enriched fraction was obtained that did contain enough repressor to yield a detectable binding of IPTG in the equilibrium dialysis.

While it was possible to obtain measurable amounts of repressor by this purification in the dark, a number of elegant techniques of genetics were also employed to raise the level of repressor to heights that made it one of the more abundant molecules in the cell. The first accomplishment was to find a mutant of the *i* gene, presumably a mutant in the promoter of the *i* gene itself, which raised the level of transcription of the gene and therefore enhanced the amount of repressor made. This mutation resulted in a fiftyfold increase in repressor level, bringing the content of repressors from its initial level of 0.002% to about 0.1%.

The second device employed to raise levels of repressor relied on the introduction of the lactose genes into a λ bacteriophage. *E. coli* is susceptible to many small viruses, known as bacteriophages, which can infect the cell, take over the protein-synthesis machinery to replicate themselves in large quantities, and lyse the cell to yield a burst of many bacteriophages for each one that has penetrated the cell. The λ phage can incorporate into its own genome the *lac* operon, and such a variant of the λ phage was used for these studies. In addition the particular bacteriophage strain employed was defective for the lysozyme used to lyse the cells at the end of infection. Therefore the λ phages replicated but remained within the cell and all of their DNA was also available for transcription and production of repressor. In this way a further twenty-five-fold increase in repressor levels was produced. This brought the total level from 0.1% to 2.5%, which then made it a very simple task to purify the repressor in large quantities.

Characterization of the isolated repressor showed it to be a tetrameric molecule with four subunits of molecular weights of about 38,000 each. Therefore the native molecule contained four sites for IPTG with a molecular weight of about 150,000. A technique was developed for studying the interaction of repressor with DNA by a nitrocellulose-filter assay. It was discovered by Riggs and Bourgeois that free DNA passes through a nitrocellulose filter, but DNA with repressor fixed in a DNA-repressor complex is retained by the filter. Therefore mixtures of DNA and repressor with one of the components labeled could be passed over the filter and from the amount of label trapped by the filter, the equilibrium constant for the repressor-DNA binding reaction could be determined. In this way a dissociation constant of $K = 10^{-13}$ M for the repressor-DNA reaction in 0.01 M magnesium was obtained. This is a very tight binding system, as 10^{-13} M is unusually low for a dissociation constant. If we return to our earlier view of the action of repressor, we see that the T state binds to DNA with an affinity of 10^{-13} M. In the R state, present when IPTG is added, the binding for the operator is diminished, in this case to a level of about 10^{-9} M as determined by the filter assay. The difference between the binding of the repressor in the R state and the T state is a factor of about 10^4. Even a dissociation constant of 10^{-9} M for the R state represents strong binding, but since the repressor is present at a concentration of only about 10^{-11} M, the change from 10^{-13} M in the T state to 10^{-9} M in the R state corresponds to a change from about 99% of the repressor in the T state bound to the operator to only about 1% of the repressor bound in the R state. These differences are enough to turn the operon off or on.

While the equilibrium or thermodynamic aspects of the repressor-DNA interaction can be accommodated in most respects, the kinetic aspects present some surprises. The equilibrium can be expressed as the quotient of the decay of the complex and its formation, or

$$K = \frac{k_d}{k_f}$$

Studies with the nitrocellulose-filter assay also permitted measurements on the individual rate constant. The rate at which the repressor is released from the DNA, or the decay constant, can be obtained simply by mixing *lac* repressor complexed with radioactively labeled DNA and uncomplexed cold DNA. As the repressor comes off the labeled DNA onto the cold DNA, the labeled DNA will pass through the filter and provide a measure of the progress of the reaction. If we assume that the binding of repressor is as fast as possible, limited only by the diffusion of the molecule itself, we would estimate a rate of 10^9 mole^{-1} sec^{-1}. From our equation we can then predict the rate of k_d, which would be $10^{-13} \times 10^9$ or 10^{-4} sec^{-1}. This means that the half-time for the reaction would be on the order of two hours. However, when the actual experiments to measure dissociation of the repressor-DNA complex were performed, a half-time of about 20 minutes was obtained, corresponding to a k_d of about 5×10^{-4} sec^{-1}.

This finding led to two unexpected conclusions. First, since induction takes only a few minutes, some process must occur whereby the inducer actually binds to the repressor on the DNA and causes it to fall off more rapidly. Such a situation was in fact found experimentally when a decay time of just a few minutes was measured in the presence of inducer. Therefore the allosteric model for repressor (Fig. 16-5) must include binding of inducer to a T-type conformation and promoting a conformational change that facilitates dissociation of the repressor-operator complex.

The second important conclusion arises from the calculation of the forward rate now using the measured decay rate. Since both k_d and K are known, we can rearrange our equation to calculate k_f, and a value of $5 \times 10^{-4} \times 10^{13}$, or 5×10^9 mole^{-1} sec^{-1} is obtained. Simple calculations show that this rate is 5 times too high for a simple diffusion-controlled process (maximum $k_f = 10^9$ mole^{-1} sec^{-1}). Therefore we are left with the conclusion that the repressor must be directed to the operator by some physical process, perhaps hitting the DNA at some point other than the operator and quickly sliding along it to the operator site. Thus long stretches of

DNA near the operator may act as a "lightning rod" to attract the repressor molecule and enhance the rate at which it can attach to the operator.

In summary, the simple view of the *lac* operon as an archetype of protein-control systems has undergone an evolution of thinking to the point where we now see the complexity of the *lac* operon itself, especially in regard to the several positive control proteins needed for its transcription. In the next chapter we will show that control systems in higher organisms are likely to involve considerably greater degrees of complexity.

PROBLEMS

1 What are the two ways that constitutive synthesis of β-galactosidase in *E. coli* can be achieved, and how can they be distinguished?

2 What positive control features of bacterial operons have been described?

3 How is an equilibrium constant for a reaction related to the forward and backward rates of the reaction?

Chapter 17 Frontispiece *Electron micrograph of a centriole, a key element in chromosomal division in animal cells. Magnification 240,000×. Courtesy J. Telford.*

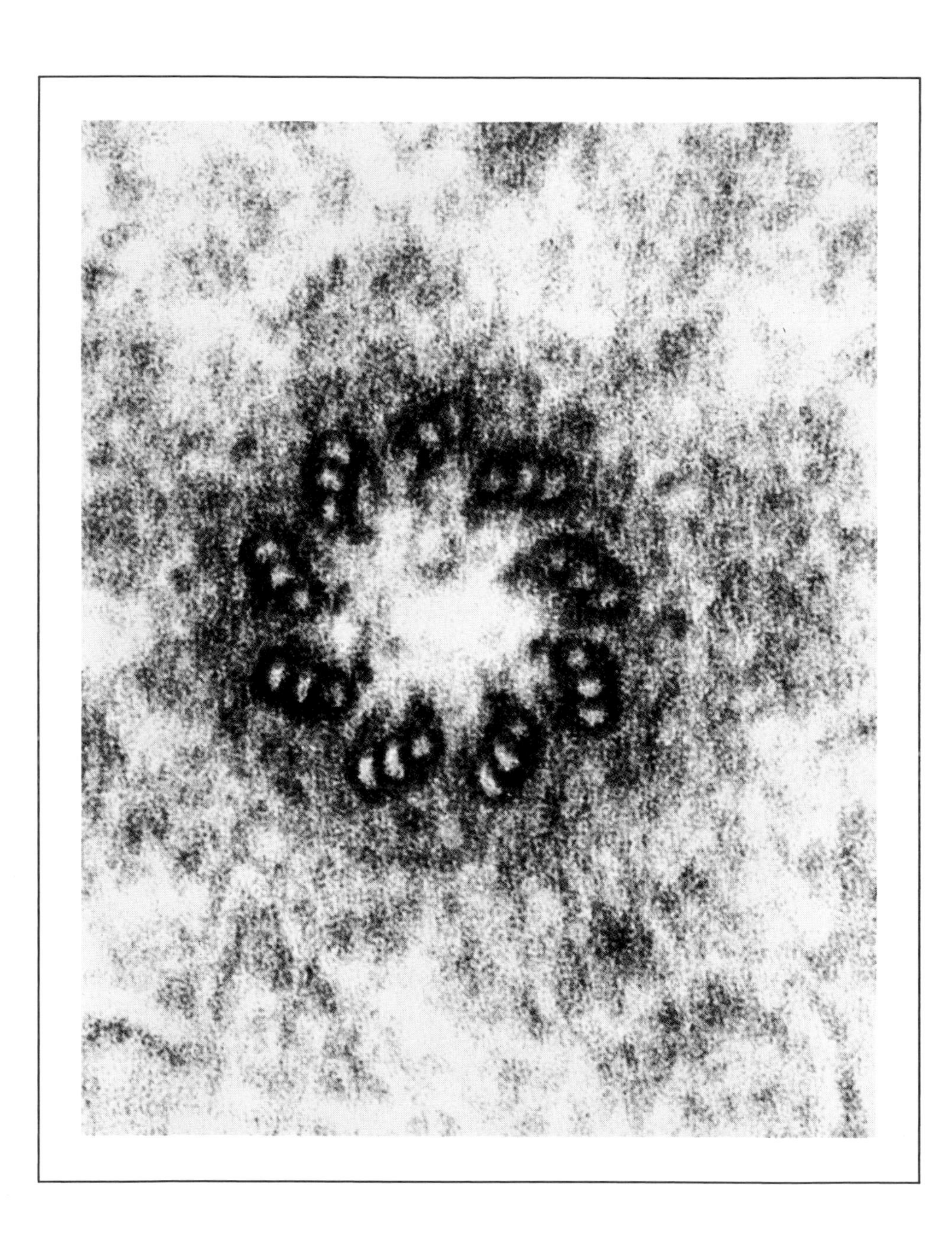

17 TOPICS IN DIFFERENTIATION

In the preceding chapters many important molecular features of life were presented, such as DNA and the genetic mechanisms, protein synthesis and folding, and the enzyme-catalyzed reactions of metabolism. The biochemical patterns of living systems were summarized in the previous chapter by discussing the interrelations of metabolism, DNA, and protein synthesis in processes of cellular regulation. The discoveries and deductions that led to this knowledge represent a tremendous accomplishment; we must recognize, however, that in some ways the achievements are just a beginning. The biologist of today is like a man who has come to an ocean for the first time. He wades in to his knees and is tremendously exhilarated by the new sensations. Yet he is hardly wet. As we will try to show in this chapter, many major challenges still confront biological science—the challenges of applying the insights of molecular biology and biochemistry to the complexities of development in higher organisms. In particular the concept of differentiation with all of its implications pinpoints the nature of these challenges.

Differentiation refers to the specialization of cells, as during animal development, in which a fertilized egg passes through the many embryonic stages on the way to forming the complete organism. At the very least, different proteins are synthesized at different times and at different levels in the cells (see, for example, changes in hemoglobin in Fig. 17-1) of the various organs and tissues of the animal. We have noted the mechanisms for control of protein synthesis in procaryotic cells in the preceding chapters. Yet the problems of the eucaryotic cell and characterization of the eucaryotic chromosome are so much more complex that a higher level of control processes must be sought. In this chapter we will try to enumerate the major questions in the molecular

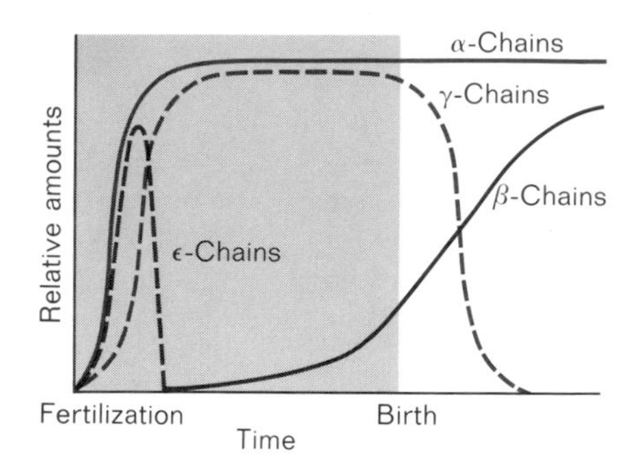

FIG. 17-1 *Changes in Human Hemoglobin During Development.* Hemoglobin is formed with two α-chains plus either two ε-chains, two γ-chains, or two β-chains, depending on the stage of development. From E. Zuckerkandl, "The Evolution of Hemoglobin," copyright © Scientific American, Inc., May 1965. All rights reserved.

aspects of differentiation, concentrating on the nature of the eucaryotic chromosome and those areas where some analogy to bacteria may be useful. We will then turn to implications in these areas for two of the more relevant topics for humans—the antibody system and cancer.

The Eucaryotic Chromosome

The increased complexity of higher organisms compared to bacteria is reflected in an increase in DNA content per cell (Fig. 17-2). Mammals possess enough DNA to code potentially for millions of proteins. Yet a need for so large a number of proteins is difficult to fathom. Most of the major metabolic pathways are already present in unicellular organisms. Humans have in fact lost many enzymatic processes available in lower animals. These observations, as well as estimates from genetic data, suggest that the bulk of the DNA in the cells of higher organisms is not coding for proteins, but is present for structural or regulatory purposes.

Further evidence in favor of a mixed function for DNA comes from microscopic studies of chromosomes, particularly the exceptionally large chromosomes in certain insects such as the fruit fly. Known as *polytene* chromosomes, these structures contain numerous strands of DNA attached side by side in a giant cable (Fig. 17-3). Bands are clearly evident on the structures, and correlations between number and position of the bands with number and mapping position of genes have been found. Britten and Davidson, Crick and others have proposed that the bands represent globular regions of DNA with some complex conformation that is recognized by control elements in differentiation. In addition this hypothesis proposes that regions between the bands, *interbands,* code for proteins (see Fig. 17-4). Crick has calculated that the average amount of DNA in an interband is enough to code for a protein in the molecular weight range of 30,000 to 40,000. Just what the control elements are which interact with the globular regions of DNA remains unknown. Both proteins and RNA molecules have been suggested as possible regulatory agents.

Another complicating factor in the eucaryotic chromosome is the presence of large quantities of protein, including histones. Histones are small, basic polypeptides, rich in lysine and arginine. They probably play a more structural than regulatory role since all histones fall into just a few chemical

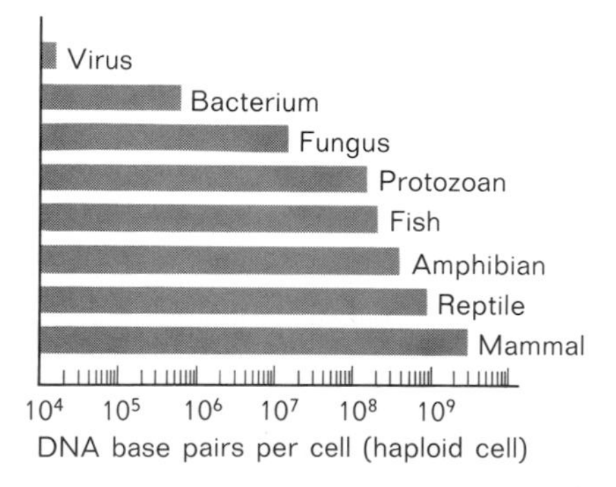

FIG. 17-2 *DNA Contents of Various Cells.* Adapted from R. J. Britten and E. H. Davidson, *Science,* **165:** 349 (1969).

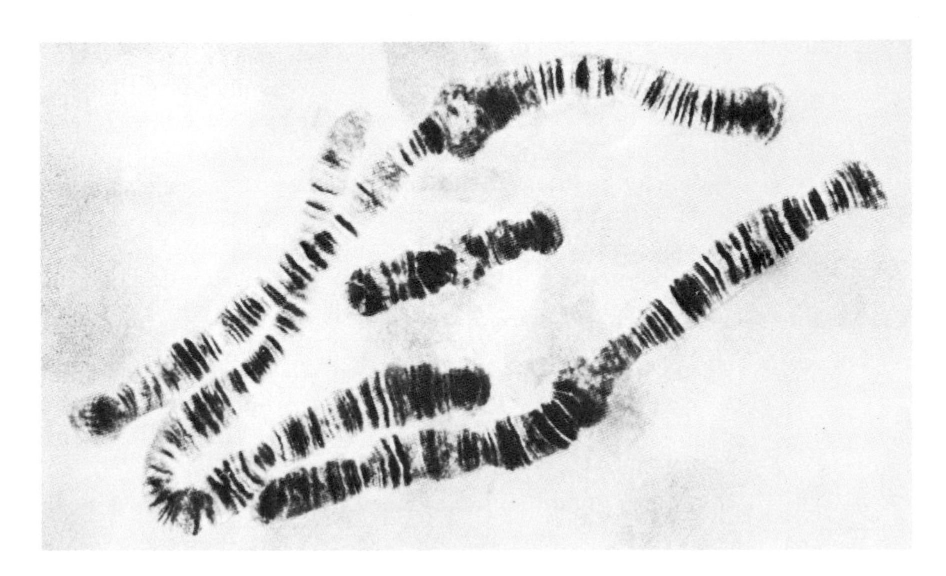

FIG. 17-3 *Giant Chromosomes.* Courtesy of W. Beerman and V. Clever. Prior publication in *Scientific American,* April 1964.

classes. Specific regulatory proteins would need much more variety to distinguish specific genes. Nevertheless, histones could be invoked as coarse timing regulators that, under the direction of other cell components, attack and repress specific genes or groups of genes.

Multiple Gene Copies

One explanation for the large amount of DNA in eucaryotic cells is redundancy of genes. If genes were present in many copies, no exotic mechanisms would be needed to account for the large quantity of DNA present. However, extensive gene redundancy could be expected to yield multiple versions of proteins. The various genes for one protein could mutate independently and lead to a multiplicity of isozymes. However, more than two or three isozymes for a given function are rarely found, suggesting that gene redundancy is not a widespread phenomenon. Nevertheless, in one case—the genes for ribosomal RNA—striking evidence for a high degree of redundancy has been found. Evidence for redundancy

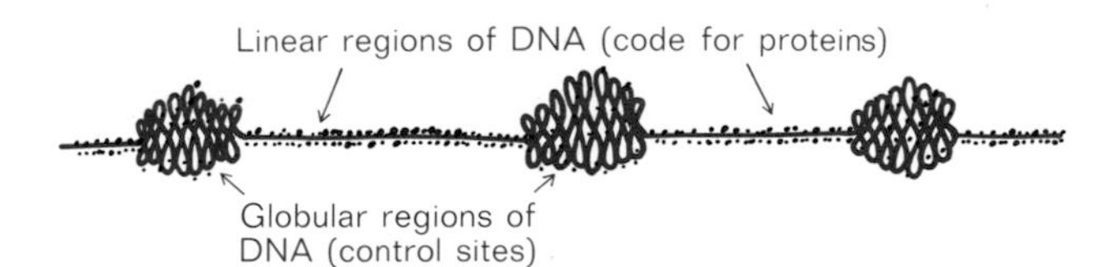

FIG. 17-4 *Hypothetical Structure for DNA in Eucaryotic Chromosome.* Histone and other proteins (in color) are attached along the entire length of DNA. Adapted from F. Crick, *Nature,* **234:** 25 (1971).

was first suggested by studies on the renaturation kinetics of DNA disrupted by one of the usual methods (see Chapter 2). The rate at which the DNA re-formed a double helix (characterized by the decrease in absorbency at 260 nm or hypochromicity associated with the ordered structure) is proportional to the size of the chromosome. The larger the amount of DNA per cell, the longer the time required for reassociation since the proper partners have more difficulty in finding one another. However, certain deviations from this size-time pattern were found, suggesting that in some cases a portion of the DNA contained sequences that were highly redundant. For these sequences, proper pairing occurred more easily, and renaturation rates prevailed that were much faster than would be predicted from the total size of the chromosome.

The existence of redundancy has been established even more clearly for the genes for ribosomal RNA. RNA-DNA hybridization techniques, as developed by Sol Spiegelman and his colleagues, permit a precise assessment of the amount of DNA responsible for a given RNA molecule. For example, radioactively labeled ribosomal RNA is isolated, mixed with the total DNA of a cell, and the mixture is "melted." As reannealing takes place, the formation of DNA-RNA hybrids between one strand of the ribosomal RNA gene and the labeled RNA can be detected by trapping the hybrids on a nitrocellulose filter. The free RNA passes through the filter, but the bound or hybridized RNA is retained along with the unlabeled DNA. In this way a saturation curve is obtained for the amount of labeled RNA trapped as a function of RNA concentration (see Fig. 17-5).

By calculating the maximal hybridizable RNA for a given amount of DNA, the gene dose of the chromosome can be calculated. Experiments of this type with the fruit fly *Drosophila* indicate that 0.27% of the DNA is devoted to ribosomal RNA. The DNA coding for ribosomal RNA is located in a small organelle of the nucleus known as the nucleolus. With a total genome size for the haploid chromosomes of about 5×10^8 nucleotide base pairs and a total chain length for ribosomal RNA of about 5000 nucleotides, one estimates that the ribosomal genes correspond to $(0.27 \times 10^{-2} \times 5 \times 10^8)/5000$, or 300 times the sequence of a single ribosomal RNA. Therefore in this case the redundancy level is 300 and presents a number of challenges to any theories of control. Redundancy poses problems of regulation and chromosome organization that will require much more elaborate theories and experiments before understanding emerges.

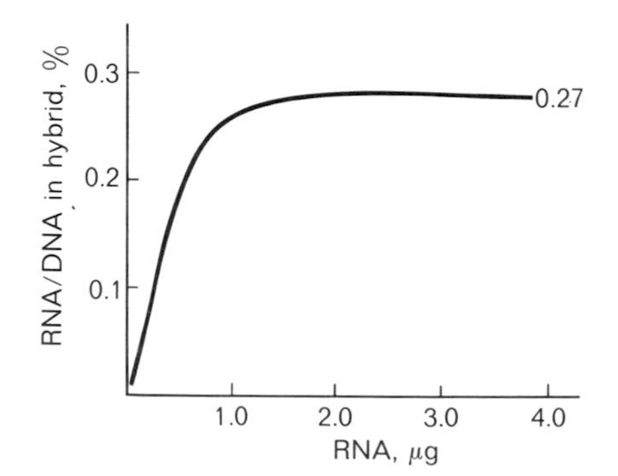

FIG. 17-5 *Hybridization Experiment.* From F. M. Ritossa and S. Spiegelman, *Proceedings of the National Academy of Science,* **53:** 737 (1965).

General Problems in Eucaryotic Control

Having described some of the structural aspects of the eucaryotic chromosome, we can now go on to discuss the detailed problems in regulation of gene expression. One problem peculiar to higher organisms, particularly in the context of differentiation, is the *timing* problem. The responses of a procaryotic cell are generally of an on or off nature. β-Galactosidase is synthesized in the presence of inducer; it is no longer synthesized in the absence of inducer. For the eucaryotic cell, the control problems are a great deal more complex. A delicate sequence of controlled events must take place to bring a cell from its primitive state as a fertilized egg through one of a variety of lines of evolution into a differentiated cell type. Even an automobile assembly line would look disorganized by comparison. The problem involves control of the transcription of DNA—calling on the proper genes at the proper time. In addition control problems at the level of RNA are known to be involved. Eggs from echinoderms, for example, contain all of the messenger RNA needed to carry the cell through several complex embryologic stages of development. However, the RNA is in a latent or blocked condition until triggered by fertilization. The presence of messenger RNA in a "repressed" form can be demonstrated by adding actinomycin to the cells. This compound inhibits DNA transcription; yet development proceeds normally for some time, so some preformed RNA must be involved.

While we have very little evidence as to how these processes of timing and control in DNA transcription and RNA translation arise, we can summarize a number of simple systems in procaryotic organisms that display some of these features. We will, at least from this discussion, arrive at some insight into how sequential timing problems can be solved and how control at the level of RNA can arise, although the precise relationship to eucaryotic organisms remains purely speculative.

As an additional note of caution against extrapolating from procaryotic to eucaryotic cells, we should note that the entire operon concept may be replaced by some other organization in higher organisms. For example, yeast, one of the simpler eucaryotic organisms, contains a battery of enzymes for the synthesis of histidine very similar to those present in *E. coli*. However, in contrast to the precise linear arrangement of the histidine genes in an operon in *E. coli*, the yeast histidine genes are widely scattered on the yeast chromosomes. Evidently little from the procaryotic cell is sacred to the eucaryotic cell, and further investigations may bring many more surprises.

Model Systems for Control of Transcription

One possible mechanism for triggering sequential events involves the synthesis of specific σ factors for RNA polymerase (Chapter 16). The factors produced at one stage trigger the genes to be transcribed at the next stage. These newly transcribed genes can also synthesize yet another specific σ factor that will go on to specify the reading of a subsequent set of genes. This cascading effect of gene sets producing a σ factor that initiates transcription for the next set of genes can provide, at least in principle, the events needed to carry a cell through a complicated sequence of differentiation. A simple system involving *E. coli* and one of its viruses, the bacteriophage T4, uses just such a mechanism.

When the T4 DNA invades the host cell, its reproduction requires a timed sequential synthesis of factors somewhat suggestive of a differentiating system. As seen in Fig. 17-6, the category of "pre-early" proteins is synthesized immediately on infection, and these enzymes arise from transcription of T4 DNA with the host σ factor. Among the proteins synthesized at that time is a new σ factor that controls the synthesis of the so-called early proteins. At the same time the host σ factor disappears. Somewhat later in the maturation process, a second new σ factor is generated, which controls the synthesis of the late proteins lysozyme and structural proteins. This σ factor displaces the first T4 σ factor. At the end of the process, the new bacteriophages are completed, assembled, and the cell is lysed and a burst of new bacteriophages appears. Thus σ factors can in principle accomplish the timed sequencing or transcriptional events needed to bring about differentiation. They also play an important role in sporulation of bacteria, a process whereby the usual cell structure is clearly modified to become enveloped in a strong, heat-resistant coat. This process as well is controlled by the appearance of sporulation-specific σ factors. Whether σ factors actually play a role in eucaryotic differentiation, however, remains to be seen.

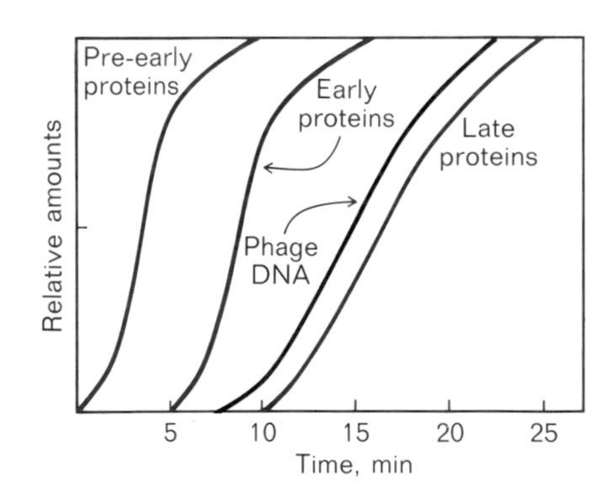

FIG. 17-6 *Synthesis of Components of Bacteriophage T4.* From J. D. Watson, *Molecular Biology of the Gene,* 2nd ed., copyright © 1970 by J. D. Watson, W. A. Benjamin, New York, 1970.

Control at the Level of Translation

Most of our attention in control has been directed at interactions involving DNA. However, from our reference to the partial differentiation of a fertilized egg with preformed messenger RNA, it is clear that control at the level of RNA must also be important. An interesting model system involving control at this level is the maturation of the small *E. coli* bacteriophage R17. This virus is extremely simple; it is composed of a molecule of single-stranded RNA encased in multiple copies of coat protein with a molecular weight of about 14,000. The genome of the RNA phage R17 is meager, containing cistrons that code for only three proteins, distributed on an RNA molecule with about 3500 nucleotides. The proteins are the A protein, involved with some maturation function; the coat protein; and an RNA-specific synthetase for replication (see Fig. 17-7).

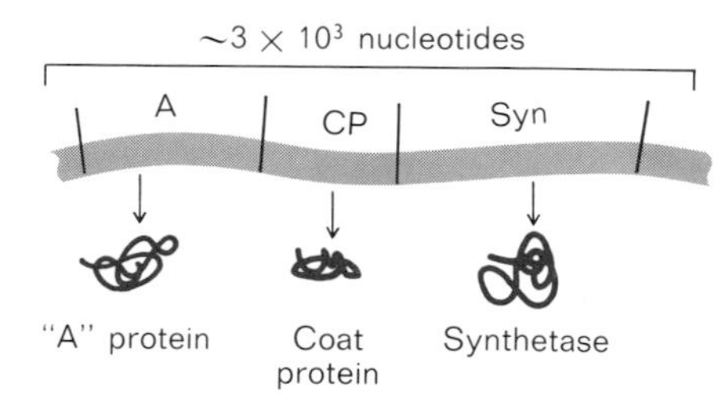

FIG. 17-7 *Bacteriophage R17 Genome.* Experiments with this bacteriophage have provided evidence for blockage of transcription as a repressionlike control process. Work with the eucaryotic cells has suggested that an apparent repression at the level of transcription may also occur by a repressor acting to make a specific messenger RNA labile to nuclease degradation (see D. W. Martin, Jr., in H. J. Vogel, ed., *Metabolic Pathways,* 3rd ed., vol. 5, Academic Press, New York, 1971).

Studies with the R17 RNA indicate that control can be exerted in two ways. The first way is blockage by secondary structure—the formation of small double-helical regions produced by the polynucleotide chain folding back on itself in a "hairpin"-like structure. Such blocks are believed to be present because when protein synthesis is initiated, a lag in the appearance of the synthetase is always noted. Presumably the lag occurs because the initiation site for the synthetase cistron is unavailable due to its participation in a hairpin at that point on the RNA molecule. As the first two cistrons are read and ribosomes traverse the RNA structure in these regions, the ribosome will eventually enter the hairpin-blocked region, thereby opening the blockage and permitting translation of the synthetase cistron. However, initiation can only occur in this way and will always give rise to a lag since translation of the first two cistrons must be accomplished first.

The second type of interaction involving R17 RNA is a specific RNA-level repression. Since synthetase molecules are needed in much lower quantity than coat protein, the R17 system has evolved two methods for ensuring this higher level. First, the rate of attachment of ribosomes for initiation is higher at the cistron for coat proteins than at the cistron for A. More important, the coat-protein molecules themselves bind to the initiation site for the synthetase gene to prevent further translation of the RNA molecule in the synthetase region. Thus the coat-protein molecules are acting in a way very analogous to the repressors of DNA operons discussed in Chapter 16. By interactions of this type, proteins synthesized from certain regions of messenger RNA could block translation by binding to other regions of the same messenger RNA or other messenger RNA. Complicated sequences of differentiation could occur by cycles of blockage and removal of blockage and could explain even the intricate RNA-dependent changes described for fertilized echinoderm eggs. However, it should also be noted that yet another type of control process may be the principal mechanism in the regulation of transcription. The work of Gordon Tomkins and his colleagues on the regulation of tyrosine aminotransferase (TAT) activity in eucaryotic cells strongly suggests that a repressor may act directly on the TAT messenger RNA to interfere with its translation and render it labile to nuclease destruction. In this case repression leads to the destruction of previously synthesized messenger RNA. Such post-transcriptional control via messenger RNA destruction has also been implicated recently in several other control phenomena and may emerge as one of the major regulatory schemes in eucaryotic cells.

The Antibody Problem

While most aspects of differentiation are "once in a lifetime" occurrences, the antibody system found in higher vertebrates continually practices a differentiation-like response. The introduction in these animals of a foreign molecule, known as an *antigen,* elicits the production of a specific protein, an *antibody,* which binds to the antigen. This immunological system provides an extremely valuable defense mechanism against foreign microbes. Antibodies in gross structure are all very similar, particularly the common type known as gamma globulin. The structure is oblate, with a molecular weight of 150,000, and is generally highly specific. When an antibody has been elicited in response to an antigen, it will bind very tightly to that antigen and rarely to any other. Antibodies are usually bivalent, having two antigen-combining sites. Bivalences lead to the potential for large lattice structures in antigen-antibody complexes. When formed, such complexes result in an insoluble precipitate. In recent years the fine structure of the antibody has been studied in detail and has been found to contain two heavy chains (MW 50,000) and

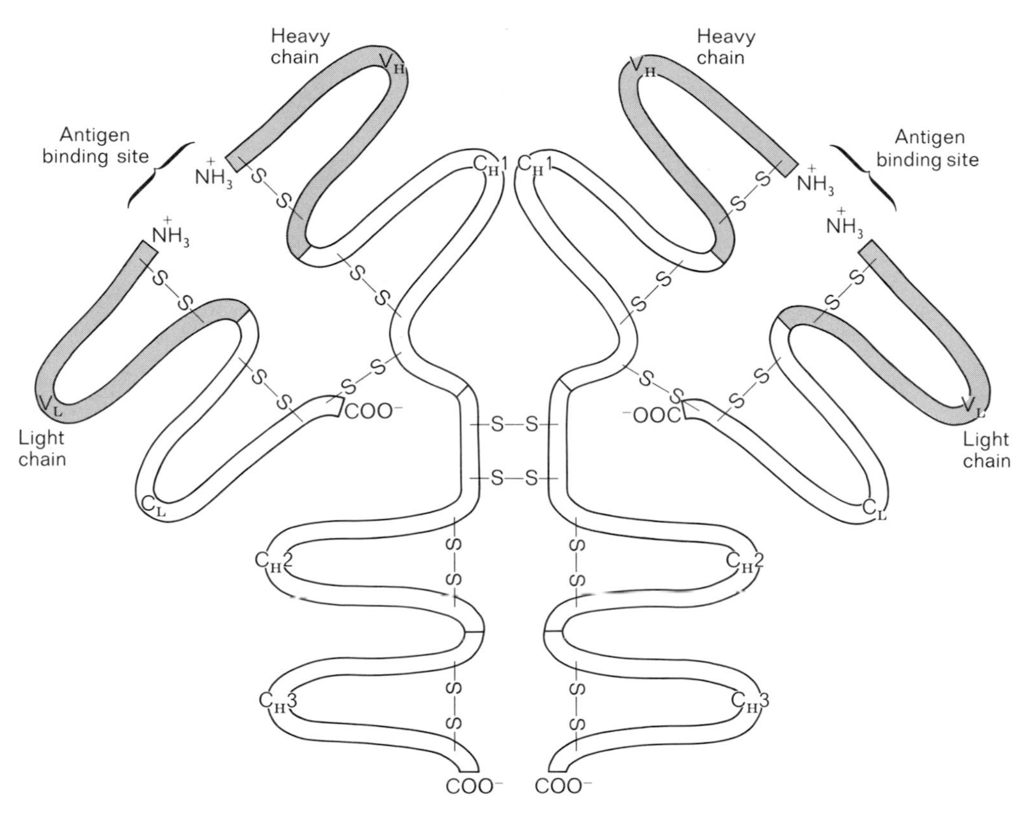

FIG. 17-8 *The Antibody Molecule.* Each antibody molecule contains two light and two heavy chains. The variable portions of the two types of chains (V_H and V_L) are homologous. The constant portion of the light chains (C_L) and the constant portions of the heavy chains (C_H1, C_H2, and C_H3) are homologous. Disulfide bonds (color) are present in each constant and variable portion and also link light chains to heavy chains and heavy chains to each other. For additional information, see G. P. Smith, L. Hood, and W. M. Fitch, *Annual Review of Biochemistry,* **40:** 969 (1971).

two light chains (MW 25,000). The chains are arranged in a specific manner, as summarized in Fig. 17-8.

Any given population of antibodies elicited in response to an antigen is generally heterogeneous. Therefore the study of antibodies in detail has been complicated by the complexity of antibody molecules in any population. A breakthrough occurred with the discovery that in the disease multiple myeloma, a specific protein is excreted, known as the Bence-Jones protein, which is a homogeneous light chain. Analysis of the amino acid sequence of Bence-Jones proteins from a variety of patients revealed a striking common pattern. Each Bence-Jones protein was composed of 214 residues. The first 108 residues contained variations from one patient to another, but the remaining part of the molecule was constant.

Many light chains produced by myeloma were sequenced and finally an entire molecule, light and heavy chains, was analyzed by Gerald Edelman and his colleagues. With the compilations from several such sources, new structural patterns were brought to light. The heavy chains were also found to contain an initial region of variability, also slightly more than 100 residues, while the remaining sequence up to residue 440 is constant. Moreover, the variable portion of the heavy chain showed similarity to the variable portion of the light chain. The constant portion of the heavy chain could be divided into three areas, each of which is homologous to the other two and to the constant region of the light chains as well. Homology is established by comparing the amino acid sequence and observing the occurrence of identical amino acids in similar places. Interestingly, each of these homologous regions contains an intraregional disulfide bond. The homology and variable versus constant regions are also summarized in Fig. 17-8. The antigen-binding site includes the variable region of both light and heavy chains.

While the protein-structure part of the antibody problem has been advanced considerably, many other aspects of antibody synthesis and stimulation by antigens remain mysterious. The mechanism of synthesis of the variable and constant regions is unknown. Possibly two or more genes are involved with the synthesis of each molecule, or fractions of it. Either the polypeptide fractions are joined together, or more probably the genes are "spliced" together in order to give the final structure. An even more fundamental question concerns how the genetic information is distributed. Evidence indicates that thousands of antigens each elicit a specific antibody. Are all the genes, for at least the variable region that binds these antigens, potentially present on the chromosome and are only one or two selected? If so, perhaps as much as 15% of the mammalian genome would be needed just for antibodies. (Nevertheless, this may be a good investment.) The alternative model involves a smaller number of genes that are amplified through somatic mutations or recombination processes to provide a larger pool of potential antibody molecules. Presumably when one cell begins synthesizing an antibody that fits an antigen in the system, a selection process, postulated as the *clonal selection theory,* stimulates the continued synthesis of the antibody and the multiplication of the cell making it. However, the process by which the synthesis is triggered and how a receptor for the antibody conveys the information to the synthesizing machinery is largely unknown. Yet another mystery is how the animal discriminates between its self-proteins and foreign matter against which antibodies are needed. This mechanism gives rise to the rejection of organs that has handicapped transplanting. Thus much remains to be learned about antibodies, and progress in this area may provide important insights into the general problems of protein synthesis and its regulation in eucaryotic organisms.

Cancer

Cancer is one of the serious general afflictions of humans that remains widespread and often incurable. Its principal characteristic is tumor formation. Cells in tumors have escaped from the normal control mechanisms and undergo continued cell division. In some cases tumor cells can be grown in the laboratory by providing proper nutrients. The cells from a tumor in one individual, Helen Lane, have been grown for many years. They are known as HeLa cells and divide about once a day. This process of laboratory growth, known as *tissue culture*, can also be applied to normal cells and has provided many of the clues now available concerning cancer cells. In particular the *transformation* of normal cells with regulated growth patterns to cancer cells with uncontrolled growth under laboratory conditions has revealed many important findings, including the role of viruses.

Tumors can be caused by a number of factors. Among them are chemical agents, as in the case of lung cancer from cigarette smoke, and genetic factors, as in the case of certain strains of mice. However, much of the excitement at the present time in cancer research concerns the role of viruses. Numerous viruses have been discovered to be involved in diseases. Some, like Rous sarcoma virus (RSV) are RNA viruses, that is, their genetic material is polyribonucleic acid. Influenza, mumps, measles, and polio in humans are caused by RNA viruses, as are leukemia and mammary tumors in several mammals other than humans.

DNA viruses cause smallpox in humans and a variety of animal tumors. Tumor-producing viruses are called *oncogenic*. DNA viruses can infect cells and either replicate themselves with the destruction of the invaded cell or transform the cell to an altered phenotype while reproduction of the virus is blocked. With RNA viruses, replication and cell multiplication can coexist while the cell is transformed. One DNA virus, SV40, has been discovered that will infect human cells in tissue culture. While no human cancer condition has been fully linked to virus infection, evidence is mounting for the involvement of a virus in human mammary cancer and a final demonstration may be forthcoming. In this case an RNA virus appears to be involved.

Although viruses can transform normal cells in tissue culture, one aspect of the biochemistry involved had been mysterious. In transformation the cell is not destroyed in a burst of virus synthesis, but rather is altered to a degenerate but stable state that propagates itself. Therefore an alteration in DNA composition would be expected, but how can such a change be effected by an RNA virus? Howard Temin proposed that a reversal of transcription must occur that passes the information in the RNA virus on to DNA that is incorporated in the chromosome. In support of this view the DNA from RSV-infected cells was found to have a small region complementary to RSV RNA. Thus reversal of the flow of information from DNA $\longrightarrow$ RNA is strongly implicated and an enzyme that catalyzes just such a process was in fact found independently by Baltimore and Temin in 1970 and is called the *reverse transcriptase*. The enzyme directs synthesis of polydeoxynucleotides complementary to an RNA template. Presence of the enzyme closely correlates with the ability of RNA viruses to transform cells. For example, viruses with diminished levels of the enzyme due to chemical inhibition with certain antibiotics are less competent at transforming cells. While a mechanism for induction of cancer with reverse transcriptase in this way has been suggested, possibly in human mammary tumors, no therapy has as yet emerged from these theories. One complication is that reverse transcriptase activity can be detected in apparently normal cells. Even virus particles and viral nucleic acids may exist in normal cells, only becoming activated under specific conditions. Many uncertainties clearly remain in this area. However, activity is at a high level and new developments are occurring rapidly.

One other aspect of cancer is relevant to our discussion of differentiation: contact inhibition. In tissue culture, normal cells placed sparsely in glass dishes multiply slowly until they reach one another. This contact appears to inhibit further growth. Although factors other than contact, such as chemicals in the growth media, also play a role, "contact inhibition" appears to be abolished in cancer cells isolated from tumors or in virus-transformed cells. An important aspect of differentiation as it occurs in higher organisms is the recognition of one cell by another. This recognition can take the form of growth stoppage when contact is made or when one cell discriminates against association with other cells of different tissues. Certain agglutins from plant sources, such as conconavalin A (con A), react with surface structures of the cell membrane in tumor cells which appear to be absent in normal cells. Thus cancer has many manifestations, all of which are aspects of a natural process gone awry. The great distance we must travel to understand the cancer condition reflects in large measure our present lack of knowledge concerning natural processes.

PROBLEMS

1 For a hypothetical hybridization experiment, how many gene copies per cell are indicated by binding of human globin messenger RNA (contains 650 nucleotides) to 0.000022% of the DNA?

2 What are three possible modes of control at the level of translation?

Chapter 18 Frontispiece *Trifid nebula. Photographed by B. Bok with the 90-inch reflector at Steward Observatory, Kitt Peak. Courtesy C. Sagan.*

18 IMPLICATIONS OF CONTEMPORARY BIOCHEMISTRY

We are all taking part in an experiment which began at least 20 billion years ago. It started with the big bang believed to have initiated this round of action in the universe. Elements of our solar system began to condense. Our early existence followed swirls of stellar dust as the planets were delineated. Eventually the earth formed in a reducing environment, dominated by the excess of hydrogen. Billions of years passed and the molecules present associated into organized structures. Somewhere a transition from chemical to biochemical processes occurred as primitive cells made their first appearance. The laws of evolution were fastidiously obeyed, cells combined into complex plants and animals, and the evolutionary tree was scaled. Then, just a moment ago on this time scale, one group—human beings—descended from the tree to start a new evolution and change the shape of earth not by alterations of chromosomes but by the power of ideas. Gradually ideas led to inquiry and science developed.

The contents of this book convey one branch of scientific progress. Humans have succeeded in discerning their very roots in the molecules of DNA and the biological reactions they control. The unique position of present-day humanity—conscious not only of itself, but of so many inner workings—is hard to appreciate. Some still struggle to comprehend the wisdom of sages spoken in simple languages thousands of years ago; yet others aspire to assimilate the complexities of life in the intricate molecular language discovered less than 20 years ago. Does biochemistry possess significance outside the areas of its immediate concern? Perhaps the achievements of modern biochemistry are only a closed endeavor, independent of other aspects of existence. However, some of the principal achievers of the current community of biochemists have attempted to relate conclusions in molecular domains to broader questions of life. Taking the view that biochemistry is what biochemists do, it seems appropriate for a book on biochemistry to include these subjects. Therefore we will briefly touch on some of the points of interest at large in the ideas of the times, such as speculations on the origin of life, views on the molecular aspects of topics such as medicine and memory, and recent reflections on the philosophical consequences of biochemical discoveries.

Speculations on the Origin of Life

All available evidence points to the origin of life in an anaerobic reducing atmosphere such as prevailed in the early history of the earth. The presence of methane, ammonia, hydrogen sulfide, and water are strongly suggested by the abundance of these compounds in the solar system and the paleological record. Many laboratory experiments have indicated that these compounds, when supplied with an energy source such as ultraviolet light and electric current or even simulated thunder, will produce amino acids in great abundance. Virtually all of the amino acids can be produced, although they are formed in equal amounts of D- and L-stereoisomers (see Chapter 8). Even the more complex nucleotide bases of nucleic acids can be generated in this way, and under the most favorable conditions, polymers of nucleic acids or amino acids can be formed. Suggestions of an anaerobic phase in biochemical evolution were also noted earlier in the text in reference to the appearance of anaerobic glycolysis as the first stage in the processing of glucose in nearly all cells. Thus the basic requirements for understanding the origins of life are present. The essential prebiotic building blocks—amino acids and nucleotides—can be synthesized from compounds believed to be present in the early stages of the earth. In addition, evidence within biochemical pathways themselves points to the existence of an anaerobic period and demonstrates the ability of cells to obtain energy under these conditions. Nevertheless, many important and as yet unanswered questions do remain, particularly concerning the origin of stereospecificity and the evolution of the machinery of protein synthesis. It is these two areas we will now discuss.

As already noted, when amino acids or other compounds with potential centers of asymmetry are synthesized spontaneously in the laboratory, both optical isomers are generated and in equal amounts. Yet when we review the occurrence of amino acids in biological systems we observe that only L-amino acids are present in all enzymes and proteins. A few instances of D-amino acids are encountered in peptide structures, especially antibiotics and cell-wall components, but these compounds are not synthesized via transcription and translation. Thus the question of how the L-amino acids became the major inheritors of the biochemical world remains a taxing one. Possibly factors of an unknown origin, such as polarized ultraviolet radiation, favored the formation or breakdown of either D- or L-amino acids at their times of initial synthesis. Alternatively, simply by chance, L-amino acids were incorporated into the first protein-like material capable of performing a stereospecific recognition. The protein system would then tend to exclude amino acids that did not have the same initial stereochemistry. The exclusion can arise from structural considerations, such as the facility of one type of amino acid (e.g., the L-amino acid) to form a continuous helix (e.g., the right-handed α-helix (Chapter 8)), whereas a combination of L- and D-amino acids in the same polymer would not readily form a regular structure. In addition, enzyme-catalyzed reactions often involve a three-point attachment and naturally tend to discriminate between stereoisomers. Therefore once the system began with L-amino acid, it could be readily propagated.

While at least a few sensible ideas can be presented for the origin of a biochemical system with only L-amino acids, current thinking is much more hard-pressed to conceive of the origin of the machinery of protein synthesis. In a modern-day version of the chicken and the egg dilemma, an important question now being raised is which came first, proteins or nucleic acids? A number of theories have been advocated from time to time favoring one or the other of these macromolecules. Recently a very convincing argument has been presented by Manfred Eigen to support the view of a simultaneous origin of proteins and nucleic acids. The hypothesis of simultaneous origin stresses the peculiar attributes of each type of macromolecule. Nucleic acids naturally tend to generate an information system since the nucleotides pair spontaneously and have built into them the capacity to replicate. Proteins, on the other hand, are ill suited for such information transfer. The amino acids do not tend to pair with one another in any regular way. However, proteins do possess the special property of spontaneously folding into three-dimensional structures that may include cavities, crevices, and catalytic sites that could concentrate nucleotides and enhance the fidelity of base pairing as well as the rates of polymer formation. The natural tendency of bases to replicate themselves according to principles of base complementarity has already been demonstrated. For example, experiments by Leslie Orgel and his colleagues have shown that free nucleotides will tend to condense into long nucleic acid polymers more readily in the presence of an existing polymer of their complementary bases.

Thus the view of simultaneous origin proposes that in some "prebiotic" stages a primitive nucleotide system provided some information for coding of a primitive protein system, and this primitive protein system in turn stabilized the coding mechanism of the nucleotide. Also at this early time an adaptor system of the type discussed in Chapter 4 would be needed and structures along the lines of a transfer RNA molecule could satisfy this requirement. The initial system may have been more simple than the current machinery, perhaps involving just the nucleotide A (which is the most readily synthesized in primitive environment situations) with its complementary nucleotide U. It is interesting to note that a genetic code based on A and U and employing the same codons as the current code would specify a reasonable number of amino acids, some charged and some hydrophobic, providing a possible first start at a system of protein synthesis and the formation of a simple enzyme. G and C and the amino acids they specify could have been incorporated at a later time. However, these theories still obviously leave an enormous amount of uncertainty and areas for both speculation and possible reconstruction experiments. The reconstruction experiments are of course difficult since the time factor present for the initial events cannot be easily duplicated. Nevertheless, there is a good deal of interest in this subject and we can undoubtedly look forward to more progress in this realm in the future.

Macromolecules in Medicine and Memory

The eminent successes of biochemical genetics have led to a growing view that given some time, a little luck, and keen analytical reasoning, virtually all aspects of life can be expected to unveil their complexities in molecular terms. More specifically the developments of molecular biology and biochemistry at the level of the gene suggest the possibility of direct chemical manipulations with the chromosomes for therapy and treatment of biochemical disease at some future time. While at present such ideas are no more than fanciful, we should also call attention to the progress that has been made in manipulation and synthesis of genes.

The first synthesis of a gene occurred in 1970 when Khorana and his colleagues applied the same techniques that had been so useful in their work with the genetic code (see Chapter 5) to synthesize the gene for alanine transfer RNA. The alanine transfer RNA gene is one of the smallest known and permitted an excellent test of Khorana's techniques. Ultimately the gene may be incorporated into a living organism and its role in synthesis of RNA by translational methods monitored directly. However, the magnitude of the accomplishment indicates the potentials to which chemical synthesis can go and might in fact lead at some distant date to rational therapy for molecular disease based on replacement of genes, perhaps through incorporation with a benign virus.

While the monumental effort of the synthesis of the alanine transfer RNA gene is an important landmark, molecular biologists, led by Shapiro and Beckwith and following the tradition of simple but ingenious experiments in which the bacteria do the dirty work, actually isolated a gene—in fact the entire operon for lactose. The demonstration that individual genes can be isolated presents perhaps a more realistic strategy for manipulating genetic information at some future date. Simply isolate proper genes—healthy genes—and somehow (that is the big uncertainty at this time) replace them for unhealthy genes. Since this method is potentially so rewarding, we will go into some detail about the actual isolation. The feat was based on the ability to incorporate lactose genes into the *E. coli* bacteriophage λ, in two different directions. λ is a small phage that infects *E. coli* and specifically incorporates the *lac* operon into its chromosome. The genes were incorporated in phage A in the usual sense and in phage B in an inverted sense. The contrast is summarized in Fig. 18-1.

With the availability of the two phages, each having the *lac* genes incorporated in the opposite direction, the isolation was made possible by separation of the strands of the phage DNA. The separation took place in the presence of added poly IG which preferentially bound one of the strands and amplified a difference in the base composition of the two strands. The strands could then be separated on a buoyant density centrifuge gradient (see Chapter 3), as shown in Fig. 18-1. When the strands were separated in this way from the two phages, the so-called heavy strands from both phages could be mixed. Since both the heavy strands contained the information for the genome in only one sense and not its complement, there would be no base pairing between strands except in the region of the *lac* operon. For here one of the strands would contain the *lac* gene in the inverted sense of the other strand and base pairing would occur. The preparation could then be treated with a nuclease and pure *lac* DNA would be obtained.

Although no immediate medical applications are likely to result from this work, the availability of the purified *lac* DNA may permit some precise sequence analysis of the material to determine the exact configuration at the operator and promoter regions of the DNA. While little likelihood of direct

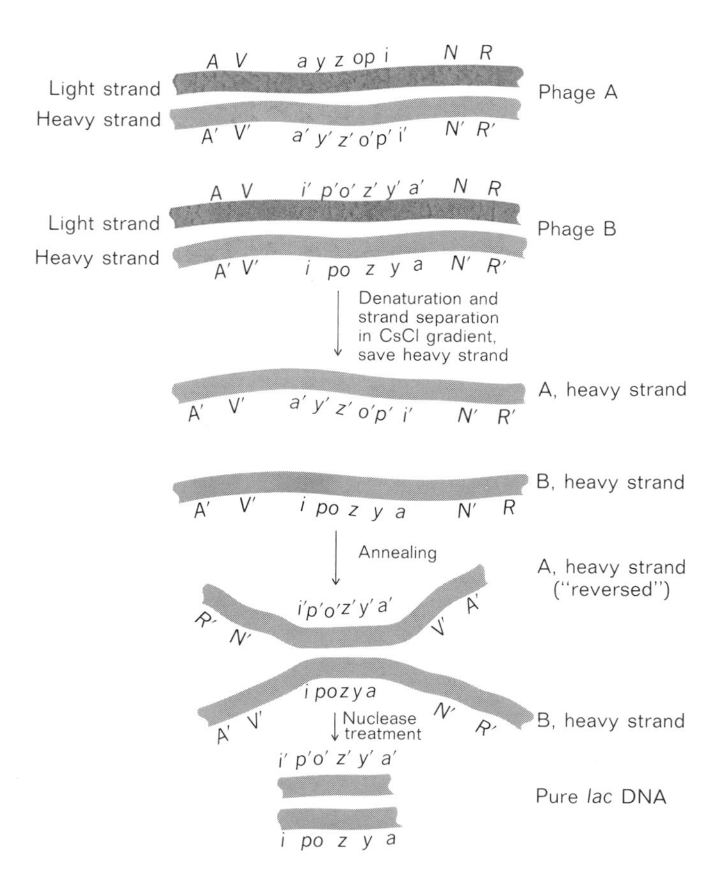

FIG. 18-1 *Purification of lac DNA*. Based on J. Shapiro et al., *Nature*, **224:** 768 (1969).

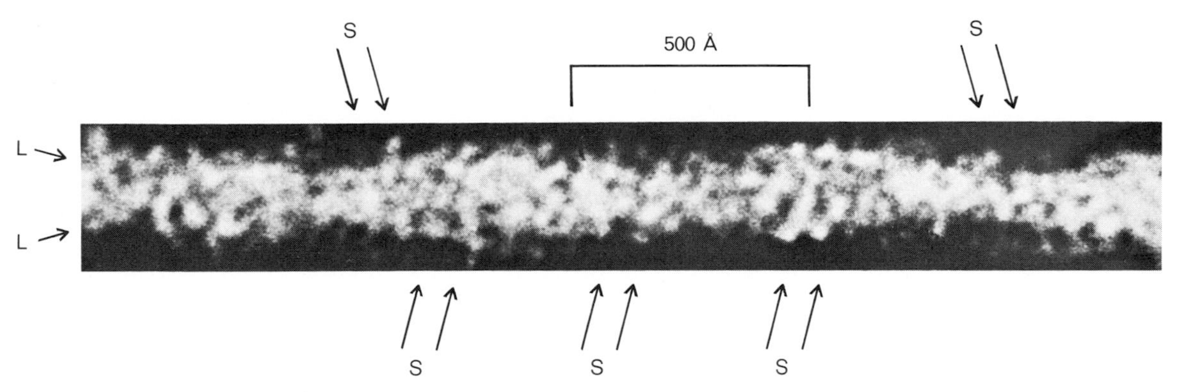

FIG. 18-2 *Electron Micrograph of Sickle Cell Hemoglobin Fiber.* The sample was prepared by injecting a gel into glutaraldehyde to stabilize the fiber by cross-linking, followed by negative staining with phosphotungstic acid. Individual globules correspond to hemoglobin tetramers which are associated in a fiber with long helical striations (L) and short helical striations (S). The pitch of the short striations indicates a double helical structure. Other measurements including observations on shadowed samples suggest that the short striations are left-handed and contain 6.4 molecules per turn. Each turn is offset from the ones above and below by about 12° to give the long striations (which form a right-handed sextuple helix). For additional details, See S. J. Edelstein, J. N. Telford, and R. H. Crepeau, *Proceedings of the National Academy of Science,* April issue, 1973.

changes in genetic information exist in the foreseeable future, alterations at the next level, proteins, are more readily imagined. One such example is in the search for a treatment of the molecular disease, sickle cell anemia. The sickle cell hemoglobin (hemoglobin S) variant (glutamate, the sixth amino acid in the β-chain, is replaced by valine) can associate with itself to form long fibers. Such fibers can distend the cell into the abnormal sickle shape in stricken individuals. The misshapen cells clog capillaries and this process leads to a very disagreeable assortment of symptoms. In addition, the cells are very fragile, leading to anemia. While no convenient treatment of sickling has yet been devised, the discovery of an anti-sickling agent with which sicklers can be treated is well within the realm of immediate possibilities. Structural details of the sickle cell fibers have recently been reported, which indicate a double helical arrangement of the individual hemoglobin molecules to give an overall appearance of a six-stranded rope (Fig. 18-2).

While therapies for biochemical disorders are a logical extension of traditional medicine, other directions of biochemical research are launching into totally new areas: the molecular aspects of physiological states, behavioral cycles, and even memory. Whether such complex processes as the events of the nervous system can be analyzed in discrete molecular terms remains to be seen. However, some progress in the area of memory has already been registered. Like so many areas of research, the findings raise more questions than they answer, but this experimental avenue may become a major research thoroughfare and deserves special attention.

One interesting set of "memory" experiments concerns the effects of inhibitors of protein synthesis and the distinction between short-term and long-term memory. Evidently the ability for immediate recall of some learned conditioned response (such as goldfish crossing a trough in response to a mild electric shock) depends on a different mechanism from the recall some days later. The important experimental finding is that inhibitors of protein synthesis such as puromycin inhibit the imprinting of the long-term memory but have little or no effect on short-term memory. Interpreted most directly these experiments imply that the "content" of memory is first held as an impermanent "engram," perhaps involving a network of neurons. Then, in a transformation to chemical information, a particular array of proteins is synthesized and stores the memory in a somewhat more permanent form. While this scheme may provide an appealing model, it leaves many serious gaps. Generally protein synthesis is directed by nucleic acids, ultimately derived from chromosomal material. If memories elicit synthesis of specific proteins, are they under control of nucleic acids? The inhibition by puromycin suggests a traditional translation mechanism since the inhibitor acts by virtue of its resemblance to a transfer RNA molecule and interferes with chain elongation on the ribosome. We are left with the view that nucleic acid–mediated protein synthesis is involved in memory storage. However, this view carries certain dilemmas of its own. To date, all nucleic acids involved in protein synthesis have been found to be under direct genetic control. Thus we are forced to the conclusion that preexisting genes exist for specific memories. The individual memories simply stimulate the appropriate gene for synthesis of a semipermanent storage protein. But, even for an analogous problem—antibodies elicited in response to one of a wide range of specific antigens—we noted (Chapter 17) that as much as 15% of the total gene library may be required for an adequate assortment of potential antibodies to permit one to be selected for each possible antigen. A memory-coding system based on a similar selection from an existing gene pool would require an absurdly large number of genes and is obviously out of the question. Thus another explanation must be sought. Perhaps memory-specific nucleic acids are synthesized in a nongenetically determined mechanism and proteins are directed by these messages. However, such a view contributes little to the understanding of the problem since it leaves us with a hypothetical model based on an unknown mechanism. A further possibility is that the puromycin inhibition reflects the interruption of some process associated with memory incorporation that requires protein synthesis, although the actual mechanism of information storage may involve an altogether different physical basis. One such possibility is that the proteins made may simply provide a "solder" for "hard-wiring" neuronal connections.

This discussion of memory, while unable to provide solid discoveries, illustrates a major assumption of current research. While mechanisms of memory and related problems such as differentiation of complex organs like the brain and nervous system are only subjects of speculation and exploratory research at present, interest and anticipation are high. The dominant view is that humans are mechanistically determined. We owe our existence to the primordial assembly of inorganic molecules into "biochemical" compounds, the spontaneous formation of cells, and evolution with a generous sprinkling of time. While generating optimism for research on the toughest questions of biology, this view also has serious consequences for humanity's self-image and philosophical outlook. We will consider the implications of these next.

Philosophical Consequences of the Mechanistic Outlook

The major philosophical consequence of modern biochemistry is the strengthening of an old theorem of scientists, the mechanistic concept of man. Stated most articulately in the recent book of Jacques Monod, *Chance and Necessity,* the human being is simply an elegant machine, an extremely sophisticated one, with elaborate macromolecular structures and exquisite capacities for abstraction and mental stimulation, but no more. Even the divine watchmaker is removed from the scene as only the haphazard occurrence of the right prebiotic molecules ("chance") coupled with their spontaneous self-assembly ("necessity") need be invoked to start evolution on the track to the present. More than just a context for research, this view leaves humanity in an existential void. The underpinnings of religious-ethical systems are removed. According to Monod, "Man is alone in the universe's unfeeling immensity." A process consisting of the evolution and selection of ideas, Monod suggests, may replace the old moral structure, but there is no certainty that the leap from knowledge to values can be executed. In fact many philosophers conclude that knowledge alone can never lead to values and any view to the contrary is a naturalistic fallacy.

According to the mechanistic doctrine, humans, by rational inquiry, have canceled the "ancient covenant" and remain alone. By penetrating to the core of DNA, biochemistry has removed the cloak from the unknown, the realm of mystery that bred every form of god-making, to reveal humans to be no more than the product of physics and chemistry. No special properties are needed. The special laws of nature that might be revealed only in living things and that many early molecular biologists sought, are not found. Having made such strides, progress itself may even be exhausted. According to another molecular biologist Gunther Stent, in *The Coming of the Golden Age,* just about everything accessible has been discovered and progress will soon fade away. Developments in cell biology, embryogenesis, and the nerve system should occupy research biochemists for a little longer according to Stent, but the drama has reached the climax and is entering the denouement. Francis Crick, whose penetrating insights have touched all areas of molecular biology, sees for the future little more of relevance to the deeper human quests than a "biochemical theology" based on the changes in endocrine secretion accompanying prayer (*Nature,* November 14, 1970).

In summary the mechanistic view states that science has penetrated the fundamental structure of biology, the gene, and found nothing special or inexplicable in terms of physical or chemical principles. Thus the last hope of many scientists to find a unique aspect of humans and other living forms has been expended and we remain—aware of our molecular heritage, optimistic that many of the remaining biological mysteries will be reduced to molecular terms, but unable to grasp any meaning or identity for the existential confrontations of life. While this view has been cogently expressed by many and is widespread among the biologists of today, there may be more to the story. To explore another side of the problem, we must consider the concept of complementarity principles.

Complementarity Principles

The special qualities of living things have always been held in awe by scientists. Even as late as 1949, Max Delbruck, one of the pioneers of molecular genetics, could hold out the possibility that "just as we find features of the atom—its stability for instance—which are not reducible to mechanics, we may find features of the living cell which are not reducible to atomic physics, but whose appearance stands in a *complementary* relation to those of atomic physics." Then a recent convert to biology from physics, Delbruck, a disciple of Niels Bohr, was very aware of the reevaluation in physics that occurred when the observation of discrete energy levels led to the development of quantum mechanics. Perhaps penetrating the gene would lead to another reevaluation requiring the development of a peculiarly biological mechanics.

That no revolutionary new physical principles are required to describe the gene is now evident—all that is needed is just a lot of fascinating chemistry. Does this situation mean that no special complementarity principles will appear? The consensus of the sources already noted is, yes, although this conclusion may be premature. Unique qualities of biological systems are more likely to be found not in penetrating the gene, but in the more complex mental qualities of the most intricate example of life, human beings. Following the pattern of physics, Delbruck and others sought new principles in the inability to *reduce* some detail of life to laws of chemistry and physics. However, the argument can be made that special relations are only to be found in biology by pointing in the direction of higher functions. The chemical interactions of genetic materials are now taken for granted, but can the composition of consciousness be viewed in the same manner? It seems much more likely for special properties to be discovered at this level—properties that might even rescue humanity from the abyss of the mechanistic conception of life detailed above and lead to the discovery that life itself possesses intrinsic value systems. If some such transcendental element of consciousness can be discovered, the molecular revelations of biochemistry and molecular biology would rest in complementary relation to these new principles.

How would such complementarity principles operate? Physical law sees planetary motion and electron motion as extremes of behavior of matter. Both are explained by quantum mechanics, but simpler newtonian mechanics can, in a *complementary* relation, represent only planetary motion. In the biological sphere the complex-simple polarity could be reversed. Consciousness and genetics are extremes of biological behavior. Both may be explained by a new consciousness factor—perhaps a component of consciousness in all matter—but simpler chemical principles can, in a complementary relation, represent DNA replication and protein synthesis. Hence the consciousness factor that may be central to understanding the mind will have a physical aspect congruent with chemistry and encompassing biological macromolecules. However, its expression at the molecular level would be slight and there would be no more hope of finding evidence for this new principle in biochemical reactions than there is of finding evidence for quantum discontinuities in the motion of the moon around the earth. In this view all matter is endowed with an element of consciousness, but the capacity for self-awareness or individual consciousness is expressed only in the most complex arrays of matter, in the human mind.

The formulation of these complementarity principles for biology in no way of course guarantees their existence. The arguments do serve, however, to provide a conceptual framework for the incorporation of new ideas. Such a conceptual framework is essential for the growth of a science, just as the implication of DNA as the genetic material had no major impact on biology until the conceptual breakthrough of the double-helix model linked DNA and the gene. Given that complementarity principles might be formulated that would coalesce with molecular genetics but also admit distinct higher mechanisms, how might these new concepts be revealed? This question is an extremely difficult one to answer without raising the specter of superstition and vitalistic dogma. Any serious challenge to the physical-chemical view of man and the universe will have to overcome the tremendous resistance of the current body of science. Physics, chemistry, and now biology and biochemistry have jelled into a mechanistic, evolutionary outlook, which constitutes, according to Thomas Kuhn (in *The Structure of Scientific Revolution*), a *paradigm*. In any epoch new knowledge is subsumed in the conglomerate paradigm of the time. Observations and theorizing inconsistent with the paradigm tend to be rejected, often ridiculed, unless the new theories generate such intellectual "pressure" that a veritable explosion or revolution occurs. Textbooks are then rewritten, and students are taught the new views. All are informed that the earth is not flat or that genes are not proteins and the transformation is completed. A new paradigm reigns, awaiting the next (if any) upheaval. If we are on the brink of such a revolutionary period, it is almost certain to express itself in the realm of consciousness. Or, if our current paradigm is stable, as Monod, Stent, Crick, and others have implied, science may truly be in a period of denouement.

The search for new scientific principles in the realm of consciousness is not new. Theories and speculations in this area are as old as recorded history. Some scientists have sought to reveal biological properties outside the laws of chemistry and physics through demonstrations of extrasensory perception (ESP). Whether such phenomena can be proven scientifically remains a very controversial topic. However, even evidence of extrasensory communication between individuals would not necessarily constitute a scientific revolution requiring new paradigms and complementarity principles. The nervous system is an extremely intricate and versatile instrument and its long spinal cord may be suitably constructed for service as an antenna in broadcasting and reception involving classic electromagnetic radiation.

Another line of research that would constitute a certain challenge to the present scientific paradigm concerns reincarnation. Recent activities, especially by Ian Stevenson, as reported in his book *Twenty Cases Suggestive of Reincarnation* and elsewhere, call attention to the spontaneous recall of past-life memories by young children, which Stevenson has verified with legalistic care. These cases are very provocative and may be the beginnings of a new paradigm, one that could have immediate application in the sphere of human values. Whether such new directions for science will be supported by time and scrutiny cannot be predicted. Nevertheless these directions are sure to be pursued and may reveal the mechanistic view of humanity to be (in the words of Schopenhauer) "the philosophy of the subject that forgets to take account of itself."

APPENDIX I
Basics of Genetics

A bridge between the descriptive areas of biology and the molecular realm of biochemistry is provided by genetics. Experiments in genetics give quantitative results that are ultimately a reflection of the molecular structure of the chromosome. When pursued in detail genetics leads to the concept of colinearity and the workings of the genetic code described at length in this book. However, the understanding of these processes rests on the simple technique known as the *genetic cross*.

The Genetic Cross

The work of Mendel, reported in 1865 and rediscovered in the early part of this century, provides a point of departure for genetics. The simple genetic cross first performed by Mendel is illustrated with garden peas. Two varieties were employed, one with round seeds and another with wrinkled seeds. When the two varieties were cross-pollinated, the first generation of offspring showed only round seeds. When these seeds were grown and crossed with themselves, some wrinkled seeds again appeared, at an occurrence of one in four. This genetic cross is summarized in Fig. I-1 and illustrates several important principles of genetics:

1 Pea shape is caused by a gene that, depending on its precise composition (*genotype*), produces either round or wrinkled seeds (*phenotype*).

2 The genes must occur in *homologous* pairs, that is, the organisms are *diploid*. The genotype for round is *dominant;* the genotype for wrinkled is *recessive*. The parents in the cross shown in Fig. I-1 each contained identical genotypes on both genes, that is, they are *homozygous*. The first filial generation

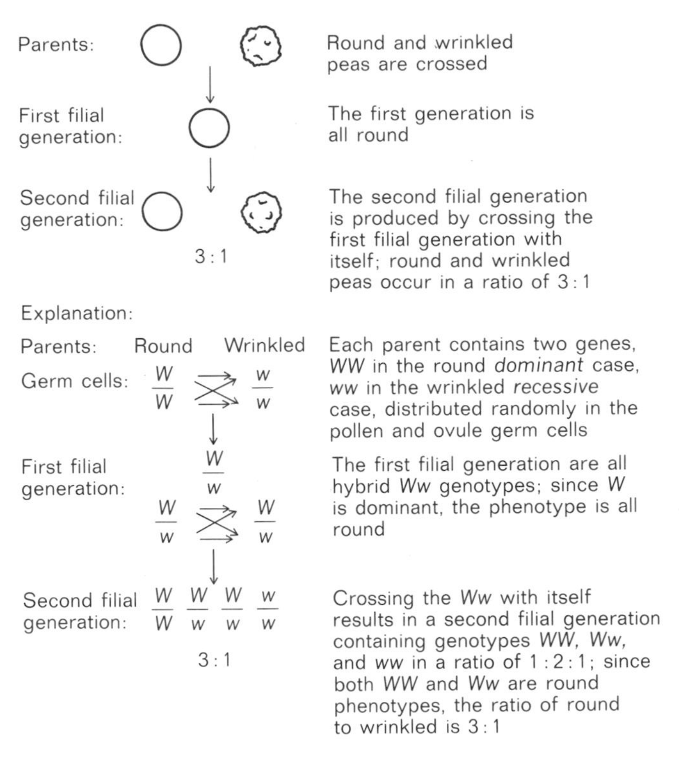

FIG. I-1 *The Basic Genetic Cross.*

contains one genotype from each parent and is *heterozygous.* Since the round genotype is dominant, the heterozygous progeny all display the round phenotype.

3 The formation of heterozygotes in the first generation and the formation of recessive homozygotes in the second filial generation suggest that the two genes of each parent assort independently. Thus the germ cells, ovule, and pollen in flowers or egg and sperm in animals possess one gene copy and are thus *haploid,* not diploid.

Linkage and Genetic Mapping

The next major advance in genetics was brought about by T. H. Morgan working with the fruit fly *Drosophila.* This species contains three pairs of homologous chromosomes per cell plus X and Y chromosomes in the male or two X chromosomes in the female. Inheritance of either an X or a Y chromosome from the male parent determines whether the sex of the offspring will be female or male. One X chromosome is also inherited from the female parent and both parents provide one member from each of the three homologous chromosome pairs. Various hereditary variants or mutants of *Drosophila* were collected with abnormalities in shape, size, and color. Crosses with one of these mutants and a normal fly confirmed Mendel's observations and were extended to crosses with normal and doubly mutant flies. In these cases the recessive phenotypes disappeared in the first filial generation and reappeared randomly in the second filial generation. However, in certain combinations of mutants, the two recessive phenotypes would always (or almost always) appear together in the second filial generation (see Fig. I-2). Such genes are said to be *linked* because they are located on the same chromosome.

Morgan noted that linkage is rarely 100% and in certain cases individuals arise in the second filial generation with genes from both parents on the same chromosome. Presumably this process is the result of *crossing-over* (Fig. I-3). If all areas of the chromosome have a similar probability for crossing-over, then the closer two genes are, the lower the probability that a crossover will take place between them. By crossing large numbers of mutants in pairs and monitoring recombination frequencies, assignments to approximate locations on specific chromosomes can be made and a *genetic map* constructed (Fig. I-4).

Fine-structure Mapping

From the point of view of molecular genetics, the primary studies concern the work of Seymour Benzer with the bacteriophage T4 of *E. coli*. T4 contains one molecule of DNA encased in a shell of proteins. It also possesses a tail and tail fibers for attachment to the cell and injection of its DNA (see Fig. I-5). Since T4 contains only one DNA molecule or chromosome it is haploid and alone exhibits no genetic recombination. However, if two mutants infect a cell at the same time, some crossing-over can occur to give a wild-type virus (see Fig. 1-8). Such experiments are readily performed with T4 and mutants in the *rII* cistron, since these mutants form altered plaques on strain B of *E. coli* but are unable to grow on strain K. This system provided a key in determining the size of the codon in protein synthesis (see Chapter 5). It also makes detailed genetic mapping relatively simple. A fine-structure map of the *rII* cistron can be constructed by crossing mutants in pairs and growing the bacteriophages produced

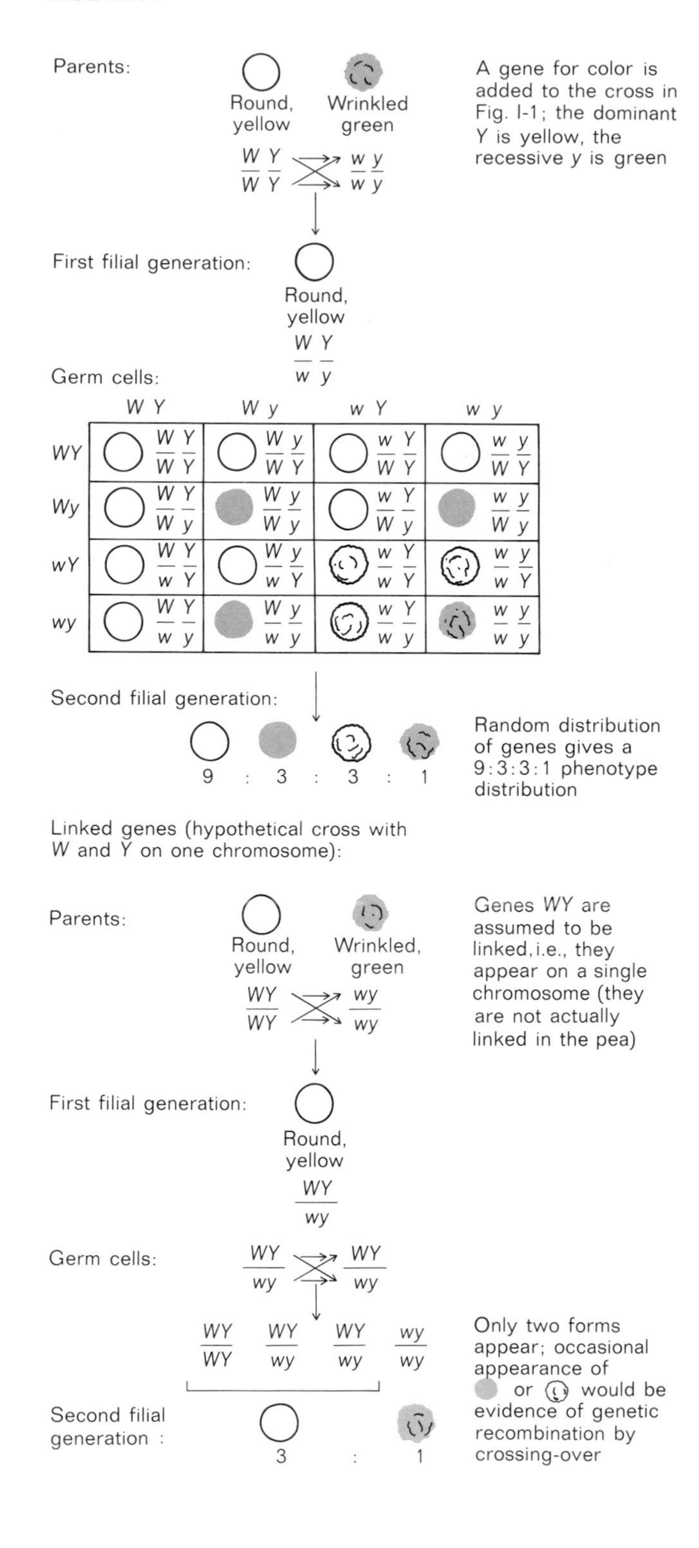

FIG. I-2 *Linked and Nonlinked Genes.*

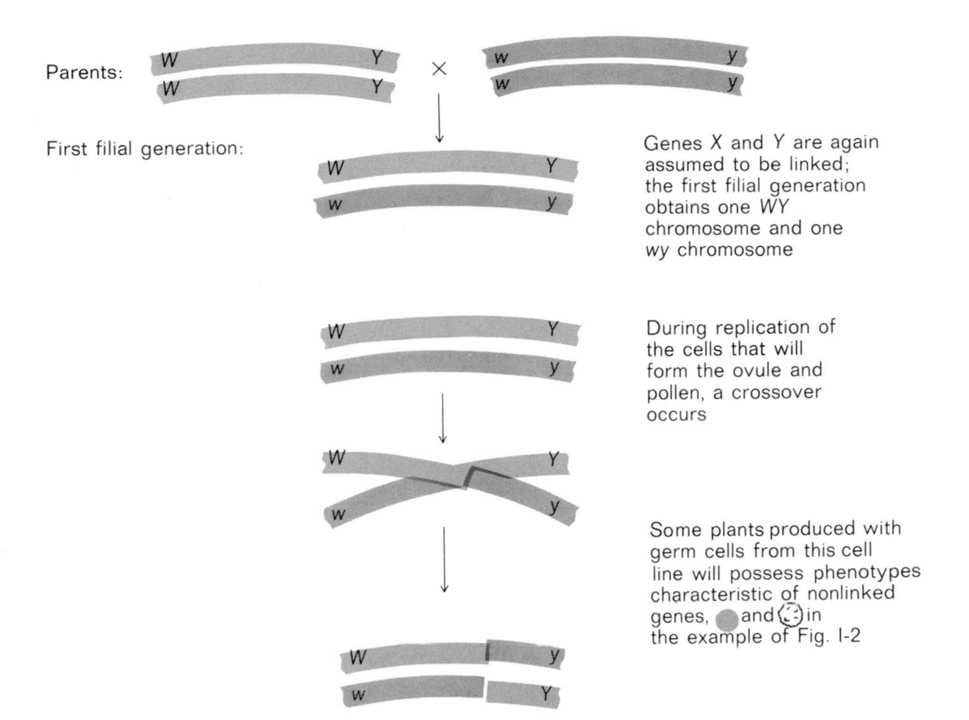

FIG. 1-3 *Crossing-over.*

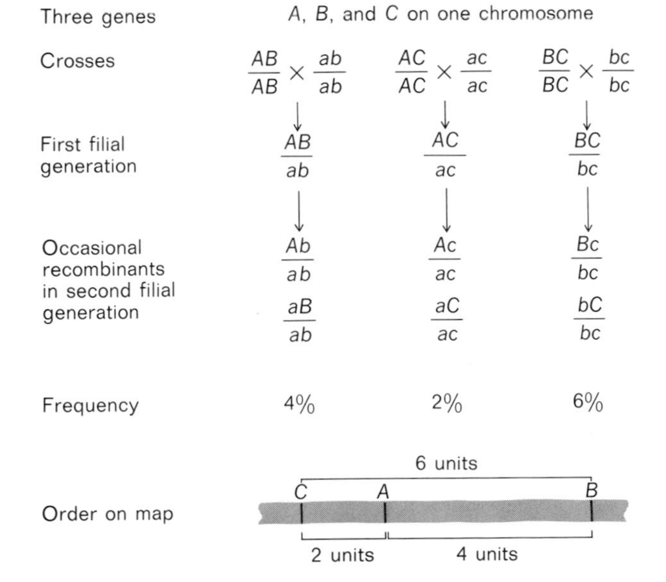

FIG. 1-4 *Genetic Mapping.* Data for constructing a genetic map are obtained from the frequency of recombinants in the second filial generation. One map unit is the distance between two mutants that give a recombination frequency of 1% in a standard cross.

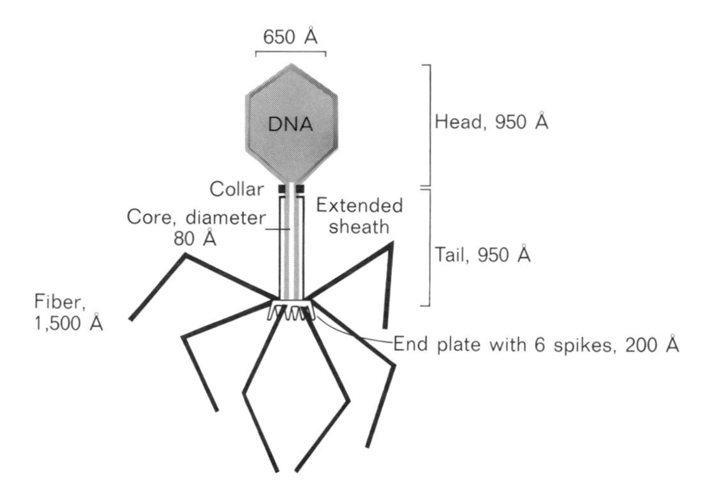

FIG. I-5 *Bacteriophage T4.*

on strains B and K. The growth on strain B gives the total number of bacteriophages produced; the growth on strain K gives the number of normal recombinants. Together the frequency of recombination can be obtained, which is related to distance on the genetic map. The technique is sufficiently sensitive to detect a recombination frequency as low as 0.0001%, although the lowest frequency actually found was 0.01%.

The correspondence between recombination frequencies and the genetic map can be estimated. The T4 DNA has about 2×10^5 nucleotide base pairs and the genetic map contains about 1500 map units, where one unit is the distance between two points that recombine at a frequency of 1%. The *rII* mutants that produce normal recombinants at the minimum frequency 0.01% are located at a distance of 0.02 map unit, since the *rII* assay on the K strain only measures half the recombinants (the normals, not the double mutants). The 0.02 map unit corresponds to $\frac{0.02}{1500}$ or 1.3×10^{-5} of the total chromosome. Since 2×10^5 nucleotide base pairs are present, the 0.02 map unit represents (1.3×10^{-5}) (2×10^5) or 2 to 3 base pairs. Thus the genetic map can distinguish sites that are virtually adjacent on the DNA molecule. The genetic concepts are thereby related to the molecular structure of the chromosome in the fundamental units of heredity, individual nucleotide bases of DNA.

APPENDIX II
Buffers and pH

Biochemistry is a wet science. Water is the major constituent of almost all living things. One of the important characteristics of water is the dissociation of the H_2O molecule to H^+ and OH^-. The ionization reaction

$$H_2O \rightleftharpoons H^+ + OH^- \qquad (1)$$

is described by the dissociation constant

$$K = \frac{[OH^-][H^+]}{[H_2O]} \qquad (2)$$

where brackets mean "concentration of." The *ion product* $[OH^-][H^+]$ can be expressed by K_w, where

$$K_w = [OH^-][H^+] = K[H_2O] \qquad (3)$$

At 25°C, K_w is very close to 10^{-14}. Since $[H_2O]$ is effectively constant, K_w is also constant. Therefore the product $[OH^-][H^+]$ will remain at 10^{-14} although the ratio $[OH^-]:[H^+]$ may vary.

The Concept of pH

As a convenient convention for expressing the proton concentration, the notation pH has been introduced, where

$$\text{pH} = \log \frac{1}{[H^+]} = -\log [H^+] \qquad (4)$$

If we define a corresponding term for hydroxyl ions,

$$\text{pOH} = -\log [OH^-] \qquad (5)$$

the equation for K_w can be rephrased as

$$-\log K_w = \text{pH} + \text{pOH} = 14 \tag{6}$$

Thus the sum of pH and pOH is 14 and the two components are related reciprocally. If pH is low, pOH is high. *Neutrality* prevails at

$$\text{pH} = \text{pOH} = 7 \tag{7}$$

Acids and Bases

A proton donor is noted as an acid; a proton acceptor as a base. Thus amino acids possess both acid and base functions. As generally described, the carboxyl group is an acid:

$$NH_2{-}\overset{\overset{\displaystyle CH_3}{|}}{\underset{\underset{\displaystyle H}{|}}{C}}{-}COOH \rightleftharpoons NH_2{-}\overset{\overset{\displaystyle CH_3}{|}}{\underset{\underset{\displaystyle H}{|}}{C}}{-}COO^- + H^+ \tag{8}$$

In contrast the amino group is a base:

$$NH_2{-}\overset{\overset{\displaystyle CH_3}{|}}{\underset{\underset{\displaystyle H}{|}}{C}}{-}COO^- + H^+ \rightleftharpoons {}^+NH_3{-}\overset{\overset{\displaystyle CH_3}{|}}{\underset{\underset{\displaystyle H}{|}}{C}}{-}COO^- \tag{9}$$

Actually the carboxylate ion can also act as a base and accept a proton. In this case it is known as a *conjugate base.* Similarly the protonated amino group can donate a proton. In this capacity it acts as an acid and is known as a *conjugate acid.* Thus every acid-base reaction involves a proton donor and acceptor. The uncharged compound is usually known as the acid or base, while the charged molecule in the reaction is known as a conjugate acid or base.

Strong acids (such as HCl) and strong bases (such as NaOH) are fully dissociated in water. Therefore the concentration of the acid or base directly determines the pH. For a millimolar solution of HCl, pH = 3; for a 0.1 mM solution of NaOH, pH = 10. Many other acids and bases, particularly those of interest in biochemistry, are weak electrolytes. In the case of a weak acid, for example, dissociation is incomplete and can be represented by an equilibrium,

$$AH \overset{K}{\rightleftharpoons} A^- + H^+ \tag{10}$$

where

$$K = \frac{[A^-][H^+]}{[AH]} \tag{11}$$

Buffers

Weak electrolytes, due to their partial dissociation, possess an important feature. When the proton donor and proton acceptor are present in roughly equal amounts, addition of a strong acid or base changes the pH less than in the absence of the weak electrolyte. This feature is known as *buffering* of pH.

Buffering is conveniently described by reformulating the equilibrium equation [Eq. (11)] for a weak electrolyte. By taking logs and rearranging, the equation becomes

$$\mathrm{pH} = \mathrm{p}K + \log \frac{[\mathrm{A}^-]}{[\mathrm{AH}]}$$

This equation is known as the Henderson-Hasselbalch equation. When $[\mathrm{A}^-] = [\mathrm{AH}]$, we obtain the relationship $\mathrm{pH} = \mathrm{p}K$, and under these conditions the system is poised for maximum buffering since addition or removal of protons is "absorbed" in the transition between A^- and AH. The buffering tendency is readily visualized by a *titration* curve (Fig. II-1). As OH^- is added to raise pH, a weak electrolyte such as acetic acid passes through a transition from acid to conjugate base. At the extremes of the transition (shown in Fig. II-1), additions of base are reflected in an appreciable increase in pH. Near the midpoint of the transition, however, OH^- is neutralized by H^+ released in the transition and the pH change is depressed. Thus the pH is *buffered* and the maximum buffering capacity is at the midpoint where $\mathrm{pH} = \mathrm{p}K$. The system can also be viewed in reverse, begin-

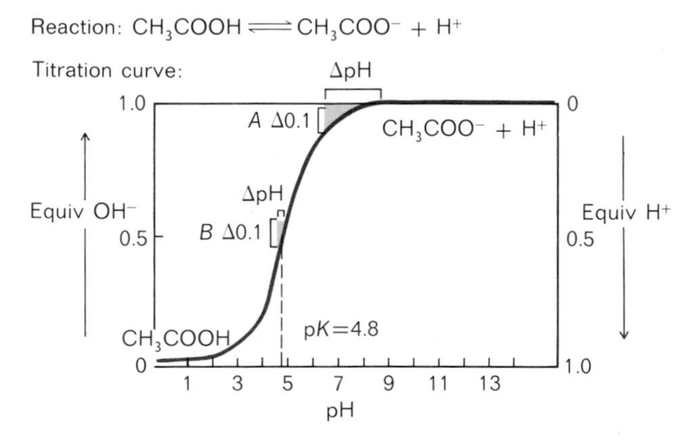

FIG. II-1 *Titration Curve.* Buffering is indicated by the small change in pH on addition of 0.1 equivalents of acid or base near the p*K* (point *B*) compared to the larger change in pH on addition of the same amount of acid or base far from the p*K* (point *A*).

TABLE II-1 Ionizable Groups in Proteins

Groups	Approximate pK
Carboxyl Groups	
(α-Carboxyl of Carboxyl-terminal Amino Acid; Side Chains of Glu, Asp)	3–4
Imidazols (His)	6.5
Amino Groups	
(α-Amino Group of Amino-terminal Amino Acid)	7.8
(ε-Amino Group of Lys)	10.2
Sulfhydryl Groups (Cys)	8.5
Hydroxyl Groups (Ser, Thr, Tyr)	9–10
Guanidino Groups (Arg)	12.5

ning with high pH and adding H^+. In this case the H^+ is neutralized by the conjugate base near the midpoint of the transition. Identical behavior is found with a weak base, except that the conjugate acid occurs at low pH and the base at high pH. The exact value of pK varies widely for different acids and bases. For example, many of the functional groups found in proteins are weak acids or bases and several are listed with their pK's in Table II-1.

The charged groups on proteins provide an overall or net charge on the molecule. Since a constellation of charges is present, some pH can be found where the net charge is 0. This pH is known as the *isoelectric point* and the protein will fail to migrate in an electric field (electrophoresis) at this pH.

Certain weak electrolytes undergo multiple dissociations. One especially important biochemical example is phosphoric

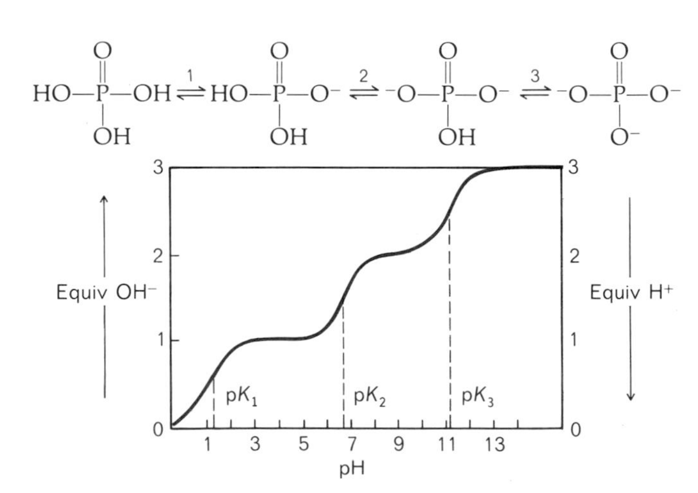

FIG. II-2 *Phosphate Ionizations.*

acid. Phosphate and its esters play an important role in nucleotide structure and provide a natural cellular buffering agent with a p*K* of about 7. Phosphate participates in three ionization reactions, as summarized in Fig. II-2, with p*K*'s in the acid, neutral, and basic range. Ionized phosphate esters also contribute to the high negative charge of DNA, since the one ionizable OH not in ester linkage has a p*K* near 1.

APPENDIX III
Chemical Principles

A key to understanding chemical reactions is the concept of charge distribution in bonding. Every student of chemistry knows that carbon can make four covalent bonds, nitrogen three, and oxygen two. However, what is not always recognized is that the atoms may attract the electrons in the bonds unequally. The attraction for electrons is known as *electronegativity* and on a scale of 1 to 4, various elements of biochemical importance can be rated (see Table III-1). Oxygen and also nitrogen are considerably more electronegative than carbon. Thus a carbonyl structure usually written as

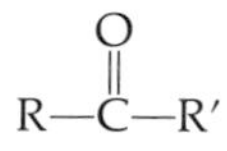

$$\mathrm{R{-}\overset{\overset{\displaystyle O}{\|}}{C}{-}R'}$$

is more accurately described by the notation

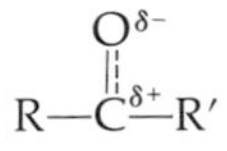

$$\mathrm{R{-}\overset{\overset{\displaystyle O^{\delta-}}{\|}}{C}^{\delta+}{-}R'}$$

The carbon experiences the *inductive* effect of the oxygen. Groups such as carboxyls, as well as atoms, can be electron-withdrawing.

The presence of a partial positive charge on the carbon renders it more susceptible to attack by an electron-rich group (a *nucleophile*). Such a reaction would be characterized as *nucleophilic* since the reaction center is deficient in electrons. A positively charged compound (an *electrophile*) attacking an electron-rich atom is characterized as an *electrophilic* reaction. Carbon can also assume a partial negative charge as in the thiazole moiety of the coenzyme thiamine pyrophosphate (see Chapter 14).

TABLE III-1 Electronegativity

Element	Value
F	4.0
O	3.5
N	3.0
C	2.5
P	2.1
H	2.1

$$\text{—}^{+}\text{N(C=C)S—}\bar{\text{C}} \longleftrightarrow \text{—}\overset{+}{\text{N}}\text{(C=C)S=C—H} \quad + \text{H}^+$$

The partial charge is drawn here in the *resonance* scheme, with two extreme structures related by a double-headed arrow. The actual structure lies somewhere in-between the two extremes. Resonance forms should be distinguished from *tautomers,* which are true chemical structures and can be isolated.

$$CH_3-\overset{O}{\overset{\|}{C}}-CH_2-COO^- \rightleftharpoons CH_3-\overset{OH}{\overset{|}{C}}=CH-COO^-$$

Tautomerism

Tautomers are of particular interest in nucleic acid structure (see Chapter 2).

Stereoisomerism

Two types of isomers are known, *structural isomers* and *stereoisomers.* Structural isomers have the same chemical composition but a different arrangement of atoms. An example of structural isomerism is found in the two amino acids leucine and isoleucine:

$$(CH_3)_2CH-CH_2-CH(\overset{+}{N}H_3)-COO^- \qquad CH_3-CH_2-CH(CH_3)-CH(\overset{+}{N}H_3)-COO^-$$

Leucine Isoleucine

Stereoisomerism can itself be divided into two types, *optical* isomerism and *geometrical* isomerism. The geometrical type concerns the arrangement of groups about a double bond, either cis or trans. An example of geometrical isomers is fumaric and maleic acids:

$$\text{H(COOH)C=C(H)HOOC} \qquad \text{H(COOH)C=C(H)COOH}$$

Fumaric acid
Trans

Maleic acid
Cis

Having distinguished the simpler forms of isomerism we can proceed to the more complex type—optical isomerism. Since carbon has a tetrahedral structure, bonding of four different groups leads to an asymmetric structure. Glyceraldehyde (D and L) illustrates this point:

$$\begin{array}{ccc} \begin{array}{c} CHO \\ H\!-\!C\!-\!OH \\ CH_2OH \end{array} & \Big| & \begin{array}{c} CHO \\ HO\!-\!C\!-\!H \\ CH_2OH \end{array} \end{array}$$

D-Glyceraldehyde | L-Glyceraldehyde

Mirror plane

Two *mirror-image* configurations can be drawn. Each is an *enantiomer* and together they form an *enantiomeric pair.* The enantiomers are identical in most chemical properties but can be distinguished by rotation of plane-polarized light. One enantiomer rotates light in a clockwise direction and is called *dextrorotatory.* The other enantiomer rotates light counterclockwise and is called *levorotatory.* A mixture of the two is optically inactive and is called a *racemic* mixture.

According to convention, D-glyceraldehyde is written with the OH group to the right, where D indicates dextrorotatory. L-Glyceraldehyde has the OH to the left in the perspective drawing and is levorotatory. Glyceraldehyde also provides the reference point for other compounds. For example, L-alanine refers to the alanine enantiomorph with a configuration similar to L-glyceraldehyde:

$$\begin{array}{cc} \begin{array}{c} COOH \\ H_2N\!-\!C\!-\!H \\ CH_3 \end{array} & \begin{array}{c} CHO \\ HO\!-\!C\!-\!H \\ CH_3 \end{array} \end{array}$$

L-(+)-Alanine L-(−)-Glyceraldehyde

The notation (+) and (−) refers to the fact that L-glyceraldehyde is levorotatory (−), but L-alanine, in the configuration corresponding to L-glyceraldehyde, is dextrorotatory (+). Thus in general usage D and L refer to configuration and cannot be taken as indications of the direction of the optical rotation.

A family of D-aldoses (sugars with an aldehyde at the first carbon) can be derived, all related to the configuration of D-glyceraldehyde (see Fig. III-1). Many of these sugars are found in prominent roles in carbohydrate metabolism and

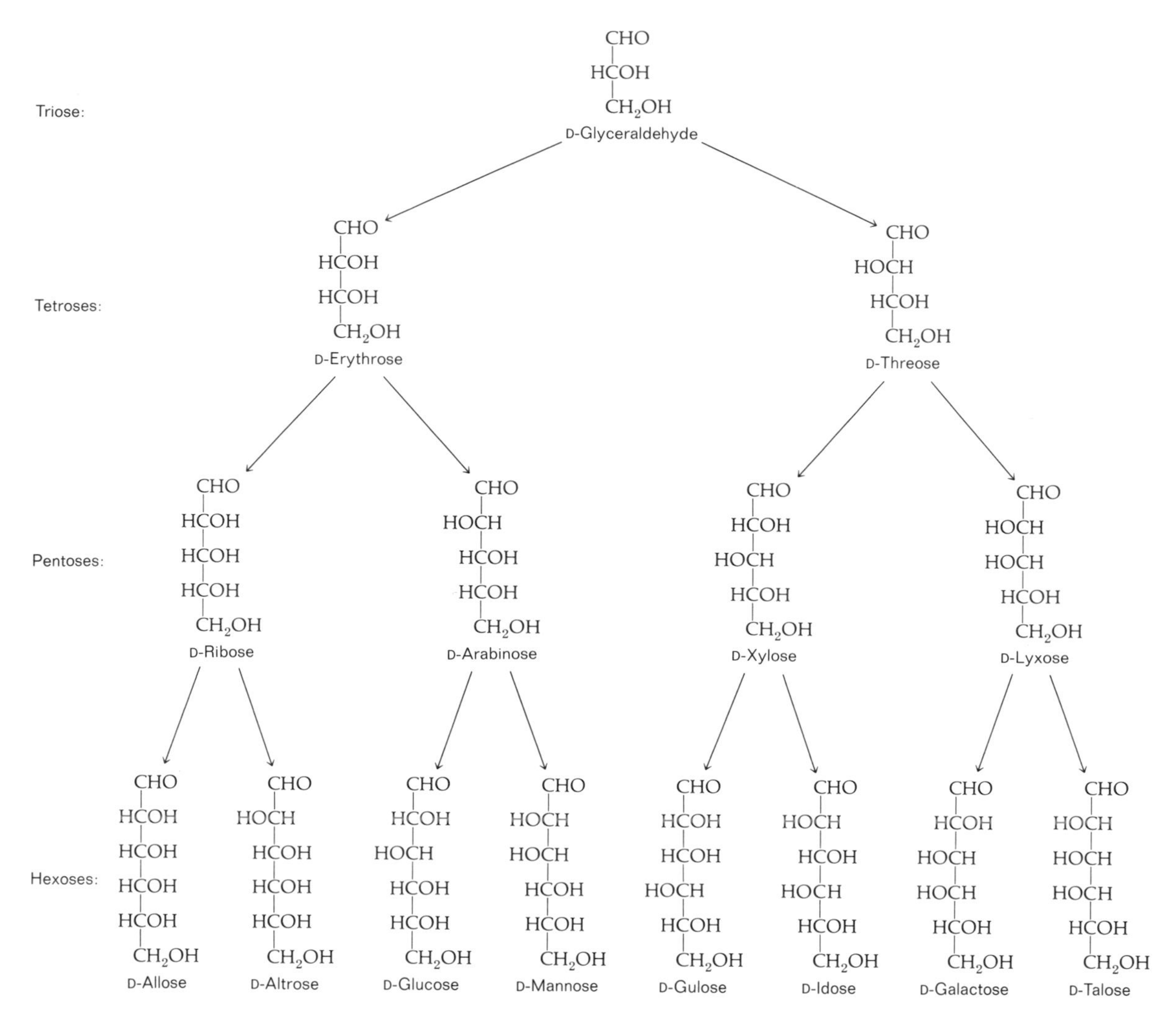

FIG. III-1 *Aldoses Related to* D-*Glyceraldehyde.*

biochemical structures. For each sugar the last asymmetric carbon atom conforms to the configuration of D-glyceraldehyde. Sugars with the same chemical composition but a different configuration at one of the other carbon atoms, for example, glucose and galactose, are *diastereoisomers*. Sugars known as ketoses, which are derived from dihydroxyacetone with the D-glyceraldehyde configuration, are also encountered in a number of biochemical processes (Fig. III-2).

Hexoses and pentoses can form ring structures with the production of an additional center of asymmetry. Aldehyde

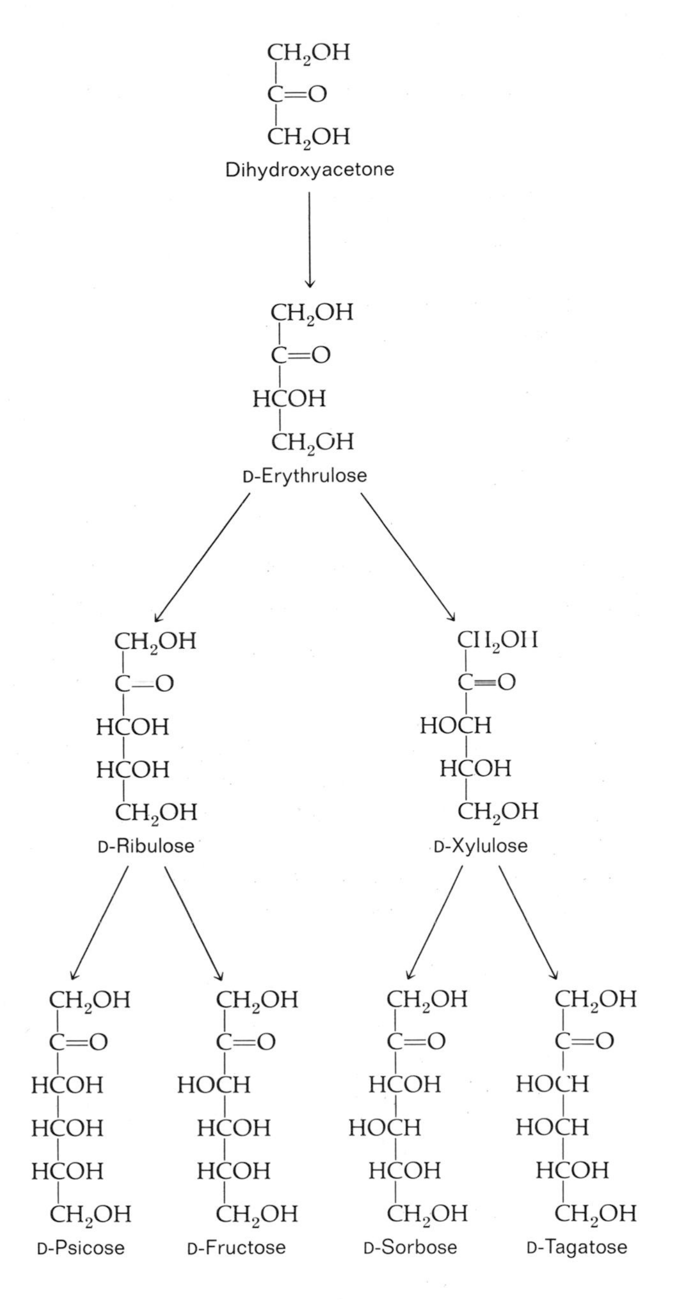

FIG. III-2 *Ketoses Derived from Dihydroxyacetone.*

and hydroxyl functions react to yield a hemiacetyl structure:

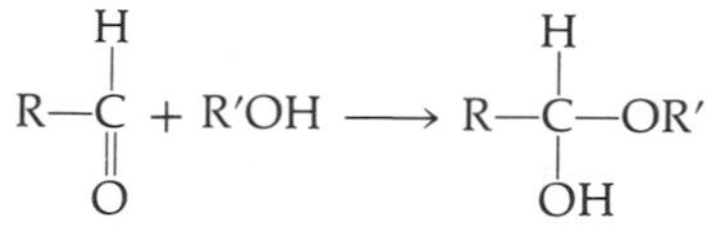

$$R{-}\overset{\displaystyle H}{\underset{\displaystyle O}{C}} + R'OH \longrightarrow R{-}\overset{\displaystyle H}{\underset{\displaystyle OH}{C}}{-}OR'$$

In this way glucose closes to give a six-membered ring. Two arrangements (α and β) are possible:

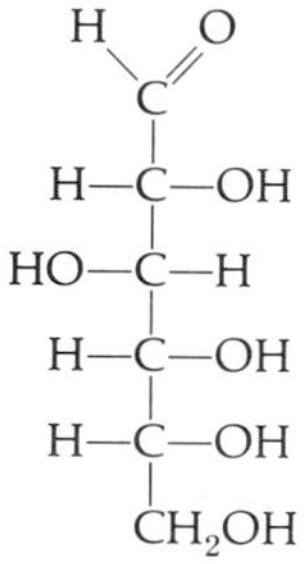

D-Glucose	α-D-Glucose	β-D-Glucose
H, O = C	H, OH — C (ring O)	HO, H — C (ring O)
H—C—OH	H—C—OH	H—C—OH
HO—C—H	HO—C—H	HO—C—H
H—C—OH	H—C—OH	H—C—OH
H—C—OH	H—C (ring O)	H—C (ring O)
CH_2OH	CH_2OH	CH_2OH

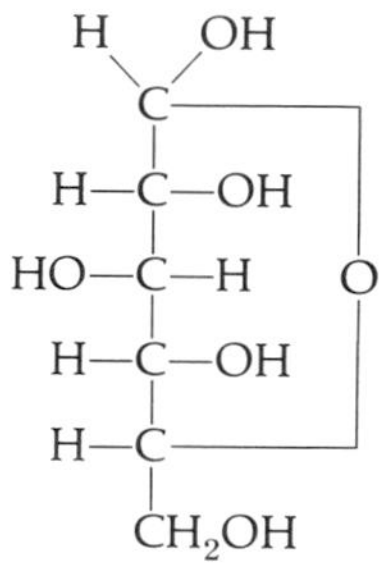

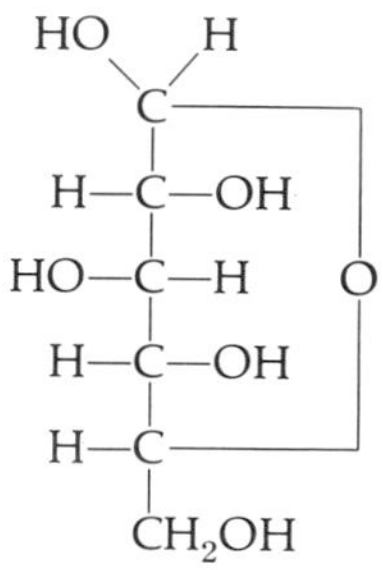

Formation of the ring structure changes the optical rotation properties of the molecule. The changes dependent on the ring formation are called *mutarotation*. The new center of asymmetry is called the anomeric carbon atom and the α- and β- forms are known as *anomers*. The ring structure here for α-D-glucose can also be written in perspective as

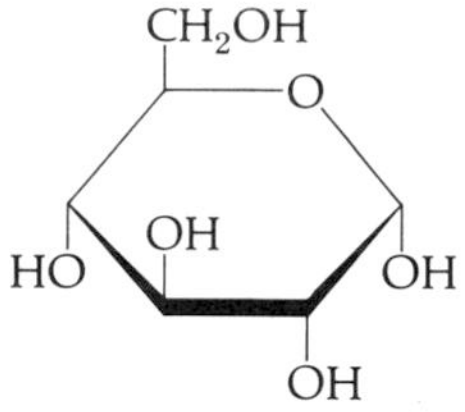

CH_2OH
O
HO OH OH
OH

or in abbreviated form as

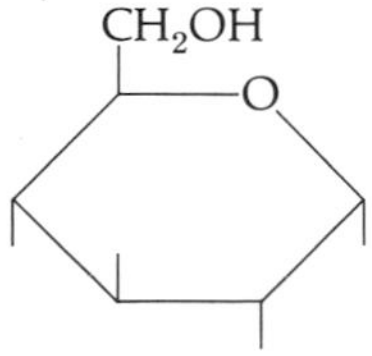

CH_2OH
O

Polysaccharides

Polymers of hexoses, polysaccharides, are important in carbohydrate storage. The polymers are formed by *glycosidic linkage.* In starch and glycogen the basic linkage is α-(1–4)-glycosidic linkage.

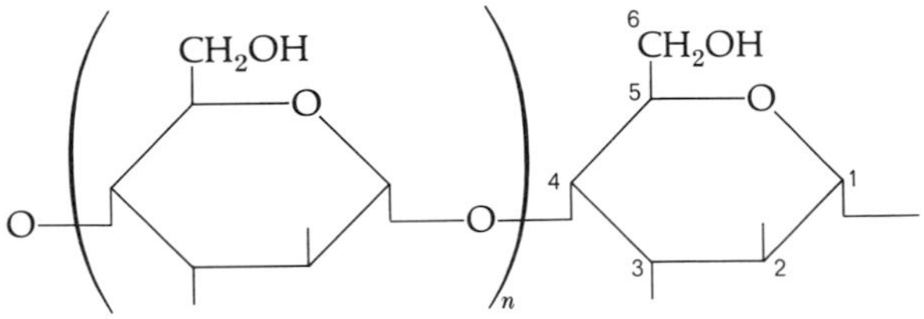

Branches also occur through α-(1–6) linkages.

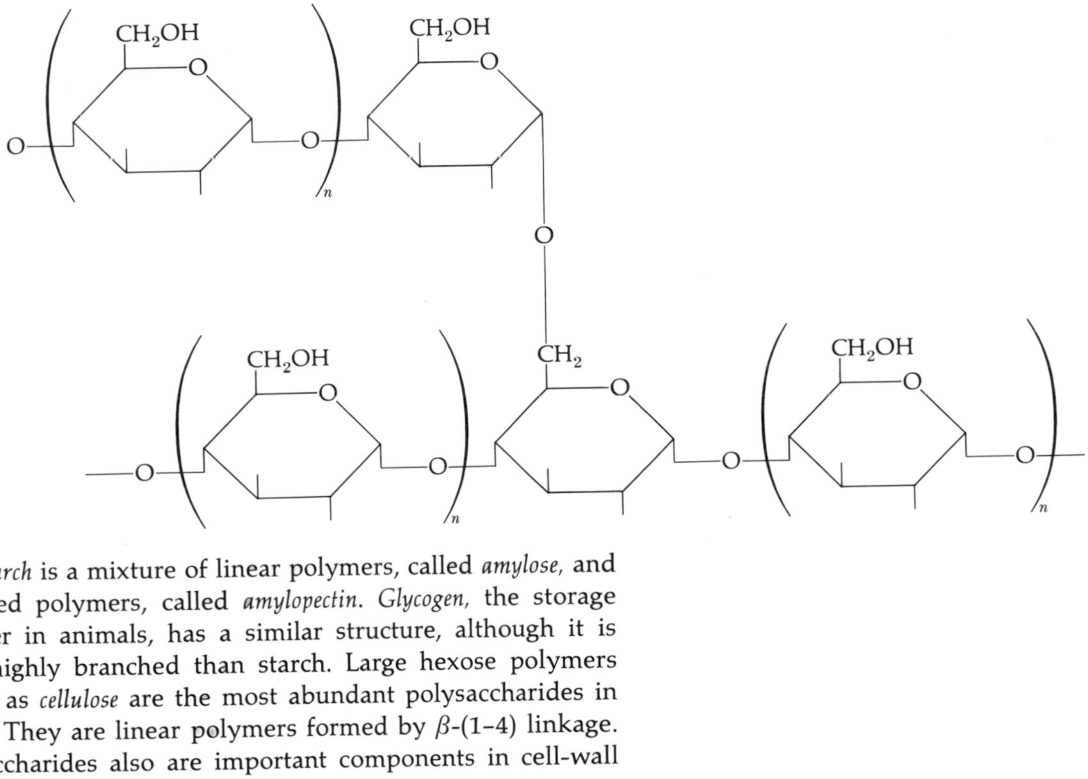

Starch is a mixture of linear polymers, called *amylose,* and branched polymers, called *amylopectin. Glycogen,* the storage polymer in animals, has a similar structure, although it is more highly branched than starch. Large hexose polymers known as *cellulose* are the most abundant polysaccharides in plants. They are linear polymers formed by β-(1–4) linkage. Polysaccharides also are important components in cell-wall structure.

APPENDIX IV
Elementary Thermodynamics

Thermodynamics studies the spontaneous tendency of matter to reach an equilibrium state. This tendency is reflected by such things as the exchange of heat by two adjacent objects at different temperatures or the diffusion of a dye added as a drop to a container of water to distribute uniformly and randomly in the container. When an imbalance exists in a system, work can be derived by tapping the tendency of the system to reach equilibrium. The energy that derives from these adjustments is known as the *free-energy change,* symbolized as ΔG. The relation of ΔG to heat and randomness is expressed in the equation

$$\Delta G = \Delta H - T\,\Delta S$$

Heat change is given as ΔH, which is also known as the *enthalpy change.* The contribution of randomness to the free-energy change is given by $T\,\Delta S$ where ΔS is the *entropy change* and T is the *absolute temperature.* While heat tends to equalize throughout a system, entropy or disorder tends to maximize throughout a system, bringing the components of the system, such as dye in water, into a random or disordered arrangement. Both principles contribute to chemical reactions and determine the equilibrium state or free-energy change. In some cases the two factors tend in opposition and the net effect determines the free-energy change.

The concept of free-energy change can now be related to chemical equilibria. For the reaction

$$a\mathrm{A} + b\mathrm{B} \rightleftharpoons c\mathrm{C} + d\mathrm{D} \qquad (1)$$

where the lowercase letters indicate moles of the reactants and products, specified by uppercase letters, the equilibrium constant K is

$$K = \frac{[C]^c[D]^d}{[A]^a[B]^b} \tag{2}$$

Thus mixing either A and B or C and D will result eventually in some distribution of all four species that balances the tendencies of heat release and entropy gain in accord with K in Eq. (2).

The free-energy change for the reaction is given by

$$\Delta G = \Delta G^\circ + RT \ln \frac{[C]^c[D]^d}{[A]^a[B]^b} \tag{3}$$

where R is the gas constant. When the reaction is at equilibrium, $\Delta G = 0$ and ΔG°, the *standard free-energy* change, is given by

$$\Delta G^\circ = -RT \ln K \tag{4}$$

At 25°C,

$$\Delta G^\circ = -1360 \log K \tag{5}$$

If ΔG° is negative, the reaction is *exergonic* and will proceed spontaneously, beginning with concentrations of all components at 1 M. A reaction characterized by a positive value of ΔG is *endergonic* and will not proceed spontaneously when all components are present at 1 M, but will reverse.

Where pH changes are involved in a reaction, values of the standard free-energy change are expressed as $\Delta G^{\circ\prime}$ to indicate that the value refers to behavior at pH 7.0. An endergonic reaction ($\Delta G^\circ < 0$) can still give rise to products if the reactants are present in sufficiently large quantities. Since ΔG refers to the extent of the reaction, not ΔG° [Eq. (3)], the concentration of the reactants is all important in determining free-energy changes. However, exergonic reactions with very large free-energy changes ($\Delta G^\circ < -7$ kcal/mole) are effectively irreversible. The hydrolysis of ATP is an example of a highly exergonic reaction (see Chapter 11).

BIBLIOGRAPHIC NOTES

A number of excellent textbooks are available and can be consulted for a more detailed treatment of the topics discussed in this book. These sources will also lead the interested reader to the original research journals.

Additional background and greater depth of coverage on the subjects of nucleic acids and protein synthesis and control can be found in:

Watson, J. D.: *Molecular Biology of the Gene,* 2nd ed., W. A. Benjamin, Inc., New York, 1970.

Stent, G. S.: *Molecular Genetics,* W. H. Freeman and Company, San Francisco, 1971.

Articles by some of the major scientific investigators can be found in a compilation of reprints from *Scientific American:*

The Molecular Basis of Life, W. H. Freeman and Company, San Francisco, 1968.

The material on proteins can be supplemented by:

Dickerson, R. E., and I. Geis: *Protein Structure and Function,* Harper & Row, Publishers, Incorporated, New York, 1969.

Many of the topics covered in the chapters on metabolism can be amplified by readings in one of the longer textbooks of biochemistry. Recommended are:

Lehninger, A. L.: *Biochemistry,* Worth Publishers, Inc., New York, 1970.

Mahler, H. R., and E. H. Cordes: *Biological Chemistry,* 2nd ed., Harper & Row, Publishers, Incorporated, New York, 1971.

White, A., P. Handler, and E. L. Smith: *Principles of Biochemistry,* 4th ed., McGraw-Hill Book Company, New York, 1968.

McGilvery, R. M.: *Biochemistry,* W. B. Saunders Company, Philadelphia, 1970.

Much of the material presented in this book concerns "current" areas of research where changes are continually taking place. The developments can be tracked by referring to the latest *Annual Review of Biochemistry,* or, for even more up-to-the-minute summaries, the "News and Views" section of the British publication *Nature* (Friday issue).

ANSWERS TO PROBLEMS

Chapter 1

1 (a) Genotype; (b) phenotype
2 (a), (c), (d) are eucaryotic; (b) is procaryotic
3 (a) Sugar, phosphate, base; (b) adenine, guanine, thymine, cytosine
4 20
5 Mutation probability of A cistron $= 3 \times 267 \times 10^{-9} = 8 \times 10^{-7}$
6 The linear sequence of amino acids in a protein is determined by a corresponding linear sequence of nucleotides in DNA.

Chapter 2

1 A with T, G with C
2 Transformation occurs with isolated DNA; virus growth correlates with the presence of ^{32}P radioactivity in DNA.
3 (a) Yes; (b) no
4 (a) 2.45×10^9 base pairs; (b) 80 cm
5 (a) $A = 0.3$; (b) $A = 0.45$
6 50% G-C
7 One phosphate group

Chapter 3

1 25% G-C
2 Two separate bands would form, corresponding to pure ^{15}N DNA and pure ^{14}N DNA.
3 Four pairs: CA and TG, GA and TC, AG and CT, AC and GT
4 No, a second set of nicks is needed to account for the θ structure observed.

Chapter 4

1 2^3 or 8
2 Arg, Asn, Asp, Cys, Glu, Gln, His, Lys, Ser, Thr, Tyr
3 Ribothymidine
4 Yes

Chapter 5

1 Any multiple of 3 ($3n$, where n is an integer)
2 (a) Phe, Leu, Ile, Tyr, Asn, Lys; (b) Ile, Tyr
3 (a) Ser, Arg, and Leu, with six codons each; (b) Trp, Met, with one codon each
4 (a) UUU, UUC; (b) Phe

Chapter 6

1 (a), (b), (c), formylmethionine; (d), methionine
2 No incorporation of 5′-deoxyadenosine; inhibition only as it competes with ATP for site on RNA polymerase
3 (a)
4 The growing polypeptide chain attached to the transfer RNA of the last amino acid residue (in site D) is donated to the next amino acid attached to its transfer RNA (in site A).
5 Such a transfer RNA might be lethal. Because of wobble the anticodon $^{2'}$AUU$^{5'}$ could recognize the three termination codons UAA, UAG, and UGA.
6 No, the triplet code is uniform, but synthetases for a given amino acid and transfer RNA in one species may charge transfer RNA molecules from other species incorrectly.

Chapter 7

1 (a) Tyr, Thr; (b) Lys, Arg; (c) Glu
2 Divergent evolution
3 Hemoglobin would be saturated with oxygen at the lungs, but the hemoglobin would not give up the oxygen at the tissues; symptoms of insufficient oxygen would result.
4 Half-time in the absence of enzyme $= 10^{18}$ msec $= 10^{15}$ sec $= 2.8 \times 10^{11}$ hr $= 1.2 \times 10^{10}$ da $= 3.2 \times 10^{7}$ yr (32 million years).

Chapter 8

1 Answer left as an exercise
2 (a) Asp, Glu, Tyr, Thr, Ser, Cys; (b) Lys, Arg, His

Chapter 9

1 10^3
2 Acid, 6.5; base, 7.5

Chapter 10

1 The results indicate four moles of Arg per mole of enzyme, consistent with the presence of four identical subunits.
2 Four bands; ratios of concentrations, 1:3:3:1
3 By protein heterogeneity, giving a mixture of binding sites with different affinities, or by negative cooperativity, in which the binding of ligand to a multisite protein diminishes the affinity of the remaining sites

Chapter 11

1 $\Delta E'_0 = +0.3 - 0 = 0.3$ volts
$\Delta G^\circ = (-2)$ (23 kcal) (0.3) = 13.8 kcal
2 Yes, in the methyl group
3 Positions 2 and 3
4 Yes
5 12

Chapter 12

1 (a) Inside; (b) outside; (c) inside
2 P:O = 2; four protons

Chapter 13

1 (a) S; (b) serine
2 All feed the TCA cycle at succinate.
3 No, one high-energy bond is consumed per urea produced.
4 (a) Ammonia, aspartate; (b) aspartate, glycine, glutamine
5 No, ribonucleoside diphosphates are involved.

Chapter 15

1 ATP shifts curve of velocity versus substrate to the right by stabilizing T and increasing L, which has the effect of reducing glycolysis when ATP levels are high.
2 More cyclic AMP → more kinase activity → more phosphorylation of phosphorylase and synthetase → more glycogen breakdown, less glycogen synthesis
3 (a) Malate shuttle to form phosphoenol pyruvate from pyruvate; (b) fructose diphosphatase; (c) glucose-6-phosphatase

Chapter 16

1 Two ways: mutation in i or mutation in o (i/i^- inducible or o/o^c constitutive)
2 Cyclic AMP, sigma factors, arabinose operon control protein
3 The equilibrium constant is the ratio of the forward and reverse rate constants.

Chapter 17

1 No. of gene copies = $(6 \times 10^9$ base pairs/cell$)(2.2 \times 10^{-7})/650 = 2$
2 (a) A messenger RNA repressor; (b) secondary structure of RNA that leads to a sequential reading of cistrons; (c) control of messenger RNA degradation

INDEX